Significant Changes

TO THE NEC® 2011

IN PARTNERSHIP WITH

This book conveys the information related to each change as of July 2, 2010, but does not reflect any subsequent appeal or action taken by the NFPA Standards Council.

1 2 3 4 5 6 7 8 9 – 10 – 9 8 7 6 5 4 3 2 1

Printed in the United States of America

Contents

Chapter 1

Contents

Chapter 2

Contents

Chapter 2 (continued)

Contents

Chapter 3

Contents

Chapter 4

Contents

Chapter 5

Contents

Chapter 6

Contents

Chapter 6 (continued)

Contents

Chapter 7

Contents

Chapter 8

Appendix

Introduction

The *National Electrical Code* (*NEC*) is a living document; as such, it is in a constant state of change. As new technologies, equipment, wiring methods, and industry needs evolve, the *NEC* must stay in step to address installation and safety requirements. The three-year revision cycle of this document facilitates a never-ending *Code*-development process. As you are reading the changes for the 2011 *NEC*, the public will be submitting proposals for change and technical committee task groups will begin to address existing issues for a revised edition in 2014. The *NEC* is the most widely used electrical installation standard in the world and, in fact, is now being adopted internationally as codes and standards evolve and are standardized. This document is used daily by electricians, contractors, maintainers, inspectors, engineers, and designers.

The *NEC* revision process begins with the submission of public proposals. The period for proposal submission to the NFPA to change the 2008 *NEC* (that is, to develop a revised 2011 edition) began following issuance of the 2008 *NEC* and continued up until November 7, 2008. A total of 5,077 proposals were submitted to revise that edition. In January 2009 the Technical Committees met to act on the proposed revisions. After the balloting was complete, the results were made available to the public on the internet and in printed form in a document called the *Report on Proposals* (*ROP*). The public then had a second opportunity to modify or reverse the actions taken by the Technical Committees, by submitting public comments directed at any proposal.

The comments for the 2011 *NEC* were accepted up until October 23, 2009. A total of 2,935 comments were submitted to revise the 2008 *NEC*, following which, in December 2009, the Technical Committees met to act on the comments. After the balloting is complete the results are made available to the public on the internet and in printed form in a document called the *Report on Comments* (*ROC*). Throughout this process, the Technical Correlating Committee (TCC) reviews the output during the *ROP* and *ROC* stages of the process to ensure that no conflicting actions occurred between the work of the nineteen Technical Committees and that all revisions conform to NFPA rules and the *NEC Style Manual*.

After the *ROC* is published, the public has an opportunity to submit a notice of intent to make a motion (NITMAM) on the floor of the NFPA annual meeting. The Standards Council reviews these requests, and all that are in order become certified amending motions (CAMs). At the NFPA annual meeting, the work of the Technical Committees in the *ROP* and *ROC* may be modified. The NFPA annual meeting for the 2011 *NEC* was held in Las Vegas in June 2010. Appeals may be made to the Standards Council if an individual or organization feels that it is necessary. The Standards Council met in Boston in August 2010 to hear those appeals. After appeals were heard and acted upon, the Standards Council issued the 2011 *NEC*.

This consensus-driven revision process provides all users of the *NEC* with an opportunity to mold the next edition through individual participation. As you read through these significant changes for the 2011 *NEC*, be sure to note your ideas for an improved *Code* and submit them as proposals for the next edition.

Features

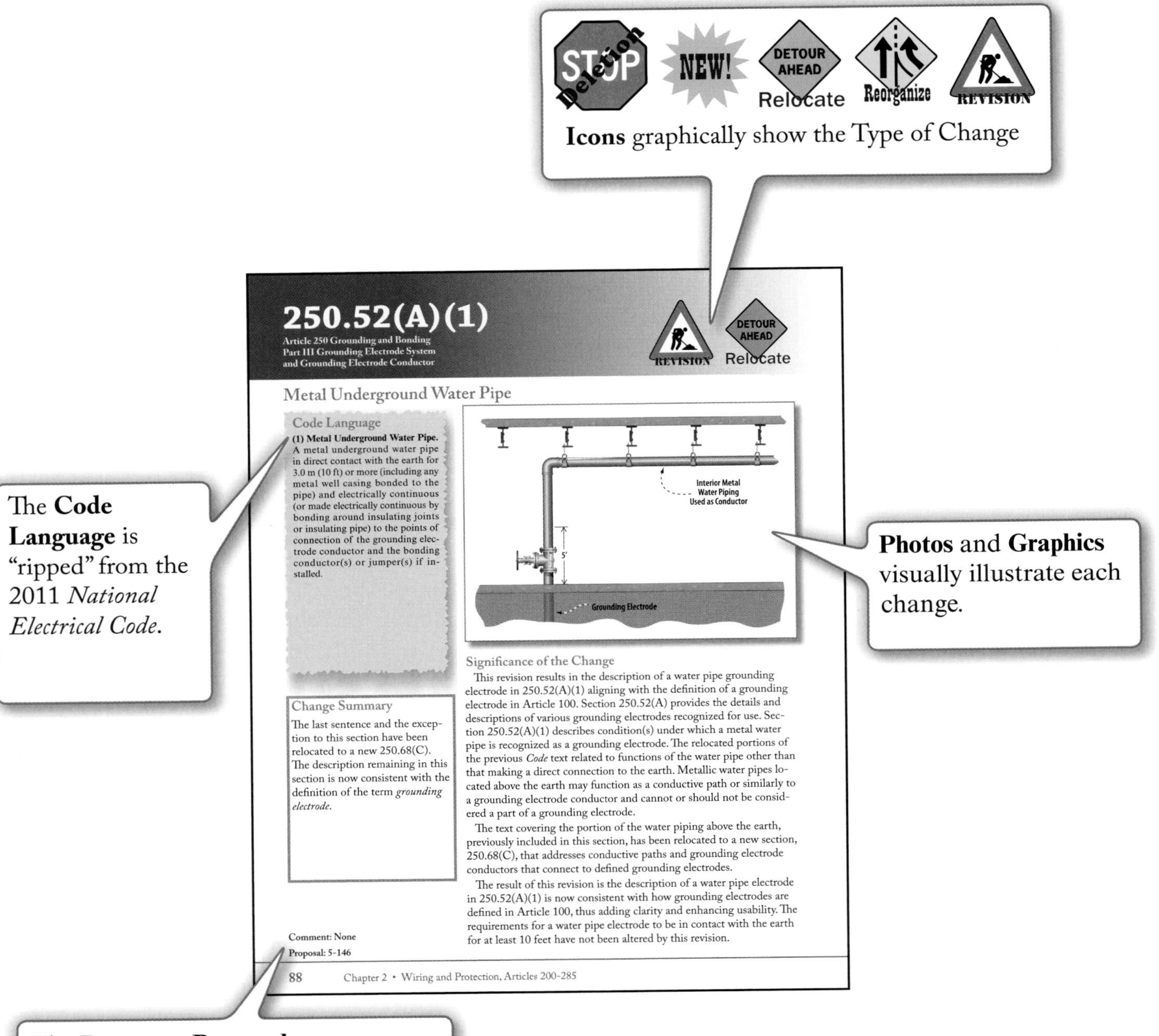

250.52(A)(1)

Article 250 Grounding and Bonding
Part III Grounding Electrode System and Grounding Electrode Conductor

REVISION
DETOUR AHEAD Relocate

Metal Underground Water Pipe

Code Language

(1) Metal Underground Water Pipe. A metal underground water pipe in direct contact with the earth for 3.0 m (10 ft) or more (including any metal well casing bonded to the pipe) and electrically continuous (or made electrically continuous by bonding around insulating joints or insulating pipe) to the points of connection of the grounding electrode conductor and the bonding conductor(s) or jumper(s) if installed.

Change Summary

The last sentence and the exception to this section have been relocated to a new 250.68(C). The description remaining in this section is now consistent with the definition of the term *grounding electrode*.

Comment: None
Proposal: 5-146

Significance of the Change

This revision results in the description of a water pipe grounding electrode in 250.52(A)(1) aligning with the definition of a grounding electrode in Article 100. Section 250.52(A) provides the details and descriptions of various grounding electrodes recognized for use. Section 250.52(A)(1) describes condition(s) under which a metal water pipe is recognized as a grounding electrode. The relocated portions of the previous *Code* text related to functions of the water pipe other than that making a direct connection to the earth. Metallic water pipes located above the earth may function as a conductive path or similarly to a grounding electrode conductor and cannot or should not be considered a part of a grounding electrode.

The text covering the portion of the water piping above the earth, previously included in this section, has been relocated to a new section, 250.68(C), that addresses conductive paths and grounding electrode conductors that connect to defined grounding electrodes.

The result of this revision is the description of a water pipe electrode in 250.52(A)(1) is now consistent with how grounding electrodes are defined in Article 100, thus adding clarity and enhancing usability. The requirements for a water pipe electrode to be in contact with the earth for at least 10 feet have not been altered by this revision.

88 Chapter 2 • Wiring and Protection, Articles 200-285

The **Report on Proposals** sequence numbers and **Report on Comments** sequence numbers are provided to allow the user to the ability to do additional research on a particular change.

Features

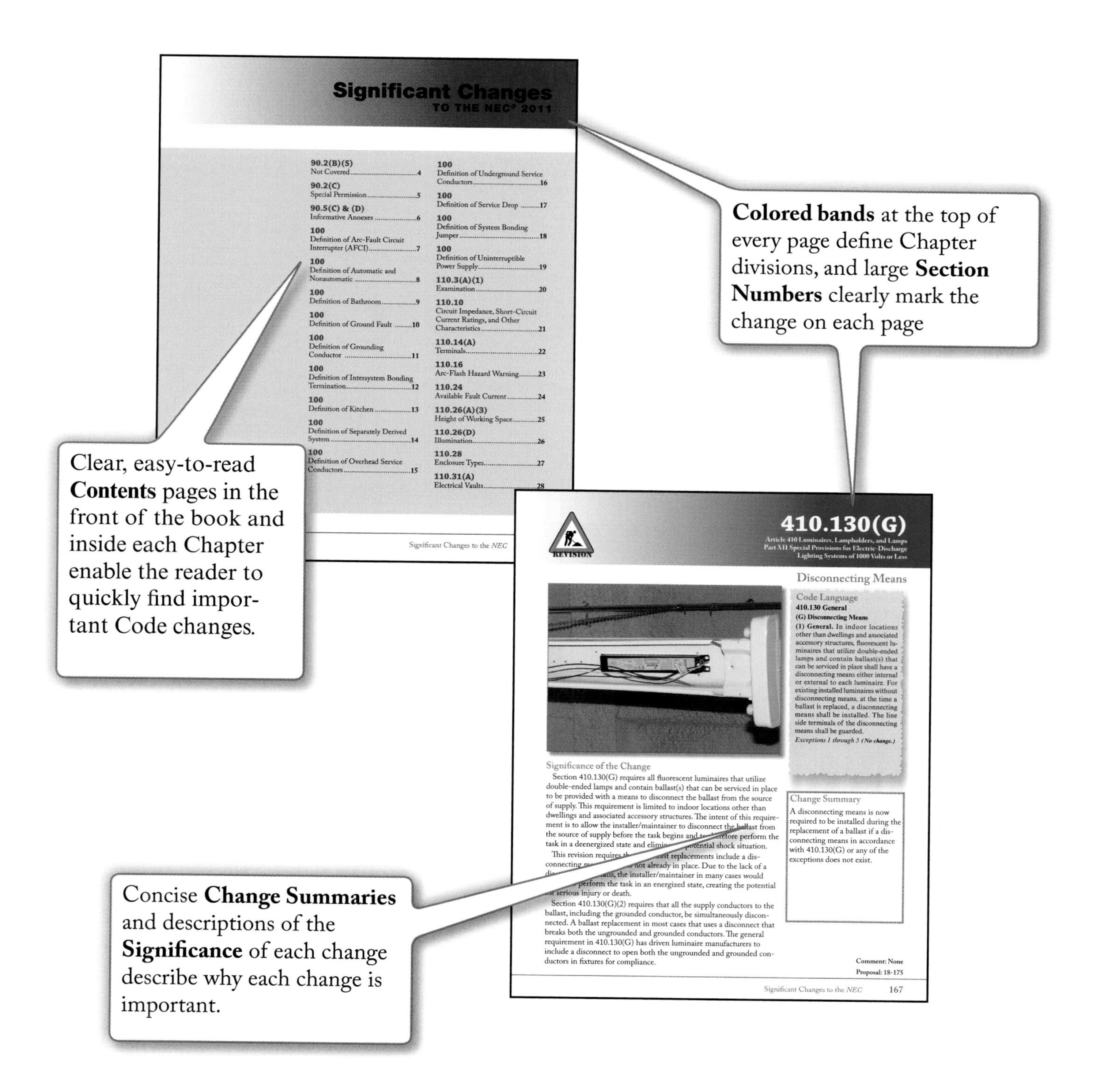

Significant Changes TO THE NEC® 2011
410.130(G)
Disconnecting Means
Colored bands at the top of every page define Chapter divisions, and large Section Numbers clearly mark the change on each page
Clear, easy-to-read Contents pages in the front of the book and inside each Chapter enable the reader to quickly find important Code changes.
Concise Change Summaries and descriptions of the Significance of each change describe why each change is important.

Acknowledgements

Jim Dollard, IBEW Local 98 Philadelphia
Michael Johnston, NECA

Chris Bayer
Cogburn Brothers, Inc.
Cooper Bussmann
Christopher Edwards, Illustrations
Eaton Corporation
ERICO International Corporation
Koninklijke Philips Electronics N. V-Lightolier
National Electric Fuse Association
NECA (I-Stock Photo & Rob Colgan)
Pass and Seymour Legrand
PHASE-A-MATIC Incorporated
Schneider Electric USA, Inc.
Shermco Industries
Tyco Thermal Controls
YESCO

About This Book

This text is written to inform electricians, contractors, maintenance personnel, inspectors, engineers, and system designers of the most significant changes to the 2011 *NEC.* This text will notify the reader of all significant changes with valuable information explaining in detail the reason behind the change and how you will be impacted by it.

Contributing Writers

Jim Dollard is a journeyman wireman in IBEW Local 98 in Philadelphia, Pennsylvania, where he currently holds the position of Safety Coordinator. He is a master OSHA 500 Instructor and works toward safe working conditions on all jobs. As a current member of the following NFPA committees, Jim plays a significant role in the development of electrical codes and standards:

— The *National Electrical Code*; Technical Correlating Committee, Code-Making Panels 10 and 13

— NFPA 70E, *Standard for Electrical Safety in the Workplace*

— NFPA 90A, *Standard for the Installation of Air-Conditioning and Ventilating Systems.*

Jim is also a member of the Underwriters Laboratories Electrical Council and a licensed electrical inspector in Pennsylvania and Delaware. Jim is the author of the NJATC *Codeology* textbook, the 2008 NJATC *NEC Significant Changes,* and co-author of the 2011 NJATC *NEC Significant Changes*. His excellent presentation skills, knowledge of the electrical industry, extensive background in the electrical construction field, and involvement in electrical safety and codes/standards allow Jim to make the most complex requirements easy to understand and apply.

Michael Johnston is NECA's Executive Director of Standards and Safety. Prior to working with NECA, Mike worked for the International Association of Electrical Inspectors as Director of Education, Codes and Standards. He also worked as an electrical inspector and electrical inspection field supervisor for the City of Phoenix, Arizona, and achieved all IAEI and ICC electrical inspector certifications. Mike earned a BS in Business Management from the University of Phoenix. He served on *NEC* Code-Making Panel 5 in the 2002, 2005, and 2008 cycles and is currently the chair of Code-Making Panel 5, representing NECA for the 2011 *NEC* cycle. He currently chairs the NFPA *NEC* Smart Grid Task Force. Among his responsibilities for managing the codes, standards, and safety functions for NECA, Mike is secretary of the NECA Codes and Standards Committee. He is also a member of the IBEW and has experience as an electrical journeyman wireman, foreman, and project superintendent. Mike is an active member of IAEI, ICC, NFPA, ASSE, the NFPA Electrical Section, Education Section, the UL Electrical Council, and National Safety Council.

Chapter 1

Articles 90, 100 and 110

Introduction, Definitions, and Requirements for Electrical Installations

Significant Changes
TO THE NEC® 2011

90.2(B)(5)

Article 90 Introduction

Not Covered

Code Language

(5) Installations under the exclusive control of an electric utility where such installations

a. Consist of service drops or service laterals, and associated metering, or

b. Are on property owned or leased by the electric utility for the purpose of communications, metering, generation, control, transformation, transmission, or distribution of electric energy, or

c. Are located in legally established easements or rights-of-way, or

d. Are located by other written agreements either designated by or recognized by public service commissions, utility commissions, or other regulatory agencies having jurisdiction for such installations. These written agreements shall be limited to installations for the purpose of communications, metering, generation, control, transformation, transmission, or distribution of electric energy where legally established easements or rights-of-way cannot be obtained. These installations shall be limited to Federal Lands, Native American Reservations through the U.S. Department of the Interior Bureau of Indian Affairs, Military bases, lands controlled by port authorities and State agencies and departments, and lands owned by railroads.

Change Summary

This section has been revised, and the former informational note has been incorporated into a new list item (d), which lists the properties to which the term *other written agreements* specifically refers.

Comments: 1-23, 1-25

Proposal: 1-29

Significance of the Change

Section 90.2(B) provides what is not covered by the requirements in the *NEC*, and 90.2(B)(5) specifically addresses installations under the exclusive control of an electric utility. The revision to this section restores the term *other agreements* in a new list item (d). The term *other written agreements* in this section significantly reduces subjectivity regarding which installations are included or excluded from *NEC* requirements by such agreements. As revised, this addition should help avoid conflict at regulatory bodies and at the state and local levels regarding jurisdictional boundaries and which codes or standards are or are not applicable. Further, this section now avoids utility interest in modifying 90.2(B)(5), by local revision of the *NEC* scope in its adoption processes, which had already occurred in some areas for the 2008 *NEC*. As such, confusion will be avoided in the field regarding installations where legally acquired easements and rights-of-way cannot be obtained. Installations not meeting the specific conditions of this section do not qualify for exclusion from the *NEC* requirements.

Special Permission

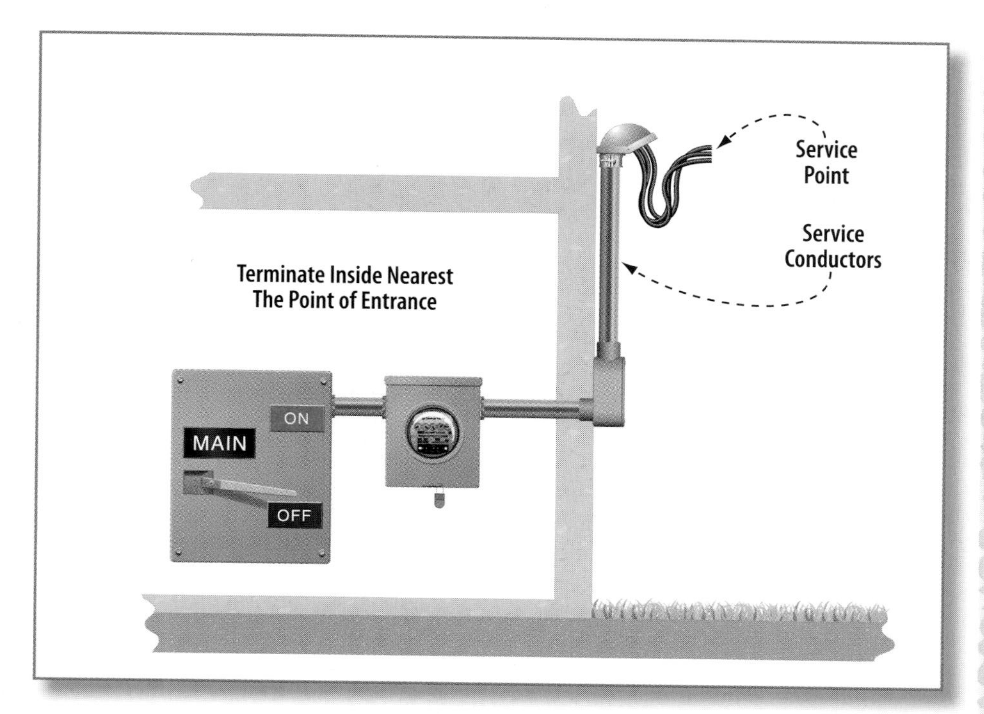

Code Language

90.2 Scope

(C) Special Permission. The authority having jurisdiction for enforcing this *Code* may grant exception for the installation of conductors and equipment that are not under the exclusive control of the electric utilities and are used to connect the electric utility supply system to the service conductors of the premises served, provided such installations are outside a building or structure, or terminate inside nearest the point of entrance of the service conductors.

Significance of the Change

This revision makes clear that the special permission addressed in this section may apply both to service conductors and equipment for an overhead system and to those from an underground system that enter a building or structure. The new text at the end of this section provides a close alignment with the provisions in 230.70(A)(1) and clarifies that service conductors can enter a building or structure in some instances under the permission afforded by this section. Under these conditions, 230.70(A)(1) regulates the location of the service disconnecting means.

The main objective of the requirement in 230.70(A)(1) is to limit the length of unprotected service conductors extended into a building or structure to the practical maximum length necessary to complete the installation. Correlating this special permission provision with the limitations of 230.70(A)(1) reduces the probability of extending excessive lengths of unprotected service conductors through buildings. This revision also promotes consistent and practical enforcement through effective correlation to applicable requirements in Article 230.

Change Summary

This section was revised by changing the term *service-entrance conductors* to *service conductors* and replacing the words "or terminate immediately inside a building wall" with the words "or terminate inside nearest the point of entrance of the service conductors" at the end of the section.

Comment: 1-32

Proposal: 1-33

90.5(C) & (D)

Article 90 Introduction

Informational Notes, Informative Annexes

Code Language

90.5 Mandatory Rules, Permissive Rules, and Explanatory Material.

(A) ***(Unchanged)***

(B) ***(Unchanged)***

(C) Explanatory Material. Explanatory material, such as references to other standards, references to related sections of this *Code*, or information related to a *Code* rule, is included in this *Code* in the form of Informational Notes. Such notes are informational only and are not enforceable as requirements of this *Code*.

Brackets containing section references to another NFPA document are for informational purposes only and are provided as a guide to indicate the source of the extracted text. These bracketed references immediately follow the extracted text.

Informational Note: The format and language used in this *Code* follows guidelines established by NFPA and published in the *NEC Style Manual*. Copies of this manual can be obtained from NFPA.

(D) Informative Annexes. Non-mandatory information relative to the use of the *NEC* is provided in informative annexes. Informative annexes are not part of the enforceable requirements of the *NEC*, but are included for information purposes only.

Change Summary

Section 90.5(C) has been revised to replace globally the term *fine print note* and the acronym (*FPN*) with the single term *Informational Note*. This revision applies in all instances where fine print notes previously appeared throughout the *NEC*. A new subdivision (D) has been added to address the informational annexes provided in the back of the *Code*.

Comment: None

Proposal: 1-37a

90.5 Mandatory Rules, Permissive Rules, and Explanatory Material.

(A) Mandatory Rules. ***(Unchanged)***

(B) Permissive Rules. ***(Unchanged)***

(C) Explanatory Material. Explanatory material, such as references to other standards, references to related sections of this *Code*, or information related to a *Code* rule, is included in this *Code* in the form of Informational Notes. Such notes are informational only and are not enforceable as requirements of this Code.

Brackets containing section references to another NFPA document are for informational purposes only and are provided as a guide to indicate the source of the extracted text. These bracketed references immediately follow the extracted text.

Informational Note: The format and language used in this *Code* follows guidelines established by NFPA and published in the *NEC Style Manual*. Copies of this manual can be obtained from NFPA.

(D) Informative Annexes. Non-mandatory information relative to the use of the *NEC* is provided in informative annexes. Informative annexes are not part of the enforceable requirements of the *NEC*, but are included for information purposes only.

Significance of the Change

The global change provides more positive differentiation between *Code* text (rules) and informational text. The term *fine print* is more applicable to type size than to portrayal of nonmandatory information as clarification for users. Action by the Technical Correlating Committee changed the proposed term *advisory notes* to *informational notes* to denote useful information and to avoid the appearance of the *Code* offering advice. This change also provides a distinction between notes to *NEC* tables, which are enforceable, and informational notes, which are not. The proposal also indicated that many standards now contain normative and informative annexes. Normative annexes are mandatory, while informative annexes, such as the annexes in the *NEC*, are not enforceable.

Definition of Arc-Fault Circuit Interrupter (AFCI)

Courtesy of Eaton Corporation

Code Language

Arc-Fault Circuit Interrupter (AFCI). A device intended to provide protection from the effects of arc faults by recognizing characteristics unique to arcing and by functioning to de-energize the circuit when an arc-fault is detected.

Significance of the Change

Section 2.2.2.1 of the *NEC Style Manual* indicates that terms appearing in more one article be defined in Article 100. Because the term *arc-fault circuit interrupter* is used in 210.12, 550.25, 760.41, and 760.121, as defined by the definition that appeared only in 210.12(A), relocating the definition to Article 100 results in conformity to the *NEC Style Manual* and in a general definition of an AFCI device that applies to the protection necessary in any present or future rule in which the term is used. Note that as technology changes and *Code* rules are revised, AFCI devices may evolve to forms other than circuit breaker types, for example as outlet-device AFCI protection, which is also provided in cord assemblies such as those required in 440.65. Revisions in Sections 210.12 and 406.4(D)(4) specify outlet-type AFCI protective devices.

Change Summary

The definition of *arc-fault circuit interrupter* has been relocated from 210.12(A) to Article 100. No editorial or technical changes have been made to this definition. This revision results in the definition of *arc-fault circuit interrupter* having global *NEC* application.

Comment: None

Proposal: 2-162

100

Article 100 Definitions
Part I General

Definitions of Automatic and Nonautomatic

Code Language

Automatic. Performing a function without the necessity of human intervention.

Nonautomatic. Requiring human intervention to perform a function.

Change Summary

The definitions of *automatic* and *nonautomatic* have been revised and simplified to correlate with the NFPA *Glossary of Terms*.

Significance of the Change

The terms *automatic* and *nonautomatic* have been revised to simplify their meaning and to provide a direct contrast between them. The revised definitions clarify and specify when human intervention is required and when it is unnecessary. Additionally, the revised definitions are now consistent with definitions of the same terms used in other NFPA codes and standards. The scope of the NFPA Glossary of Terms Technical Advisory Committee is to facilitate the use of accurately defined terms and to minimize the number of terms that carry differing definitions in various NFPA codes and standards. Similar proposals are being submitted for NFPA 70E, 96, 99, 101, 101B, 550, and 901.

Comment: None
Proposals: 1-54, 1-110a

Definition of Bathroom

Code Language

Bathroom. An area including a basin with one or more of the following: a toilet, a urinal, a tub, a shower, a bidet, or similar plumbing fixtures.

Significance of the Change

This revision clarifies that other plumbing fixtures, in addition to a tub, shower, toilet, and basin, provide the features and functions that qualify a room or area as a bathroom. Bathrooms are used primarily for human waste excretion and sanitation of the body. Including urinals, and bidets is appropriate because of how they function and where they are typically installed in occupancies. The revision also addresses other "similar plumbing fixtures," to allow the definition to remain open-ended in coverage of rooms that would qualify as bathrooms when constructed with fixtures typically used in bathrooms. The revisions to this definition enhance enforceability and clarify applicability of requirements that include references to the term *bathrooms*.

Change Summary

The definition of the term *bathroom* has been revised by including the words "a urinal" and "a bidet, or similar plumbing fixtures."

Comment: 2-1
Proposal: 2-5

Definition of Ground Fault

Code Language

Ground Fault. An unintentional, electrically conducting connection between an ungrounded conductor of an electrical circuit and the normally non-current-carrying conductors, metallic enclosures, metallic raceways, metallic equipment, or earth.

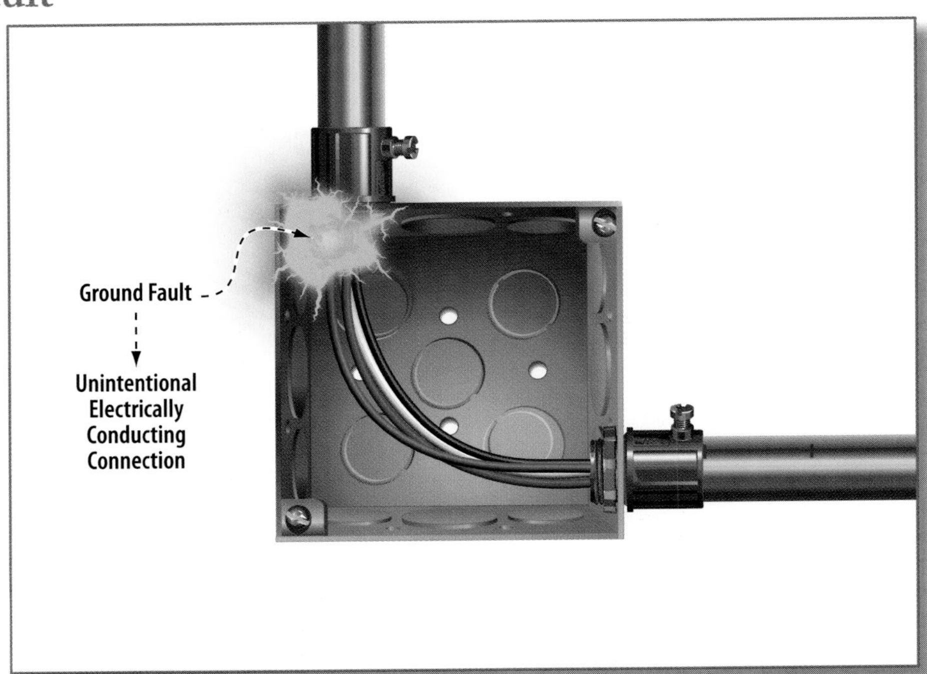

Change Summary

The definition of the term *ground fault* has been relocated from 250.2 to Article 100. No technical changes to the revision result from this change. This revision results in the definition of ground fault having global *NEC* application.

Significance of the Change

The term *ground fault* appears in more than two articles of the *Code*, thus qualifying it for definition in Article 100, per Section 2.2.2.1 of the *NEC Style Manual*. This change does not affect the current definition of the term. It should be understood that the words *ground fault* are used as part of larger terms throughout the *NEC* and that this definition of *ground fault* is not intended to apply to those terms. For example, the term *ground-fault circuit interrupter*, which is a type of protective device that provides protection against electric shock, is defined on its own within Article 100. Defining the term *ground fault* in Article 100 improves clarity and enhances usability of the requirements in which the term appears throughout the *NEC*, rather than just within Article 250.

Comment: 5-39

Proposal: 5-10

Definition of Grounding Conductor

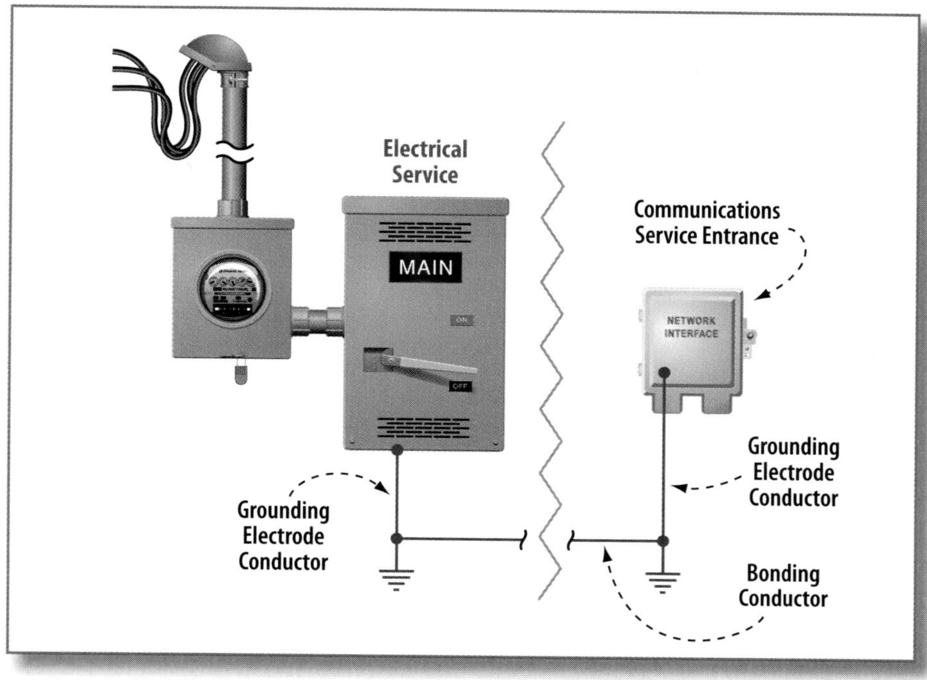

Code Language

(Delete the definition of grounding conductor.)

~~**Grounding Conductor.** A conductor used to connect equipment or the grounded circuit of a wiring system to a grounding electrode or electrodes.~~

Significance of the Change

Deletion of this term and its definition is a continuation of work begun in the 2008 *NEC* development process by a task group assigned to grounding and bonding words and terms. Having two defined terms for a single component of the grounding and bonding scheme is unnecessary and leads to misapplication of requirements. The term *grounding conductor* is used primarily in the limited energy articles in Chapters 7 and 8, and this component of the grounding scheme performs the same functions as a grounding electrode conductor. *Grounding conductor* has been replaced by *grounding electrode conductor* or by *bonding conductor* within Chapters 7 and 8, as a coordinated set of revisions to correlate with the deletion of *grounding conductor*. It should be understood that the specific requirements for sizing and installation of grounding electrode conductors are contained in the Chapter 7 and 8 articles. Sizing and installation rules for grounding electrode conductors used for power systems and services are as specified in Chapter 2, with no changes made.

Change Summary

The definition of the term *grounding conductor* has been deleted from Article 100.

Comments: 5-7, 5-8, 5-9, 5-10, 5-11, 5-12, 5-13

Proposals: 5-13, 5-18

Definition of Intersystem Bonding Termination

Code Language

Intersystem Bonding Termination. A device that provides a means for connecting bonding conductors for communications systems to the grounding electrode system.

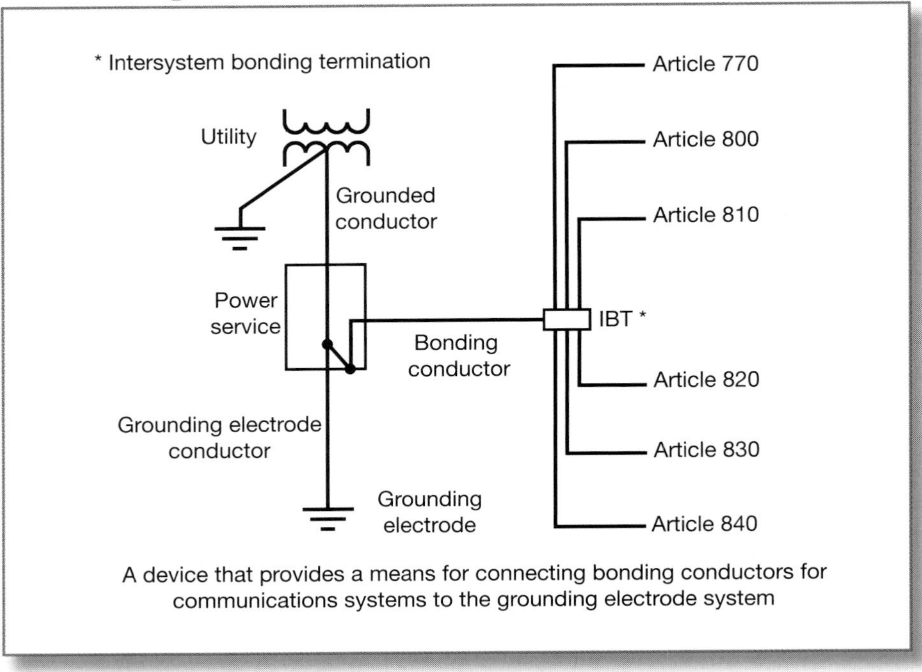

A device that provides a means for connecting bonding conductors for communications systems to the grounding electrode system

Change Summary

The definition of *intersystem bonding termination* has been revised by deleting the words "grounding conductor(s) and bonding conductor(s) at or near the service equipment, or metering equipment, or at the disconnecting means for buildings or structures supplied by a feeder or branch circuit" and adding the words "to the grounding electrode system."

Significance of the Change

This revision simplifies the definition and no longer specifies where such terminations can be installed. Intersystem bonding terminations are required to be installed at the locations specified in 250.94, including both service equipment and metering equipment locations. The change means that metering equipment is now an acceptable location for the installation of the intersystem bonding termination. The revision to this definition also deleted the word *near*, to reduce subjectivity and promote conformity to the *NEC Style Manual*, which recommends avoiding use of *near* and other words or terms that could result in vague and unenforceable requirements.

The rules using this term provide the performance criteria intended for installations involving intersystem bonding terminations. These connection points provide a specific means for connecting bonding conductors of limited energy systems covered by Article 770 and Chapter 8. Section 250.94 covers the general requirements for installing an intersystem bonding termination device and where it is required to be located.

Comment: 5-19

Proposal: 5-21

Definition of Kitchen

I-Stock Photo, Courtesy of NECA

Code Language

Kitchen. An area with a sink and permanent provisions for food preparation and cooking.

Significance of the Change

This revision brings the definition of the term *kitchen* into harmony with the definition of *dwelling unit*, which includes the words "permanent provisions for cooking." The use of the word "provisions" in the definition recognizes that not all kitchen appliances are permanent. For example, electric ranges or fixed microwave ovens that are cord-and-plug-connected occupy a dedicated location in the kitchen. When kitchens are constructed, a dedicated space and outlet are provided and thereby constitute "provisions." The word "provisions" is more appropriate for consistency and actual meaning in this definition, as it relates to cooking means that qualify the room or area as a kitchen.

Change Summary

The definition of *kitchen* has been revised by replacing the word "facilities" with the word "provisions."

Comment: None

Proposal: 2-23

Definition of Separately Derived System

Code Language

Separately Derived System. A premises wiring system whose power is derived from a source of electric energy or equipment other than a service. Such systems have no direct connection from circuit conductors of one system to circuit conductors of another system, other than connections through the earth, metal enclosures, metallic raceways, or equipment grounding conductors.

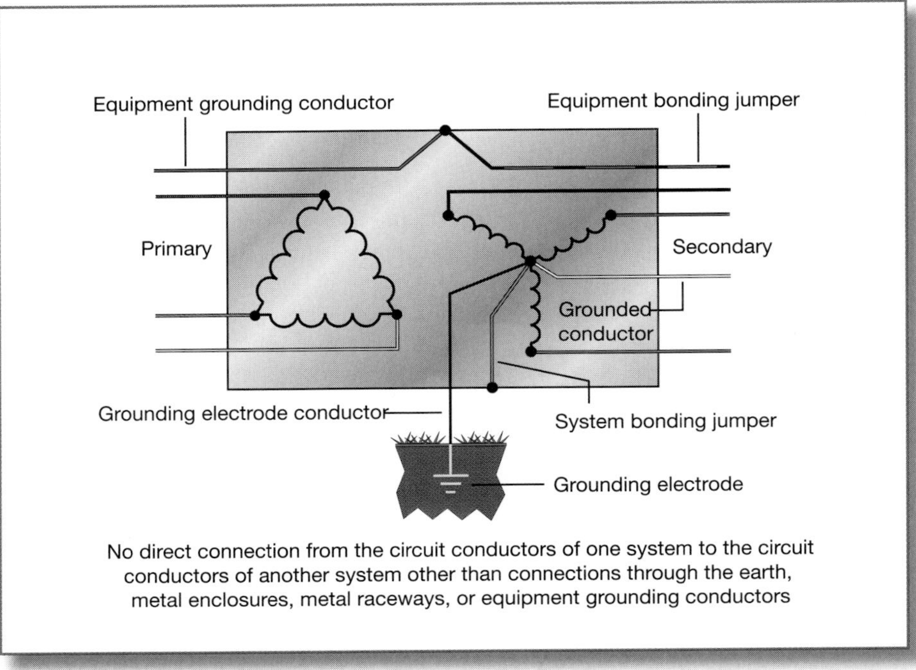

No direct connection from the circuit conductors of one system to the circuit conductors of another system other than connections through the earth, metal enclosures, metal raceways, or equipment grounding conductors

Change Summary

The second sentence of this definition has been revised to read as follows: "Such systems have no direct connection from circuit conductors of one system to circuit conductors of another system, other than connections through the earth, metal enclosures, metallic raceways, or equipment grounding conductors."

Significance of the Change

The definition of *separately derived system* has been revised to clarify what constitutes a separate system specifically as it relates to the direct connection between system conductors. The second sentence clarifies that, although there can be direct connections between equipment grounding conductors and equipment bonding conductors, the circuit conductors of other systems cannot be directly connected other than through metal enclosures, earth, metallic raceways, or equipment grounding conductors. The revision clarifies when a system would qualify as "separately derived" in the determination of applicable grounding and bonding rules. An example of a separately derived system that has no direct connection between the circuit conductors of another system is a transformer that has a 480-volt delta primary and a 208Y/120-volt secondary. While there is a direct connection between grounding and bonding conductors of the primary and secondary, there is no direct connection between circuit conductors on the primary and secondary.

Comment: 5-21
Proposal: 5-28

Definition of Overhead Service Conductors

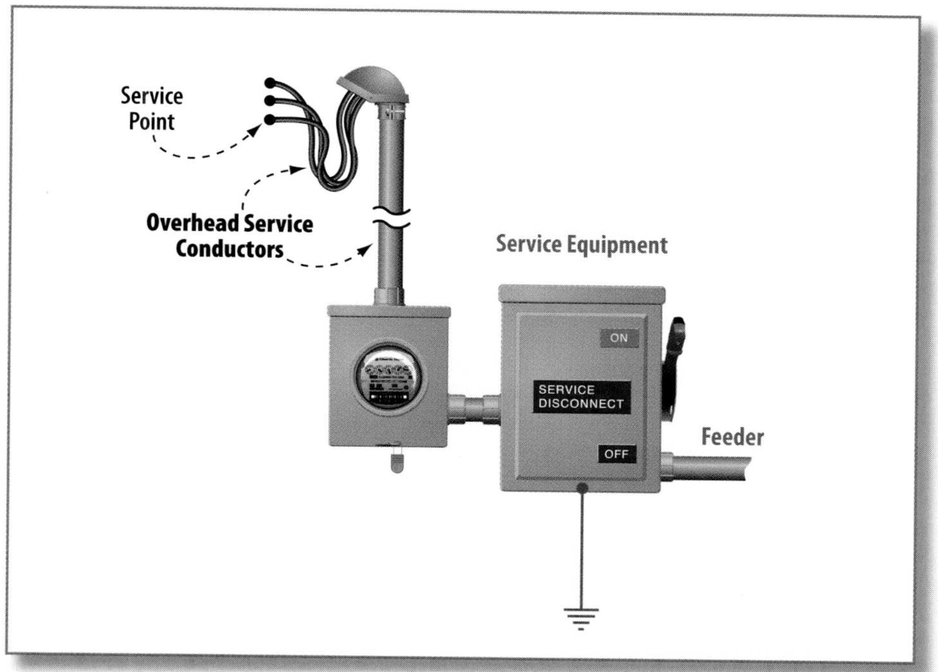

Code Language

Service Conductors, Overhead. The overhead conductors between the service point and the first point of connection to the service-entrance conductors at the building or other structure.

Significance of the Change

This definition is new and applies to the service conductors in an overhead service arrangement. This definition clarifies which portion of the installation includes the overhead service conductors. This is one of several coordinated changes to help differentiate between the service drop conductors and the overhead service conductors on the load side of the service point. The service conductors in an overhead system are typically routed from the service point through a service head (weather head) and through a riser that terminates within the service equipment. The revision to the definition of the term *service drop* clarifies that service drops are overhead conductors between the utility distribution system and the service point. In many cases, the service drop is owned and under the control of the serving utility. The new definition increases clarity and enables users to more effectively apply *Code* requirements that pertain to overhead service conductors on the load side of the service point.

Change Summary

A definition of the term *Service Conductors, Overhead* has been added to Article 100.

Comment: None
Proposal: 4-3

Definition of Underground Service Conductors

Code Language

Service Conductors, Underground. The underground conductors between the service point and the first point of connection to the service-entrance conductors in a terminal box, meter, or other enclosure, inside or outside the building wall.

Informational Note: Where there is no terminal box, meter, or other enclosure, the point of connection is considered to be the point of entrance of the service conductors into the building.

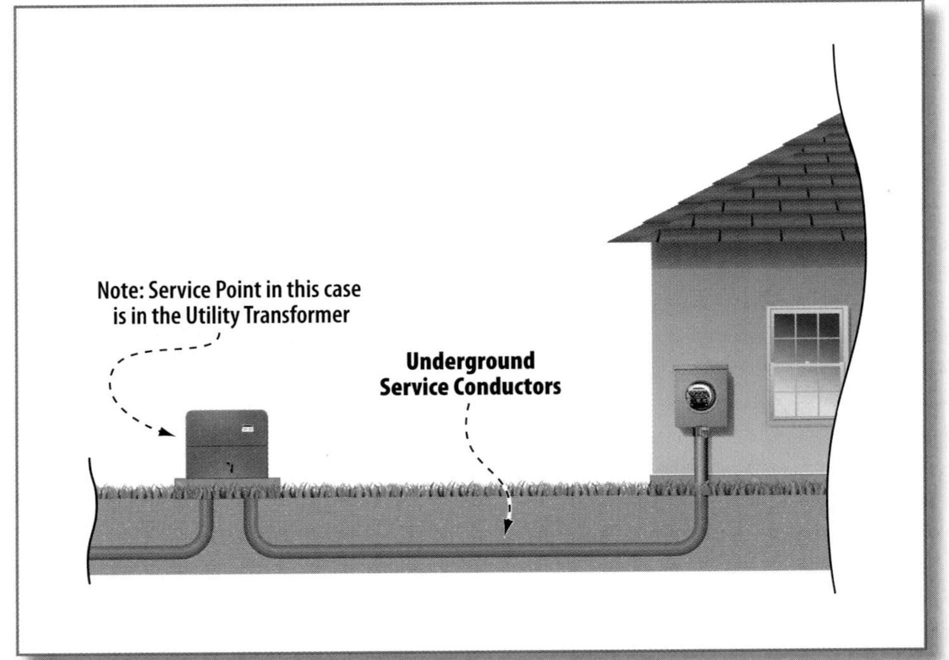

Change Summary

A new definition, for the term *service conductors, underground*, and an informational note have been added to Article 100.

Significance of the Change

This new definition applies to service conductors in an underground arrangement and clarifies which portion of an installation comprises the underground service conductors. During the 2008 *NEC* development process, concerns were expressed over the term *service laterals* and whether they were always the responsibility of the utility. Underground service conductors are not always under the exclusive control of the serving utility.

This revision is intended to bring differentiation between the underground service conductors and the utility conductors on the line side of the service point. The service conductors in an underground system are typically routed from the service point through an underground conduit system or direct burial cables that terminate within the service equipment located on or within a building. The definition applies where the underground service conductors begin at the service point.

The new informational note clarifies that where no enclosure exists, such as a meter or terminal box outside the building, the point of connection is considered to be that point at which the service conductors enter the building.

Comment: 4-2
Proposal: 4-15

Definition of Service Drop

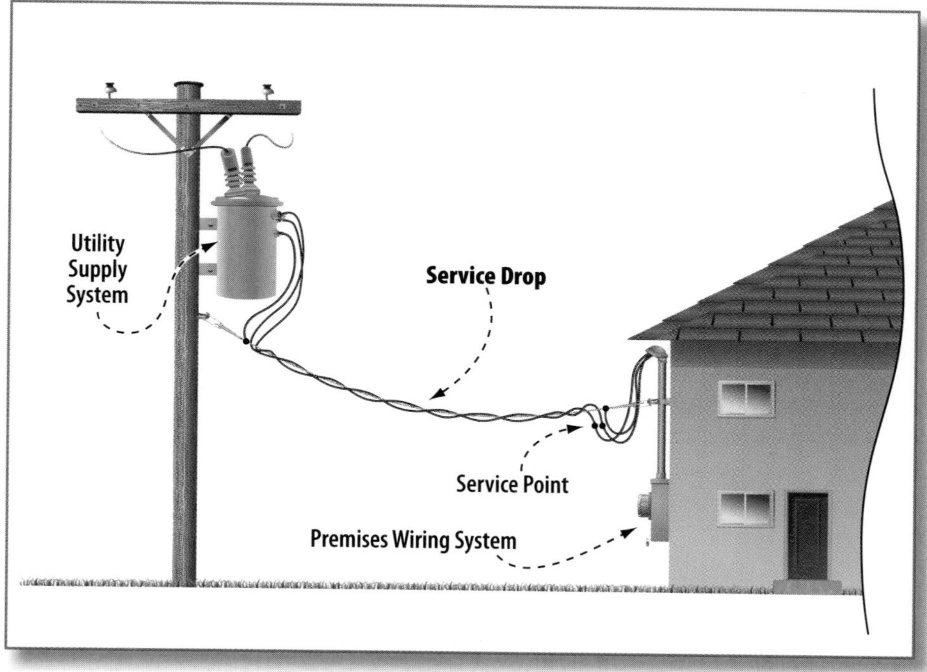

Code Language

Service Drop. The overhead conductors between the utility electric supply system and the service point.

Significance of the Change

Revision of the term *service drop* is one of several coordinated changes to clarify where utility responsibilities and applicable regulations end and where the *NEC* application begins relative to overhead service installations. This revision provides users with a more precise description of which conductors make up the service drop in an overhead service arrangement. The revised definition uses the term *service point* and makes clear that service drops are typically under the utility responsibility and are typically governed by the requirements in the *National Electrical Safety Code* (*NESC*) or other utility regulations. The service drop typically originates at the pole-mounted utility transformer and extends over a property so as to connect to the overhead service conductors at the service point. The overhead service conductors extend from the load side of the service point to the service equipment and service-entrance conductors where they enter a building or structure.

Change Summary

The definition of the term *service drop* has been simplified and now includes the words "service point."

Comment: 4-3

Proposal: 4-8

Definition of System Bonding Jumper

Code Language

Bonding Jumper, System. The connection between the grounded circuit conductor and the supply-side bonding jumper, or the equipment grounding conductor, or both, at a separately derived system.

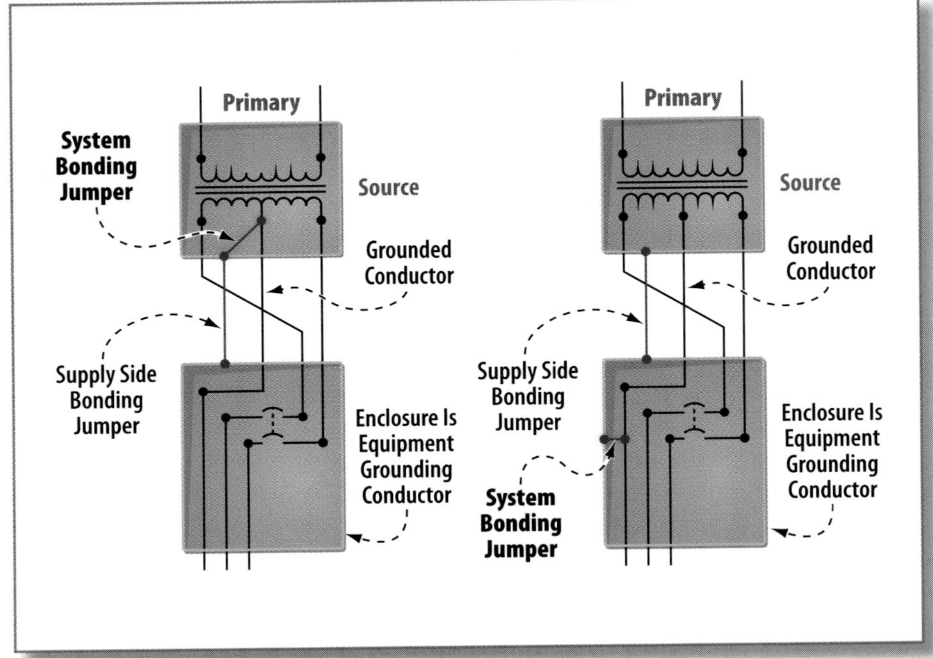

Change Summary

The definition of the term *system bonding jumper* has been revised and relocated from Section 250.2 to Article 100.

Significance of the Change

Because the term *system bonding jumper* appears in two or more articles of the *NEC*, it qualifies for inclusion in Article 100, in accordance with 2.2.2.1 of the *NEC Style Manual.* While most uses of this term appear in 250.30 covering separately derived systems, the term is also found in Part III of Article 708, covering power sources for critical operations power systems.

The definition has been revised to reference the supply-side bonding jumper in addition to the equipment grounding conductor at a separately derived system. This revision clarifies that on the secondary of a transformer derived system, a supply-side bonding jumper is often used to which the grounded conductor is also connected. The previous definition only referenced a connection between the grounded conductor and the equipment grounding conductor of a system.

Comments: 5-37, 5-4a

Proposals: 5-52, 5-6

Definition of Uninterruptible Power Supply

Code Language

Uninterruptible Power Supply. A power supply used to provide alternating current power to a load for some period of time in the event of a power failure.

Informational Note: In addition, it may provide a more constant voltage and frequency supply to the load, reducing the effects of voltage and frequency variations.

Significance of the Change

The term *uninterruptible power supply* (UPS) is used in various *Code* articles, qualifying it as a definition in Article 100. This new definition is consistent with national product safety standards covering uninterruptible power supply equipment and describes the functions of uninterruptible power supplies, including serving as a backup power source. The new informational note indicates that UPS systems or equipment may include such functions as power filtration, voltage stabilization, and frequency regulation. It should be noted that the new definition applies to an uninterruptible power supply that could be in the form of an assembly packaged as one piece of equipment or to a system that serves to provide a backup source power during normal power interruption. Typical UPS systems use batteries, fly wheels, and other alternative sources of power and a means to transfer the load during a normal power system failure or interruption. UPS equipment and systems are typically automatic in function and maintain a steady state of power to equipment during a power failure in such a manner that the equipment or facility supplied does not experience a power loss.

Change Summary

A new definition of the term *uninterruptible power supply* has been added to Article 100, and a new informational note now follows this new definition.

Comments: 12-1, 12-2, 12-3

Proposal: 12-3

110.3(A)(1)

Article 110 Requirements for Electrical Installations
Part I General

Examination, Identification, Installation, and Use

Code Language

110.3 Examination, Identification, Installation, and Use of Equipment.

(A) Examination. In judging equipment, considerations such as the following shall be evaluated:

(1) Suitability for installation and use in conformity with the provisions of this *Code*.

Informational Note: Suitability of equipment use may be identified by a description marked on or provided with a product to identify the suitability of the product for a specific purpose, environment, or application. Special conditions of use or other limitations and other pertinent information may be marked on the equipment, included in the product instructions, or included in the appropriate listing and labeling information. Suitability of equipment may be evidenced by listing or labeling.

Change Summary

The informational note following 110.3(A)(1) has been revised by adding a new second sentence that reads as follows: "Special conditions of use or other limitations and other pertinent information may be marked on the equipment, included in the product instructions, or included in the appropriate listing and labeling information."

Significance of the Change

The informational note following 110.3(A)(1) has been revised to address special conditions of use and other important and pertinent information marked on a product or instructions provided with a product. The existing informational note addressed suitability of use but did not alert the user to any special conditions of use that may be essential to safe use or proper functioning of equipment. Examples of special conditions of use provided in the substantiation included elevated or reduced ambient temperatures, stringent power quality requirements, and specific overcurrent protective devices.

In the 2008 *NEC* development process, a new informational note was accepted to 500.8(A)(3) that addresses product certificates or other pertinent information marked on a product that demonstrates compliance. Inspection authorities typically use listing and labeling as a basis for approvals. Section 110.3(B) requires listed and labeled equipment to be installed and used according to instructions included in the listing or labeling. Specific information about use may now be included in a product certificate, in pertinent markings on products, as well as in installation instructions.

Comment: 1-80

Proposal: 1-111

Circuit Impedance, Short-Circuit Current Ratings, and Other Characteristics

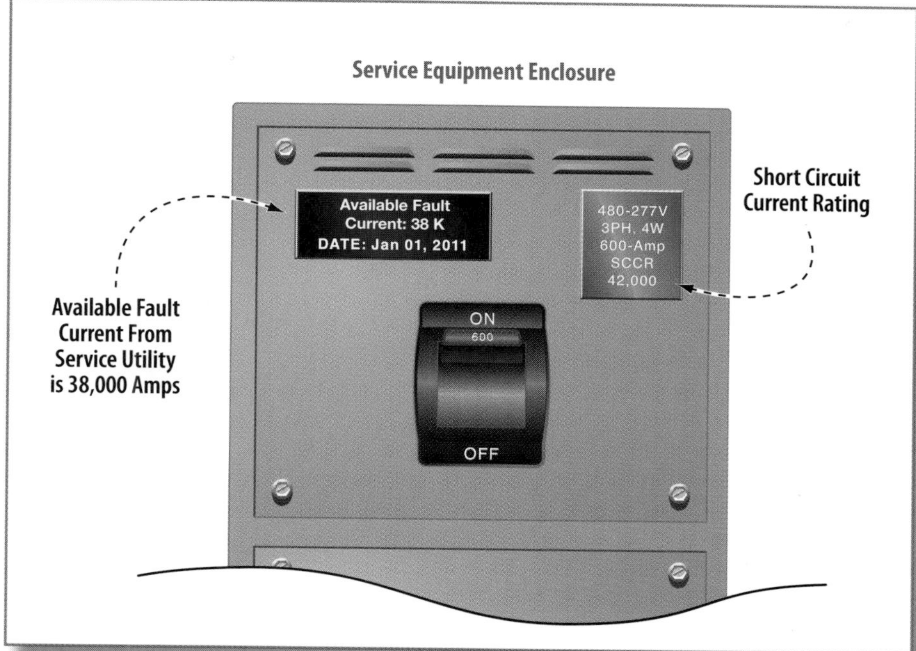

Code Language

110.10 Circuit Impedance, Short-Circuit Current Ratings, and Other Characteristics. The overcurrent protective devices, the total impedance, the equipment short-circuit current ratings, and other characteristics of the circuit to be protected shall be selected and coordinated to permit the circuit protective devices used to clear a fault to do so without extensive damage to the electrical equipment of the circuit. This fault shall be assumed to be either between two or more of the circuit conductors or between any circuit conductor and the equipment grounding conductor(s) permitted in 250.118. Listed equipment applied in accordance with their listing shall be considered to meet the requirements of this section.

Significance of the Change

Section 110.10 requires the installation and use of equipment that has short-circuit current ratings equal to or greater than the amount of available short-circuit current. The revision to the title is appropriate because short-circuit current ratings are one aspect of compliance with this rule. The word *equipment* was added before the term *grounding conductor* to correlate with the deletion of the term and its associated definition. The ground fault described in the second sentence is between any of the ungrounded conductors of the circuit and the equipment grounding conductor of the wire type or any type permitted in 250.118, including any enclosing metal raceway that is part of the grounded equipment used in the system.

The revisions to this section promote accuracy in the use of terms and enhance clarity. The words *component* and *product* may be interpreted as requiring evaluation of individual internal components of equipment even when such equipment is marked with an overall short-circuit current rating as required by other articles or product standards. Changing to the defined term *equipment* provides needed clarity and consistency.

Change Summary

The title to this section has been revised by adding the words *Short-Circuit Current Ratings*, and the second sentence has been revised by adding the word *equipment* before the term *grounding conductor* and adding the words "permitted in 250.118." The word *equipment* has replaced the words *components* and *products* within this section.

Comments: 1-88, 1-90
Proposals: 1-129, 1-130

110.14

Article 110 Requirements for Electrical Installations
Part I General

Electrical Connections

Code Language

110.14 Electrical Connections

Because of different characteristics of dissimilar metals, devices such as pressure terminal or pressure splicing connectors and soldering lugs shall be identified for the material...

...will not adversely affect the conductors, installation, or equipment.

Connectors and terminals for conductors more finely stranded than Class B and Class C stranding as shown in Chapter 9, Table 10, shall be identified for the specific conductor class or classes.

Change Summary

A new last sentence was added to 110.14 that reads as follows: "Connectors and terminals for conductors more finely stranded than Class B and Class C stranding as shown in Chapter 9, Table 10, shall be identified for the specific conductor class or classes."

Comment: 1-101

Proposal: 1-149

Significance of the Change

The revision to 110.14 requires that terminals for fine-stranded conductors be identified for the specific conductor or class of stranding. Fine-stranded conductors behave differently in conventional terminations for standard stranded conductors; thus specific termination means must be used for terminating fine-stranded conductors. Requirements for terminating fine-stranded cables were added to Article 690 in the 2008 cycle. Listed fine-stranded conductors are now available for installations covered by the *NEC* rules. Fine-stranded conductors are used in applications other than photovoltaic equipment and installations, for example, DC conductors installed in battery rooms and dock wiring.

This revision incorporates a general requirement in Article 110 that applies to all installations involving fine-stranded conductors, even though specific conductor termination rules may exist within various articles of the *Code*. The new text references a new Table 10, Conductor Stranding, in Chapter 9, which provides information about conductors with stranding finer than Class B or C stranding. These conductors should be identified for the specific conductor class. The origin of this new table is from UL 486A-B, Table 14.

110.16

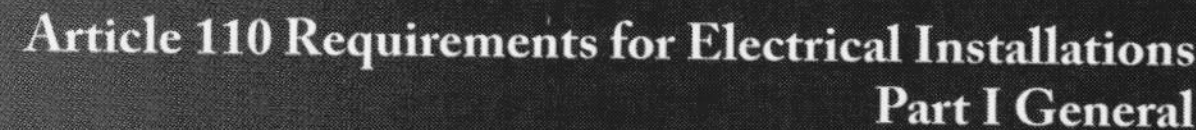

Arc-Flash Hazard Warning

Code Language

Arc-Flash Hazard Warning. Electrical equipment, such as switchboards, panelboards, industrial control panels, meter socket enclosures, and motor control centers, that are in other than dwelling units, and are likely to require examination, adjustment, servicing, or maintenance while energized shall be field marked to warn qualified persons of potential electric arc flash hazards. The marking shall be located so as to be clearly visible to qualified persons before examination, adjustment, servicing, or maintenance of the equipment.

Significance of the Change

The title of Section 110.16 has been changed to Arc-Flash Warning Label. This requirement addresses a field-applied warning label, but it does not provide any requirements dealing with flash protection. Flash protection requirements are provided in NFPA 70E, *Standard for Electrical Safety in the Workplace.*

The term *dwelling occupancies* has been changed to *dwelling unit* to clarify the exemption from an arc-flash warning label requirement. In the *NEC*, the term *dwelling occupancies* encompasses multifamily dwellings. A large multifamily dwelling could have larger electrical services with high levels of available fault current similar to those in other than dwelling occupancies. For example, the equipment in an office building would require arc-flash warning labels, while the equipment within a dwelling unit of a multi-occupancy building multifamily dwelling would not. The lack of this additional reminder and warning label could lead to a catastrophe for the electrical worker in the multifamily dwelling, while the electrician in the office building would have the benefit of seeing the warning sign and being reminded of the danger before commencing work on any equipment.

Change Summary

The title of this section has been changed from Flash Protection to Arc-Flash Hazard Warning. The word *occupancies* has been replaced by the word *units* in the first sentence of this section.

Comments: 1-105, 1-106

Proposal: 1-162

Available Fault Current

Code Language

110.24 Available Fault Current.

(A) Field Marking. Service equipment in other than dwelling units shall be legibly marked in the field with the maximum available fault current. The field marking(s) shall include the date the fault current calculation was performed and be of sufficient durability to withstand the environment involved.

(B) Modifications. When modifications to the electrical installation occur that affect the maximum available fault current at the service, the maximum available fault current shall be verified or recalculated as necessary to ensure the service equipment ratings are sufficient for the maximum available fault current at the line terminals of the equipment. The required field marking(s) in 110.24(A) shall be adjusted to reflect the new level of maximum available fault current.

Exception: The field marking requirements in 110.24(A) and 110.24(B) shall not be required in industrial installations where conditions of maintenance and supervision ensure that only qualified persons service the equipment.

Change Summary

A new section 110.24 titled Available Fault Current has been included in Part I of Article 110.

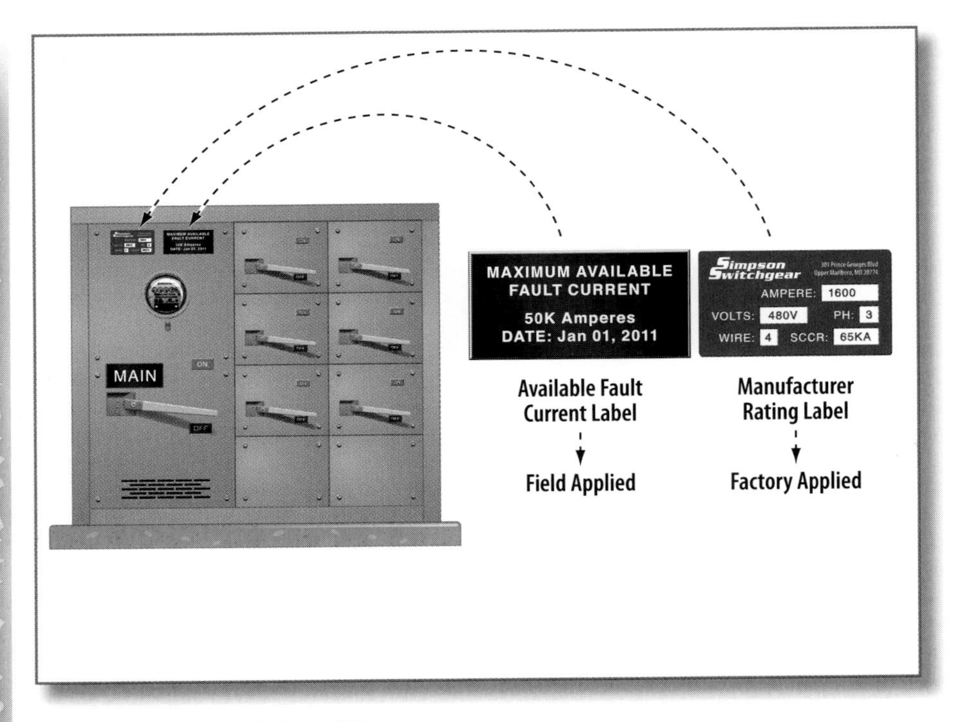

Significance of the Change

This new requirement calls for a field-applied marking indicating the maximum available short-circuit current being supplied at the line terminals of the service equipment. It does not require anything new from designers or engineers, given that this information must be known to specify and install service equipment within its ratings. The only new requirement is a marking providing the maximum level of available short-circuit current. The modifications addressed in subdivision (B) addressing installations are also often experienced by the system and equipment over time and can result in equipment with inadequate short-circuit current ratings. The exception relaxes the marking requirement for industrial establishments and is applicable only under the controlled conditions described in the exception.

This new requirement will assist engineers, designers, contractors, workers, and inspectors in attaining compliance with 110.9 not only for new installations but for any activity or modifications to the electrical installation that causes the amount of available fault current to exceed the rating of equipment.

Comment: 1-115
Proposals: 1-183, 10-72

110.26(A)(3)

Article 110 Requirements for Electrical Installations
Part II 600 Volts, Nominal, or Less

Height of Working Space

Code Language

110.26(A)(3) Height of Working Space. The work space shall be clear and extend from the grade, floor, or platform to a height of 2.0 m (6 ½ ft) or the height of the equipment, whichever is greater. Within the height requirements of this section, other equipment that is associated with the electrical installation and is located above or below the electrical equipment shall be permitted to extend not more than 150 mm (6 in.) beyond the front of the electrical equipment.

Exception No. 1: In existing dwelling units, service equipment or panelboards that do not exceed 200 amperes shall be permitted in spaces where the height of the working space is less than 2.0 m (6 ½ ft).

Exception No. 2: Meters that are installed in meter sockets shall be permitted to extend beyond the other equipment. The meter socket shall be required to follow the rules of this section.

Significance of the Change

The requirements of 110.26(A)(3) and 110.26(E) of the 2008 *NEC* were effectively the same. This revision combines the two sections to improve clarity and usability. The "headroom" requirements of 110.26(E) were limited to service equipment, switchboards, panelboards, and motor control centers. The revision also clarifies that the minimum height requirements are applicable to all equipment that qualifies for minimum working spaces governed by this section. The reference to "headroom" has been deleted to reduce confusion between what constitutes "headroom" and what qualifies as "minimum heights of the working space" about equipment.

This relocation removes the redundancy whereby two rules address the same requirements. A new Exception No. 2 has been added following 110.26(A)(3), to specifically address electric meters and meter socket enclosures. Meters are permitted to extend beyond the maximum distance of 6 inches as provided in this rule, but meter socket enclosures must meet all other requirements in this section. The new exception recognizes that many installed electric meters extend distances greater than 6 inches from the face of the meter socket enclosure.

Change Summary

The provisions in 110.26(E) and the accompanying exception have been incorporated into 110.26(A)(3). A new Exception No. 2 addressing meter installations has been added.

Comment: 1-132

Proposals: 1-207, 1-208

Illumination

Code Language

(D) Illumination. Illumination shall be provided for all working spaces about service equipment, switchboards, panelboards, or motor control centers installed indoors and shall not be controlled by automatic means only. Additional lighting outlets shall not be required where the work space is illuminated by an adjacent light source or as permitted by 210.70(A)(1), Exception No. 1, for switched receptacles.

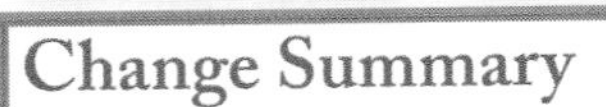

Change Summary

The words "and shall not be controlled by automatic means only" in the last sentence have been relocated to the first sentence of 110.26(D). The words "in electrical equipment rooms" have been removed from this section.

Significance of the Change

Room use and designation are often determined by a building facilities management and owner. If a room containing electrical equipment is designated as a storage room, the requirements of this section become more difficult to apply. The revision clarifies that the illumination is for all working spaces about service equipment, switchboards, panelboards, or motor control centers installed indoors, regardless of room designation or location. This revision ensures that all areas referred to in this requirement have the necessary lighting for safe servicing, examination, adjustment, maintenance, and other operations associated with the type of equipment covered in this requirement. The requirements in this section apply to work spaces and areas about electrical equipment, and not just to rooms identified as "electrical equipment rooms."

Comment: None

Proposals: 1-227, 1-231

110.28

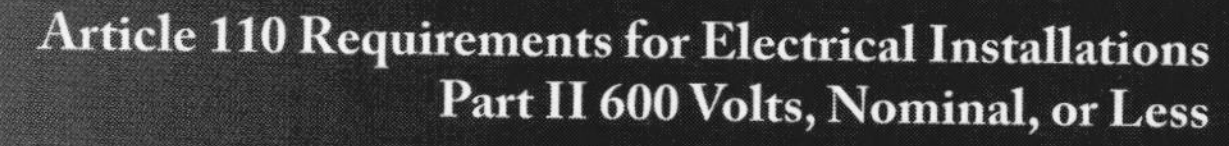

Enclosure Types

Code Language

110.28 Enclosure Types. Enclosures (other than surrounding fences or walls) of switchboards, panelboards, industrial control panels, motor control centers, meter sockets, enclosed switches, transfer switches, power outlets, circuit breakers, adjustable-speed drive systems, pullout switches, portable power distribution equipment, termination boxes, general purpose transformers, fire pump controllers, fire pump motors, and motor controllers, rated not over 600 volts nominal and intended for such locations, shall be marked with an enclosure-type number as shown in Table 110.28.

[Table 110.28 (former Table 110.20) unchanged.]

Significance of the Change

The requirements in this section and table are not applicable to equipment over 600 volts. Section 110.30 specifically incorporates all coverage in Part I except as supplemented or modified by Part III provisions, and no provision in Part III modifies NEMA enclosure types. This relocation resolves this issue and places the rules where they will apply only to installations operating at 600 volts or less. Additionally, the list of equipment enclosures has been expanded to include most types of equipment, and all are now required to be marked with an enclosure-identifying number as indicated in the applicable product standards. These additional enclosure markings will greatly assist installers in determining enclosure and equipment suitability for various types of environments and conditions.

Change Summary

Section 110.20 and Table 110.20 have been relocated to Part II of Article 110 and renumbered as Section 110.28 and Table 110.28. The list of equipment enclosures covered by this requirement has been expanded to include multiple types of equipment enclosures.

Comment: None
Proposals: 1-171, 1-172

Electrical Vaults

Code Language

110.31(A) Electrical Vaults. Where an electrical vault is required or specified for conductors and equipment operating at over 600 volts, nominal, the following shall apply.

(1) Walls and Roof. The walls and roof shall be constructed of materials that have adequate structural strength for the conditions, with a minimum fire rating of 3 hours. For the purpose of this section, studs and wallboard construction shall not be permitted.

(2) Floors. ***(See actual NEC text for the remainder.)***

(3) Doors. ***(See actual NEC text for the remainder.)***

(4) Locks. ***(See actual NEC text for the remainder.)***

(5) Transformers. ***(See actual NEC text for the remainder.)***

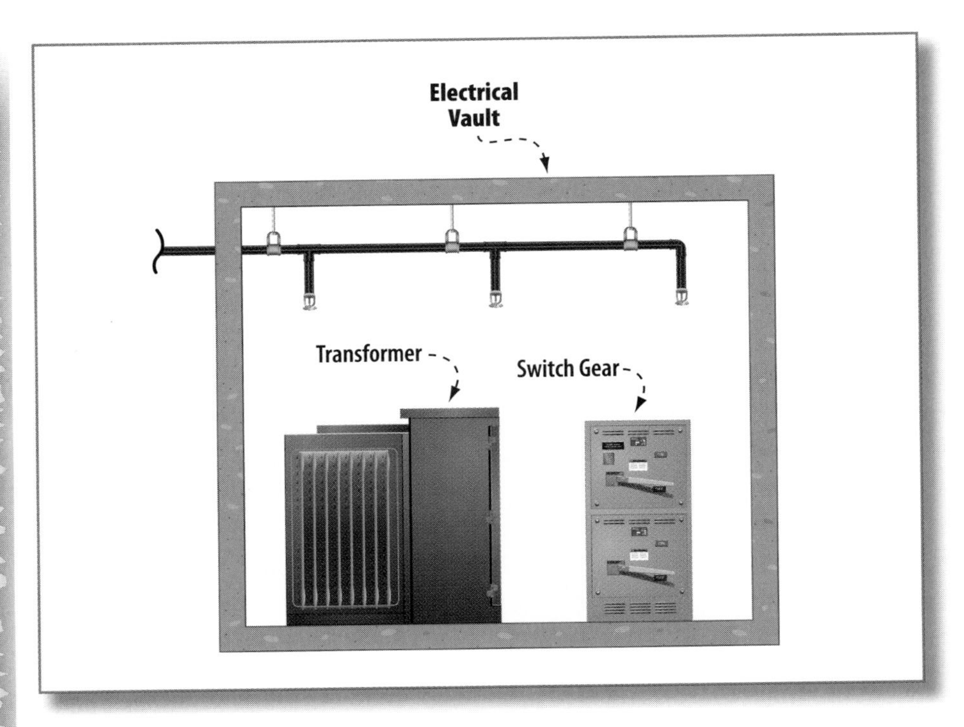

Significance of the Change

This section has been restructured and expanded in a list format to provide specific requirements for electrical vaults. The revision increases clarity and usability and provides more effective correlation with existing provisions in Article 450 that apply to transformer vaults. This section continues to address floors, roofs, walls, and locks, but it also includes provisions for transformers located in such vaults and an appropriate reference to Part III of Article 450. No additional technical requirements have been added under this revision; however the change brings more effective driving language into the rules, making the provisions more useful and enforceable. The exception to (1), (2), and (3) relaxes the construction fire rating from 3 hours to 1 hour if the vault is provided with an automatic fire-suppression system as specified in the exception.

Change Summary

The title has been changed to Electrical Vaults because it is proposed that the section cover more than just the fire resistance rating. Driving language has been added in the main paragraph to indicate that the section applies when a vault is required or specified, and the section has been split into a number of subsections as follows:

— Item (1) applies to walls and roofs and contains the requirement currently in 110.31(A). The last sentence of list item (1) has been revised as follows: "For the purpose of this section, studs and wallboard construction shall not be permitted."

— Item (2) applies to floors and contains the current provisions in 110.31(A) for floors. The sentence regarding studs and wallboard has been moved into item (1) because it would not apply to the floor.

— Item (3) has been added to apply to doors and is taken from 450.43(A).

— Item (4) has been added to specify the locking requirements for the doors on the vault. These requirements were taken from 450.43(C).

— Item (5) has been added to make it clear that any vault that is required due to the requirements of Article 450 must be constructed in accordance with Article 450, Part III. Although the language in Article 450 is similar to that in this proposal, it contains requirements for door sills and ventilation that would not be applicable in an equipment/conductor vault. As such, it makes more sense to simply defer to Article 450 where the vault includes a transformer that is required by Article 450 to be in a vault.

— Finally, a new exception to the construction requirements, taken from 450.42 and 450.43, allows for 1-hour construction when a vault is protected by a fire suppression system. If a transformer vault can be reduced to 1 hour by adding fire suppression, having similar permission for a general electrical vault would be acceptable since the transformer fires are likely more severe than what would occur in an equipment room without a transformer.

Two new informational notes have been added that parallel the existing informational notes in 450.42 and 450.53. New Informational Note No. 1 is a combination of the informational notes from 450.42 and 450.53. The new Informational Note No. 2 is taken from 450.42 Informational Note No. 2.

Comments: 1-141, 1-142

Proposal: 1-252

Chapter 2

Articles 200-285

Wiring and Protection

Significant Changes
TO THE NEC® 2011

200.2(A) & (B)

Article 200 Use and Identification of Grounded Conductors

General

Code Language

200.2 General. Grounded conductors shall comply with 200.2(A) and (B).

(A) Installation. *(No changes)*

(B) Continuity. The continuity of a grounded conductor shall not depend on a connection to a metallic enclosure, raceway, or cable armor.

Informational Note: See 300.13(B) for the continuity of grounded conductors used in multiwire branch circuits.

Change Summary

Section 200.2 has been revised by removing all but the last sentence. A new informational note has been added following 200.2(B).

Comment: 5-25

Proposals: 5-31, 5-32

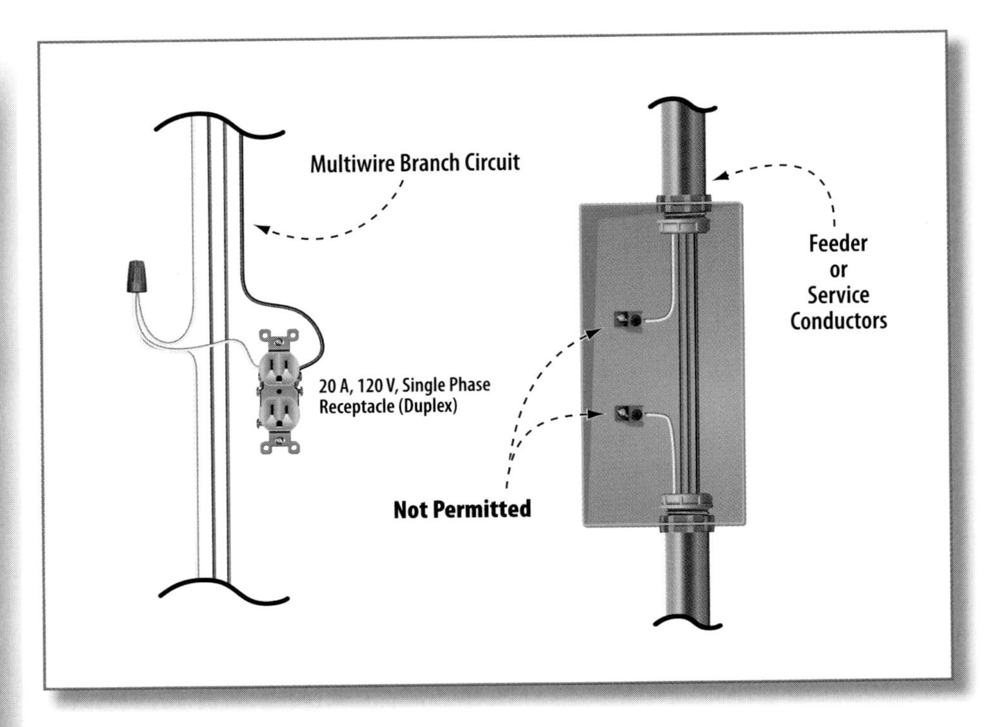

Significance of the Change

Section 200.2 has been revised to introduce the requirements in new subdivisions (A) and (B) and to remove the references to multiple *Code* sections that exempted circuits including a grounded conductor. Referencing these sections is unnecessary, as these exemptions are dealt with specifically within those rules. The requirement for inclusion of a grounded conductor with ungrounded conductors of a service or separately derived system is dealt with sufficiently in 250.24(C) and 250.30. Section 200.2(B) was incorporated into Article 200 in the 2008 *NEC* development process. Subdivision 200.2(B) requires that the continuity of any grounded conductor not depend on a connection to conductive enclosures, raceways, or cable armor.

The new informational note following 200.2(B) provides useful correlation and reference to a requirement for grounded conductor continuity as provided in 300.13(B). Although using a wiring device such as a receptacle to maintain continuity of an ungrounded circuit conductor is not prohibited, a neutral (grounded conductor) of a multiwire branch circuit is not permitted for this use, as restricted by 300.13(B). This new informational note provides users with clear direction to provisions that address grounded conductor continuity requirements specifically at the branch circuit outlet level of the system.

Neutral Conductors

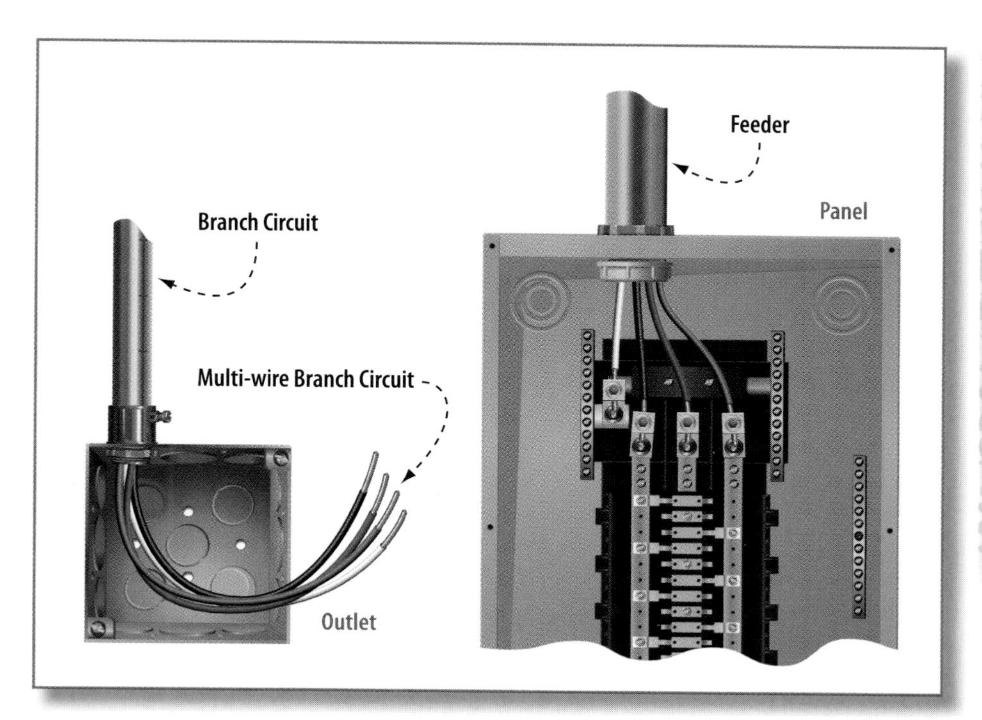

Code Language

200.4 Neutral Conductors. Neutral conductors shall not be used for more than one branch circuit, for more than one multiwire branch circuit, or for more than one set of ungrounded feeder conductors unless specifically permitted elsewhere in this *Code*.

Significance of the Change

Section 200.4 is new and now prohibits the use of a neutral conductor with more than one branch circuit, more than one multiwire branch circuit, or more than one set of ungrounded feeder conductors unless specifically permitted by other provisions in the *NEC*. Two examples of rules addressing feeders using a shared neutral conductor can be found in 215.4 and 225.7(B). Substantiation for the revision indicated that prohibited use of common neutrals was not addressed directly, that because this practice was permitted in some provisions of the *Code* but not in others meant that it was restricted to the provisions of just those sections.

This new rule generally prohibits using a common neutral for more than a single branch circuit, multiwire branch circuit, or for more than a single set of ungrounded feeder conductors. Its language clarifies that the *NEC* generally prohibits this type of shared neutral use, except for specific installations that are covered in the *Code*, and, importantly, it complements the definitions of *neutral conductor* and *neutral point*, introduced in the 2008 *NEC* cycle.

Change Summary

A new Section 200.4 has been added to Article 200. This new text clarifies that in general, neutral conductors are not permitted to be used for more than one branch circuit, for more than one multiwire branch circuit, or for more than one set of ungrounded feeder conductors unless specifically permitted elsewhere in this *Code*.

Comment: 5-33
Proposal: 5-49

200.6(A) & (D)

Article 200 Use and Identification of Grounded Conductors

Means of Identifying Grounded Conductors

Code Language

200.6 Means of Identifying Grounded Conductors.

(A) Sizes 6 AWG or Smaller. An insulated grounded conductor of 6 AWG or smaller shall be identified by one of the following means:

(1) A continuous white outer finish.

(2) A continuous gray outer finish.

(3) Three continuous white stripes along the conductor's entire length on other than green insulation.

(4) Wires that have their outer covering finished to show a white or gray color but have colored tracer threads in the braid identifying the source of manufacture shall be considered as meeting the provisions of this section.

(D) Grounded Conductors of Different Systems. ***(Revised last sentence as follows):*** The means of identification shall be documented in a manner that is readily available or shall be permanently posted where the conductors of different systems originate.

Change Summary

Section 200.6(A) has been restructured to improve clarity and increase usability. The last sentence of 200.6(D) has been revised.

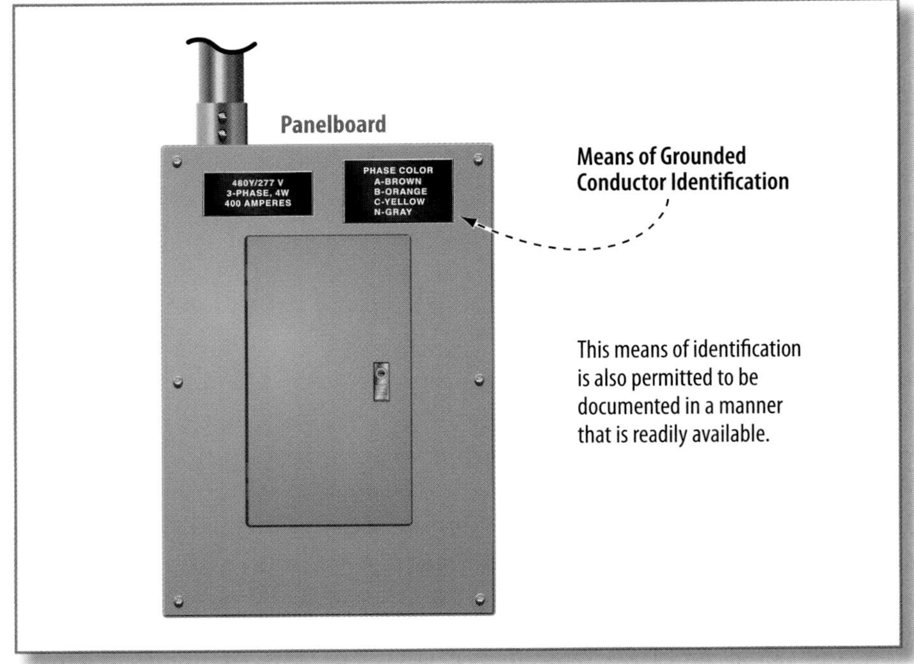

Significance of the Change

This revision introduces no additional technical requirements, but instead provides another alternative for documentation. Subdivision (A) now provides acceptable current identification methods in clear list format, in accordance with 3.3.2 of the *NEC Style Manual.* This change enhances usability while maintaining the current provisions of this section. Subdivision (D) was revised to address documentation of grounded conductor identification required in this section, in addition to promoting a more consistent correlation with identification methods already provided in 210.5(C) and 215.12(C).

The substantiation clearly indicated that the means of identification for grounded conductors must also be provided at "feeder" panelboards and switchboards in addition to "branch-circuit" panelboards as previously worded. The means of posting identification now applies to equipment where the branch circuit originates and includes readily available documentation as an alternative identification method. Readily available documentation for identification of conductors is achievable by industrial facilities and other facilities having and employing controlled conditions of use and operation.

Comment: None

Proposals: 5-34, 5-51

210.4(B)

Article 210 Branch Circuits
Part I General Provisions

Disconnecting Means

Code Language

210.4 Multiwire Branch Circuits.

(B) Disconnecting Means. Each multiwire branch circuit shall be provided with a means that will simultaneously disconnect all ungrounded conductors at the point where the branch circuit originates.

Informational Note: See 240.15(B) for information on the use of single pole circuit breakers as the disconnecting means.

Change Summary

This new informational note refers the *Code* user to the significant requirements for the application of identified handle ties on individual single-pole circuit breakers in Article 240.

Significance of the Change

This informational note provides the *Code* user with essential information as follows:

— 240.15(B)(1) permits the use of single-pole circuit breakers with an identified handle tie where only single-phase line-to-neutral loads are served.

— 240.15(B)(2) permits individual single-pole circuit breakers with identified handle ties to serve line-to-line loads, provided the circuit breakers are rated at 120/240 V AC.

— 240.15(B)(3) addresses two- and three-phase systems and permits the use of individual single-pole circuit breakers with identified handle ties where the circuit breakers are rated at 120/240 V AC and the system has a grounded neutral point and the voltage to ground does not exceed 120 volts.

A "means to simultaneously disconnect" does not require a common trip-type circuit breaker. A common trip circuit breaker is permitted in the preceding applications but is not required. Only the systems and ratings listed in 240.15(B) may utilize identified handle ties. For example, two individual single-pole circuit breakers rated 277/480 V would not be permitted to have a handle tie in any installation. A two-pole common trip 277/480 V circuit breaker would be required instead.

Comment: 2-17
Proposal: 2-37

210.5(C)

Article 210 Branch Circuits
Part I General Provisions

Identification of Ungrounded Conductors

Code Language

210.5 Identification for Branch Circuits

(C) Identification of Ungrounded Conductors. Ungrounded conductors shall be identified in accordance with 210.5(C)(1), (2), and (3).

(1) Application. Where the premises wiring system has branch circuits supplied from more than one nominal voltage system, each ungrounded conductor of a branch circuit shall be identified by phase or line and system at all termination, connection, and splice points.

(2) Means of Identification. The means of identification shall be permitted to be by separate color coding, marking tape, tagging, or other approved means.

(3) Posting of Identification Means. The method utilized for conductors originating within each branch-circuit panelboard...***(Remainder unchanged)***

Change Summary

The requirements for the identification of ungrounded branch circuit conductors have been logically separated into three second-level subdivisions for clarity and usability.

Significance of the Change

As written in the 2008 *NEC*, 210.5(C) contained three different requirements, which this revision logically now separates. First level subdivision 210.5(C) has been retitled to clarify that the requirements contained are for "identification" purposes, and its text separated into three second-level subdivisions with titles for clarity and usability as follows:

(1) Application. This requirement mandates identification of all ungrounded branch-circuit conductors where more than one nominal voltage system is employed on the premises. If a structure is supplied at 208Y/120 volts and no other nominal voltage system exists, no marking is required.

(2) Means of Identification. Permitted methods include separate color coding, marking tape, tagging, or other approved means.

(3) Posting of Identification Means. Permanent posting at each branch-circuit panelboard or similar branch-circuit distribution equipment is the most common method, although posting may be achieved by documentation in a readily available manner.

Comment: 2-27

Proposal: 2-52

210.6(C)

Article 210 Branch Circuits
Part I General Provisions

LEDs

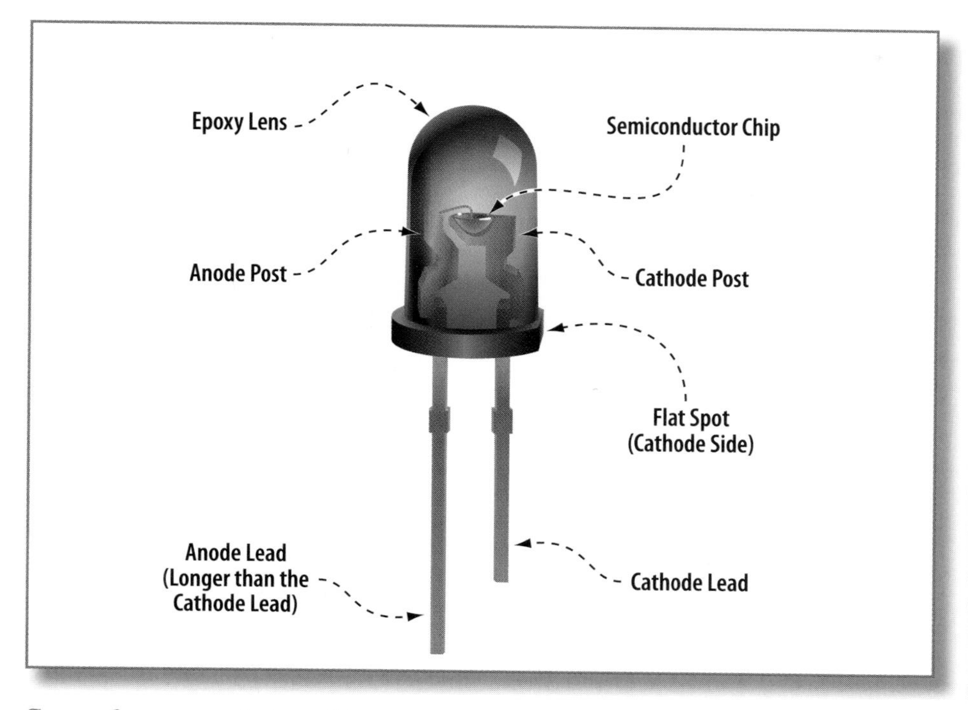

Code Language

210.6 Branch-Circuit Voltage Limitations

(C) 277 Volts to Ground. Circuits exceeding 120 volts, nominal, between conductors and not exceeding 277 volts, nominal, to ground shall be permitted to supply the following:

(1) Listed electric-discharge or listed light-emitting diode type luminaires

(List items (2) through (6) are unchanged.)

Significance of the Change

This global revision incorporates LED technology into the *NEC*. The multitude of changes includes those in Article 410 that require incorporating LEDs into existing requirements. While this technology is not entirely new, the worldwide emphasis on energy savings and "going green" is now beginning to embrace this technology.

Light-emitting diodes (LEDs) are solid-state devices that convert electric energy directly into light of a single color. These devices employ "cold" light-generation technology, which results in the production of visible spectrum in a majority of light. This technology is extremely efficient and loses no energy in the form of heat. For example, heat, the primary output of an incandescent lamp, wastes a tremendous amount of energy, whereas LEDs are extremely efficient and result in greatly reduced energy costs over their life.

Change Summary

Throughout the entire *NEC*, listed light-emitting diode-type luminaires — known more commonly as LEDs — are now recognized. This new lighting technology has been incorporated into existing Code requirements addressing luminaires.

Comment: None

Proposal: 2-58

210.8

Article 210 Branch Circuits
Part I General Provisions

Ground-Fault Circuit-Interrupter Protection for Personnel

Code Language

210.8 Ground-Fault Circuit-Interrupter Protection for Personnel. Ground-fault circuit-interruption for personnel shall be provided as required in 210.8(A) through (C). The ground-fault circuit-interrupter shall be installed in a readily accessible location.

Informational Note: See 215.9 for ground-fault circuit-interrupter protection for personnel on feeders.

(A) Dwelling Units.

(B) Other Than Dwelling Units.

(C) Boat Hoists.

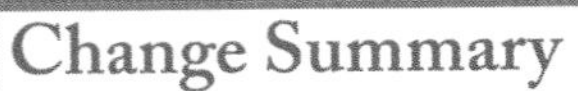

Change Summary

New parent text has been added to mandate that all ground-fault circuit-interrupter devices (GFCIs) installed to comply with the requirements of 210.8 must be located in a readily accessible location.

Comment: 2-29

Proposal: 2-77

Significance of the Change

Two sentences have been added to the text of 210.8, the first to clarify that GFCI protection for personnel must be provided as required in subdivisions (A), (B) and (C). The second sentence now requires that all GFCI devices installed as required by 210.8 be "readily accessible," which is defined in Article 100 as follows:

Accessible, Readily (Readily Accessible). Capable of being reached quickly for operation, renewal, or inspections without requiring those to whom ready access is requisite to climb over or remove obstacles or to resort to portable ladders, and so forth.

This new requirement mandates that installers and inspectors locate GFCI devices in spaces that will remain "readily accessible." GFCI devices installed on kitchen countertops and in bathrooms within 3 feet of the basin must remain "readily accessible" throughout the life of the structure. However, where GFCI devices are required in garages, in boathouses, outdoors, or in unfinished basements consideration must now be given to the location of the device with respect to its remaining "readily accessible."

All occupancies are transient to some degree, in that a change of ownership will occur. Thus, placement of GFCI devices required by 210.8 must be in spaces that will remain, to a reasonable degree of certainty, "readily accessible."

210.8(B)(6) & (7)

Article 210 Branch Circuits, Part I General Provisions

Indoor Wet Locations and Locker Rooms with Associated Showering Facilities

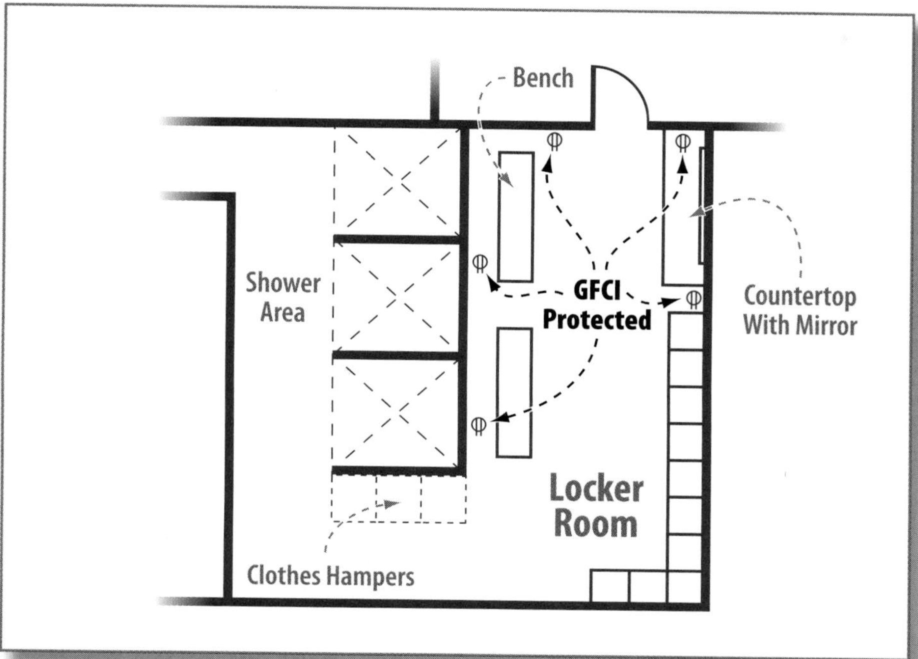

Code Language

210.8 Ground-Fault Circuit-Interrupter Protection for Personnel.

(B) Other Than Dwelling Units. All 125-volt, single-phase, 15- and 20-ampere receptacles installed in the locations specified in 210.8(B)(1) through (8) shall have ground-fault circuit-interrupter protection for personnel.

...

(6) Indoor wet locations

(7) Locker rooms with associated showering facilities

Change Summary

The safety-driven GFCI requirements for other than dwelling units have been expanded to include indoor wet locations and locker rooms with adjacent showering facilities.

Significance of the Change

New list item (6), "indoor wet locations," requires GFCI protection for all 125-volt, single-phase, 15- and 20-ampere receptacles installed in areas such as, but not limited to, car washes, food-processing areas of facilities that make food products, meat packing plants, and all other indoor wet areas that are washed down daily.

The addition of list item (7) requires that all 125-volt, single-phase, 15- and 20-ampere receptacles installed in "locker rooms with associated showering facilities" be GFCI protected. Note that no distance limitation applies — this requirement applies to all locker rooms regardless of size. For example, for a very large locker room for a college or high school sports team, with lockers in only a portion of the room but with associated showering facilities, GFCI protection of receptacle outlets is required by 210.8(B)(7).

Extending GFCI protection into areas other than dwelling units where persons are exposed to wet environments and will utilize power tools or appliances such as hair dryers is prudent. Installers and designing engineers have provided GFCI protection in these areas for many years, but GFCI protection for these areas is now required.

Comment: 2-52

Proposals: 2-105, 2-110

Garages, Service Bays, and Similar Areas

Code Language

210.8 Ground-Fault Circuit-Interrupter Protection for Personnel.

(B) Other Than Dwelling Units. All 125-volt, single-phase, 15- and 20-ampere receptacles installed in the locations specified in 210.8(B)(1) through (8) shall have ground-fault circuit-interrupter protection for personnel.

(8) Garages, service bays, and similar areas where electrical diagnostic equipment, electrical hand tools, or portable lighting equipment are to be used.

Change Summary

The safety driven GFCI requirements for other than dwelling units has been expanded to include "garages, service bays, and similar areas." This revision is a significant expansion of the present GFCI requirement for similar areas in 511.12.

Significance of the Change

All 125-volt, single-phase, 15- and 20-ampere receptacles installed in "garages, service bays, and similar areas where electrical diagnostic equipment, electrical hand tools, or portable lighting equipment are to be used" are now required to be GFCI protected.

The present GFCI requirement for these areas exists in Article 511, Commercial Garages, Repair and Storage. However, the application of Article 511 is limited in 511.1 Scope to areas "...in which volatile flammable liquids or flammable gases are used for fuel or power." This eliminates areas that will service, repair, or store diesel, propane, electric, or other types of vehicles. This revision is limited in a similar fashion as 511.12 to "...areas where electrical diagnostic equipment, electrical hand tools, or portable lighting equipment are to be used..." A parking garage would not require GFCI protection because a receptacle there would not be intended for areas where electrical diagnostic equipment, electrical hand tools, or portable lighting equipment are to be used.

This new requirement applies regardless of the type of fuel being used or the task being performed.

Comment: 2-45

Proposal: 2-122

210.12(A) (Exc. Nos.1 & 3)

Article 210 Branch Circuits
Part I General Provisions

Arc-Fault Circuit-Interrupter Protection

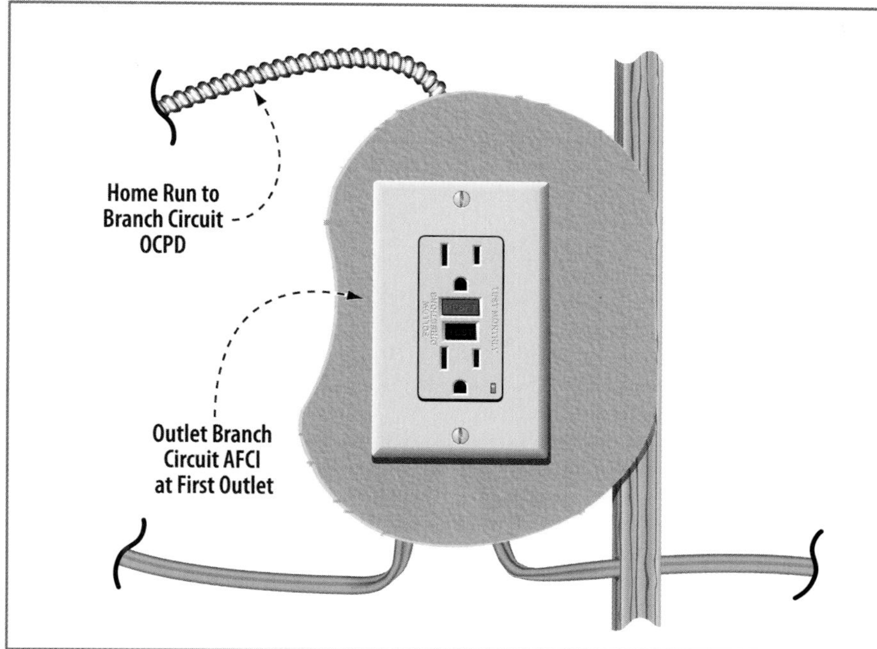

Code Language

210.12 Arc-Fault Circuit-Interrupter Protection.

(A) Dwelling Units. All 120-volt, single-phase, 15- and 20-ampere branch circuits supplying outlets installed in dwelling unit family rooms, dining rooms, living rooms, parlors, libraries, dens, bedrooms, sunrooms, recreation rooms, closets, hallways, or similar rooms or areas shall be protected by a listed arc-fault circuit interrupter, combination-type, installed to provide protection of the branch circuit.

(No change to Informational Notes)

Exception No. 1: If RMC, IMC, EMT, Type MC, or steel armored Type AC cables meeting the requirements of 250.118 and metal outlet and junction boxes are installed for the portion of the branch circuit between the branch-circuit overcurrent device and the first outlet, it shall be permitted to install an outlet branch circuit type AFCI, at the first outlet to provide protection for the remaining portion of the branch circuit.

Exception No. 3. Where an individual branch circuit to a fire alarm system......Type MC now permitted...

Significance of the Change

Section 210.12 has been editorially modified by relocating the definition of *Arc-Fault Circuit Interrupter* to Article 100, in accordance with the *NEC Style Manual* requirement that where a defined term is located in more than one article (AFCI is used in several articles), it be defined in Article 100. The text previously located in 210.12(B) has been relocated to 210.12(A).

Exception No. 1 has been revised to recognize the outlet branch circuit Type AFCI, which, in essence, is an "AFCI receptacle." These devices provide both parallel and series protection downstream and series protection upstream. This exception permits the use of an outlet branch circuit AFCI at the first outlet where specified wiring methods and metal outlet and junction boxes are used for the portion of the branch circuit between the branch-circuit overcurrent device and the first outlet. Type MC cable has been added to the list of permitted wiring methods along with RMC, IMC, EMT, or steel armored Type AC cables meeting the requirements of 250.118.

Type MC cable has also been added to the list of permitted wiring methods where Exception No. 3 (note that in the 2008 *NEC* this was Exception No.2) is employed. Exception No. 3 now applies only where a fire alarm system is supplied with an individual branch circuit.

Change Summary

The definition of *Arc-Fault Circuit Interrupter* has been relocated to Article 100. The use of Type MC cable is now permitted where Exceptions No.1 and No. 3 are applied. Exception No.1 has been modified to recognize the use of an outlet branch circuit Type AFCI at the first outlet. Exception No. 3 now requires an individual branch circuit.

Comment: 2-78
Proposals: 2-156, 162, 182

Conduit/Tubing in Concrete

Code Language

210.12 Arc-Fault Circuit-Interrupter Protection.

(A) Dwelling Units. All 120-volt, single phase, 15- and 20-ampere branch circuits supplying outlets installed in dwelling unit family rooms, dining rooms, living rooms, parlors, libraries, dens, bedrooms, sunrooms, recreation rooms, closets, hallways, or similar rooms or areas shall be protected by a listed arc-fault circuit interrupter, combination-type, installed to provide protection of the branch circuit.

(No change to Informational Notes No. 1, 2 and 3.)

Exception No. 2: Where a listed metal or nonmetallic conduit or tubing is encased in not less than 50 mm (2 in.) of concrete for the portion of the branch circuit between the branch-circuit overcurrent device and the first outlet, it shall be permitted to install an outlet branch circuit type AFCI at the first outlet to provide protection for the remaining portion of the branch circuit.

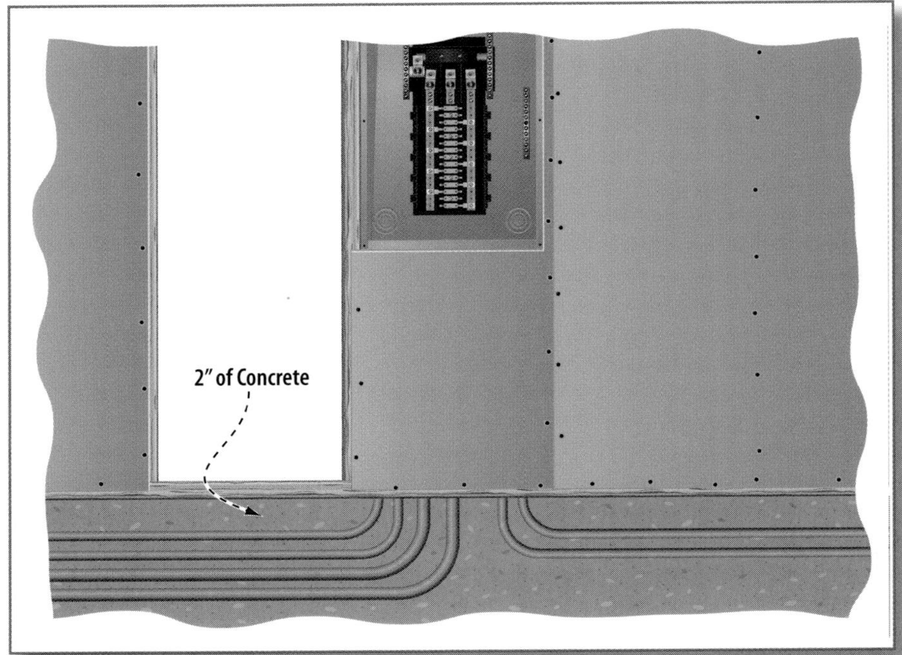

Change Summary

A new exception has been added to 210.12 to permit the use of an outlet branch circuit type AFCI at the first outlet where listed metal or nonmetallic conduit or tubing is encased in not less 2 inches of concrete.

Significance of the Change

This new exception mirrors the permissive requirements of existing Exception No. 1 to 210.12(A). This exception recognizes that, where raceways are enclosed in concrete, the branch circuit conductors within them from the branch-circuit overcurrent device to the first outlet do not require AFCI protection. This exception will permit the use of an outlet branch circuit type AFCI at the first outlet instead of a combination-type AFCI circuit breaker in the panelboard. An "outlet branch circuit type AFCI" is in essence an "AFCI receptacle." These devices provide both parallel and series protection downstream and series protection upstream. This exception requires the use of type EMT, PVC, ENT, or other listed metal or nonmetallic conduit or tubing installed in not less than 2 inches of concrete for the portion of the branch circuit between the branch-circuit overcurrent device and the first outlet.

Comment: 2-64

Proposals: 2-156, 162, 182

210.12(B)

Article 210 Branch Circuits
Part I General Provisions

Branch Circuit Extensions or Modifications — Dwelling Units

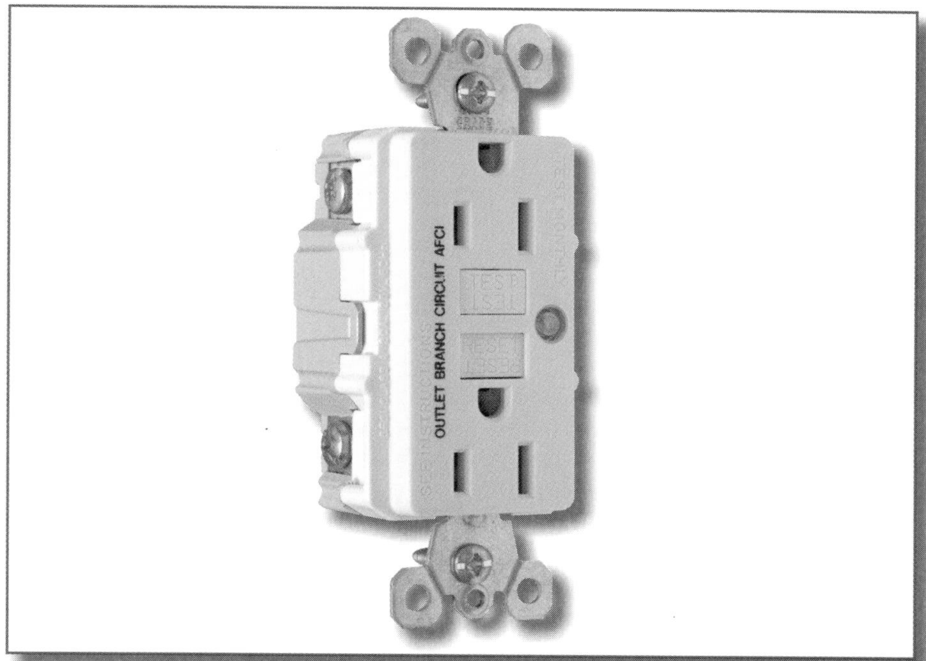

Prototype Courtesy of Pass and Seymour Legrand

Code Language

210.12 Arc-Fault Circuit-Interrupter Protection for Dwelling Units.

(B) Branch Circuit Extensions or Modifications — Dwelling Units. In any of the areas specified in 210.12(A), where branch circuit wiring is modified, replaced, or extended, the branch circuit shall be protected by one of the following:

(1) A listed combination-type AFCI located at the origin of the branch circuit

(2) A listed outlet branch-circuit type AFCI located at the first receptacle outlet of the existing branch circuit

Change Summary

New AFCI requirements have been added to 210.12(B) to address branch-circuit modifications, replacements, or extensions. A listed outlet branch-circuit type AFCI is now permitted to be installed at the first receptacle outlet of an existing branch circuit to protect any modifications, replacements, or extensions.

Significance of the Change

This revision is extremely significant, in that we now have clarity with respect to the applicability of 210.12 for branch-circuit modifications, replacements, and extensions. This revision requires the use of either a listed combination AFCI circuit breaker at the origin of the branch circuit or a listed outlet branch circuit type AFCI at the first receptacle outlet of the existing branch circuit. An "outlet branch circuit AFCI" is in essence an "AFCI receptacle." These devices provide both parallel and series protection downstream and series protection upstream.

In many areas of the country, municipalities and AHJs have struggled with AFCI requirements for branch-circuit modifications, replacements, or extensions and service upgrades. Many AHJs have cited technical committee statements that claimed it was not their intent to include existing branch circuits or service upgrades in this requirement. However, installers and inspectors are bound by the printed text in the *NEC*, wherein previous editions did not exclude branch circuit modifications, replacements, or extensions in existing dwelling units. The text of 210.12 in the 2008 and earlier editions of the *NEC* literally applied to all branch-circuit installations, modifications, replacements, or extensions. There is no qualifier in Article 90 or in Article 210 informing the *Code* user that the provisions of 210.12 apply only to new construction.

Comments: 2-63, 64
Proposals: 2-156, 162, 182

210.52(A)(2)

Article 210 Branch Circuits
Part III Required Outlets

Wall Space

Code Language

210.52 Dwelling Unit Receptacle Outlets.

(A) General Provisions

(2) Wall Space. As used in this section, a wall space shall include the following:

(1) Any space 600 mm (2 ft) or more in width (including space measured around corners) and unbroken along the floor line by doorways and similar openings, fireplaces, and fixed cabinets.

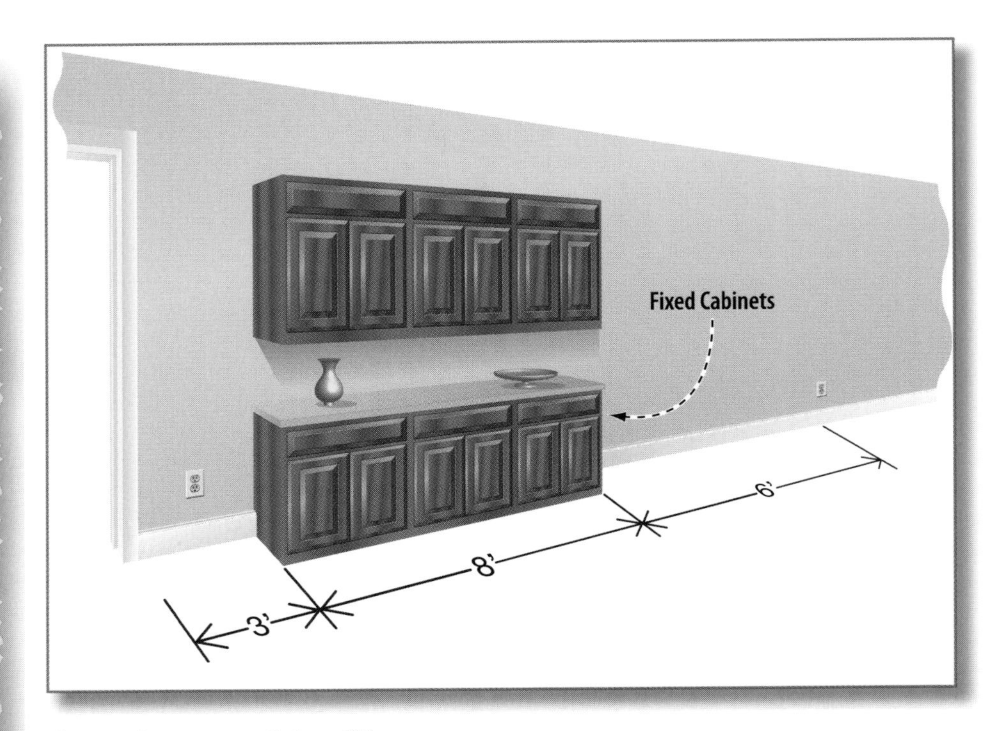

Change Summary

New text has been added to 210.52(A)(1) to clarify that where fixed cabinets are installed, that portion of the wall space is excluded from this receptacle outlet requirement. However the requirements of 210.52(C) may supercede this requirement.

Significance of the Change

This revision is the result of the addition of a new 210.52(A)(4) that added text to clarify that receptacles installed for countertop surfaces as specified in 210.52(C) shall not be considered as the receptacles required by 210.52(A).

The general requirement for the location of receptacles, located in 210.52(A)(1), requires that they be installed such that no point measured horizontally along the floor line of any wall space is more than 1.8 m (6 ft) from a receptacle outlet. To clarify where this rule applies, 210.52(A)(2) addresses wall spaces that fall under the rule in 210.52(A)(1). This requirement includes any space 2 feet or more in width, including space measured around corners unbroken along the floor line by doorways and similar openings, fireplaces, and fixed cabinets. The addition of "fixed cabinets" clearly removes these spaces from the requirements of 210.52(A)(1). This revision was intended to address a situation where an installer attempted to count a kitchen counter receptacle installed per 210.52(C) as a receptacle required by 210.52(A)(1). This revision will however address all "fixed cabinets."

Comment: 2-120

Proposal: 2-228

210.52(A)(4)

Article 210 Branch Circuits
Part III Required Outlets

Countertop Receptacles

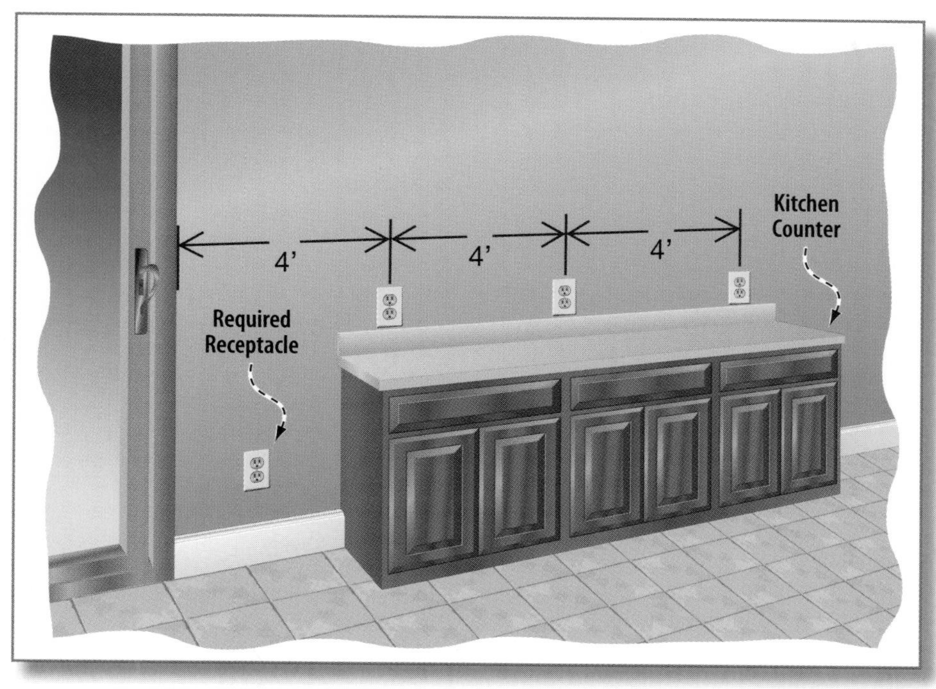

Code Language

210.52 Dwelling Unit Receptacle Outlets.

(A) General Provisions

(4) Countertop Receptacles. Receptacles installed for countertop surfaces as specified in 210.52(C) shall not be considered as the receptacles required by 210.52(A).

Significance of the Change

Location requirements for dwelling unit receptacles are covered in 210.52 by nine first-level subdivisions, each dedicated to specific requirements in the dwelling unit. Section 210.52(A) governs receptacle location in all general areas, including every kitchen, family room, dining room, living room, parlor, library, den, sunroom, bedroom, recreation room, or similar spaces. The general requirement is that a receptacle be installed such that no point measured horizontally along the floor line of any wall space is more than 1.8 m (6 ft) from a receptacle outlet.

This revision clearly prohibits the installer from including as part of that measurement a receptacle installed in accordance with 210.52(C). For example, consider a receptacle outlet mounted at the end of a kitchen countertop. At the end of the countertop the wall continues for 3 feet and ends at a doorway. A physical measurement from the counter-mounted receptacle would seem to satisfy the general 6-foot requirement given that the measurement from the receptacle to the door frame is 4 feet. However, using a counter-mounted receptacle to comply with the general rule in 210.52(A) is now clearly prohibited. A receptacle is now required in the wall space next to the counter to comply with 210.52(A).

Change Summary

Receptacles installed to serve countertops in kitchens, pantries, breakfast rooms, dining rooms, and similar areas of dwelling units as required by 210.52(C) may not be counted as a receptacle when applying the rule in 210.52(A) for general wall spaces.

Comment: None

Proposal: 2-228

210.52(C)(5)

Article 210 Branch Circuits
Part III Required Outlets

Receptacles On or In Countertops

Code Language

210.52 Dwelling Unit Receptacle Outlets

(C) Countertops

(5) Receptacle Outlet Location. Receptacle outlets shall be located on, or above, but not more than 500 mm (20 in.) above, the countertop. Receptacle outlet assemblies listed for the application shall be permitted to be installed in countertops. Receptacle outlets rendered not readily accessible by appliances fastened in place, appliance garages, sinks, or rangetops as covered in 210.52(C)(1), Exception, or appliances occupying dedicated space shall not be considered as these required outlets.

Informational Note: See 406.5(E) for requirements for installation of receptacles in countertops.

Change Summary

This revision will now permit a countertop receptacle to be installed on the countertop. Receptacle outlet assemblies that are listed for the application are now permitted to be installed in countertops. A new informational note has been added to reference 406.5(E).

Courtesy of Thomas and Betts

Significance of the Change

The previous text of 210.52(C)(5) only permitted the required receptacles to be installed "above, but not more than 500 mm (20 in.) above, the countertop," thus eliminating the possibility of using a "tombstone" or other style assembly placed in or on the countertop surface. The reason for this limitation is water on the kitchen countertop. This revision now permits "receptacle outlet assemblies listed for the application" to be installed in countertops. This means that installers cannot design or build their own installation of a receptacle in or on a countertop. A listed assembly is now required.

Several types of receptacle outlet assemblies listed for use on kitchen countertops are available on the market today. This revision now permits the development and listing of new receptacle assemblies for kitchen countertops, including the tombstone, the pop-up, and other styles.

Comment: 2-132

Proposal: 2-253

210.52(D)

Article 210 Branch Circuits
Part III Required Outlets

Bathrooms

Code Language

210.52 Dwelling Unit Receptacle Outlets

(D) Bathrooms. In dwelling units, at least one receptacle outlet shall be installed in bathrooms within 900 mm (3 ft) of the outside edge of each basin. The receptacle outlet shall be located on a wall or partition that is adjacent to the basin or basin countertop, located on the countertop, or installed on the side or face of the basin cabinet not more than 300 mm (12 in.) below the countertop. Receptacle outlet assemblies listed for the application shall be permitted to be installed in the countertop.

Informational Note: See 406.5 for requirements for installation of receptacles in countertops.

Significance of the Change

The general rule of 210.52(D) requires that all dwelling unit bathrooms have a receptacle outlet installed not more than 3 feet from the outside edge of each basin. The 2008 *NEC* requirement limited the location of the required receptacle outlet to either of the following:

(1) A wall or partition adjacent to the basin or basin countertop

(2) The side or face of the basin cabinet not more than 12 inches below the countertop

The permitted location of the bathroom receptacle outlet has been expanded, so that it is now permissible to locate the receptacle on or in the countertop, provided that the receptacle outlet assemblies used are listed for the application. Installation can be achieved through the use of a tombstone-type device or other means that would prevent water on the countertop from entering the receptacle. Other methods may include items similar to an appliance garage to contain the receptacle as well as to store hair dryers and other electrical appliances used in the bathroom.

Change Summary

The permitted location of receptacles in dwelling unit bathrooms has been expanded to include a tombstone type of receptacle or other method mounted directly on or in the basin countertop provided the receptacle outlet assemblies are listed for the application.

Comment: 2-135
Proposal: 2-258

210.52(E)(3)

Article 210 Branch Circuits
Part III Required Outlets

Balconies, Decks, and Porches

Code Language

210.52 Dwelling Unit Receptacle Outlets

(E) Outdoor Outlets

(3) Balconies, Decks, and Porches. Balconies, decks, and porches that are accessible from inside the dwelling unit shall have at least one receptacle outlet installed within the perimeter of the balcony, deck, or porch. The receptacle shall not be located more than 2.0 m (6 ½ ft) above the balcony, deck, or porch surface.

The exception for small balconies (smaller than 20 square feet) is deleted.

Change Summary

All balconies, decks, and porches that are accessible from inside a dwelling unit are now required to have at least one receptacle outlet installed within the perimeter of the balcony, deck, or porch. This rule now applies to all balconies, decks, and porches, regardless of size.

Comment: None

Proposal: 2-266

Significance of the Change

The 2008 *NEC* included a new requirement to mandate that all balconies, decks, and porches accessible from inside a dwelling unit have at least one receptacle outlet installed within the perimeter of the balcony, deck, or porch at a height of not more than 6 ½ feet above the surface of these structures. However an exception was included for such structures with a usable area of less than 20 square feet.

The exception has been deleted to recognize that where access is available from inside a dwelling, occupants often find a reason or need to supply electrical equipment or holiday lighting on balconies, decks, or porches. The purpose of this requirement is to eliminate the use of cords passing through doorways to supply electrical equipment or holiday lighting at these locations and to avoid the potential for electric shock and fire due to damage that can easily occur to a cord where it passes through a doorway. In virtually all situations where a cord is used in this manner to supply holiday lighting, a radio, or other equipment on the balcony, deck, or porch, GFCI protection is not provided because the cord is supplied from the nearest receptacle outlet in the dwelling unit. The required outlets in this second-level subdivision are required by 210.8(A) to be GFCI protected.

210.52(G)

Basements, Garages, and Accessory Buildings

Code Language

210.52 Dwelling Unit Receptacle Outlets

(G) Basements, Garages, and Accessory Buildings. For a one-family dwelling, the following provisions shall apply:

(1) At least one receptacle outlet, in addition to those for specific equipment, shall be installed in each basement, in each attached garage, and in each detached garage or accessory building with electric power.

(2) ***(No Change)***

Significance of the Change

This expansion of 210.52(G) now requires at least one receptacle outlet, in addition to those for specific equipment, in all accessory buildings associated with a one-family dwelling unit.

The 2008 *NEC* requirement was limited to garages and unfinished basements, yet the same electrical hazards exist in sheds, greenhouses, pool houses, pole barns, and other accessory structures. When a general-purpose receptacle outlet is not installed, homeowners will do whatever it takes to obtain a source of supply for power tools and other electrical appliances in a structure. Or the home owner may resort to installing a device into an incandescent lampholder to gain such access. The required outlets in this first-level subdivision are required by 210.8(A) to be GFCI protected.

Change Summary

All accessory buildings for a one-family dwelling unit such as sheds, greenhouses, pool houses, or pole barns that are supplied with electricity are now required to have at least one receptacle outlet, in addition to those for the specific equipment installed.

Comment: None

Proposal: 2-270

210.52(I)

Article 210 Branch Circuits
Part III Required Outlets

Foyers

Code Language

210.52 Dwelling Unit Receptacle Outlets

(I) Foyers. Foyers that are not part of a hallway in accordance with 210.52(H) and that have an area that is greater than 5.6 m² (60 ft²) shall have a receptacle(s) located in each wall space 900 mm (3 ft) or more in width and unbroken by doorways, floor-to-ceiling windows, and similar openings.

Change Summary

Dwelling unit foyers 60 square feet or larger that are not part of a hallway will now be required to have a receptacle outlet installed in all wall spaces over 3 feet in length.

Comments: 2-114, 119

Proposal: 2-223

Significance of the Change

Where a large foyer is not connected to a hallway, this new first-level subdivision requires that a receptacle be installed in any foyer spaces 3 feet or more in width (including space measured around corners) and unbroken along the floor line by doorways, floor-to-ceiling windows, or other similar openings. Many large foyers contain furniture with electrical appliances and/or lamps. This new requirement is intended to provide receptacles for the home owner and to eliminate the use of extension cords.

Receptacle outlets now required in dwelling units are listed in 210.52. Foyers were not previously addressed because they were considered to be a hallway and were covered by 210.52(H). This new requirement specifically addresses foyers greater than 60 square feet in size, whereas the previous requirement mandated that a space 10 feet or more in length required a single receptacle.

Section 210.52(I) treats a foyer as a separate space, provided it is not connected to a hallway, whereas the previous requirement was limited to a single receptacle. All large dwelling unit foyers not connected to a hallway must now be treated individually, and all wall spaces 3 feet or more in width require the installation of a receptacle.

220.43(B)

Article 220 Branch Circuit, Feeder, and Service Calculations
Part III Feeder and Service Load Calculations

Track Lighting

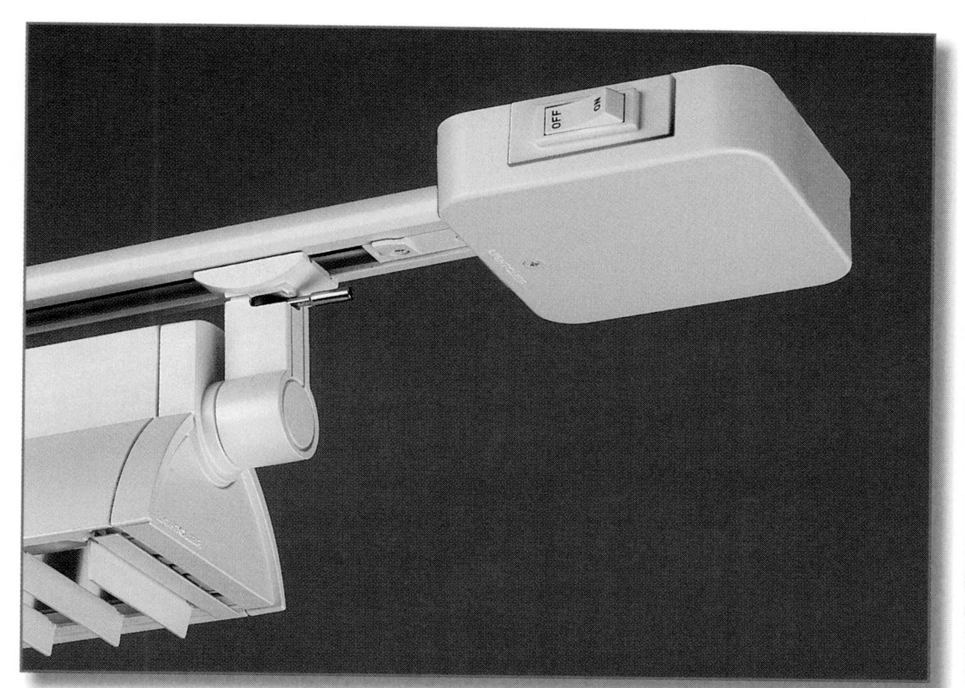

Courtesy of Koninklijke Philips Electronics N. V-Lightolier

Code Language

220.43 Show-Window and Track Lighting

(B) Track Lighting. For track lighting in other than dwelling units or guest rooms or guest suites of hotels or motels, an additional load of 150 volt-amperes shall be included for every 600 mm (2 ft) of lighting track or fraction thereof. Where multicircuit track is installed, the load shall be considered to be divided equally between the track circuits.

Exception: If the track lighting is supplied through a device that limits the current to the track, the load shall be permitted to be calculated based on the rating of the device used to limit the current.

Significance of the Change

A new exception permitting an alternative method of calculating feeder/service load for track lighting recognizes existing energy codes that in some cases require much lower values than the *NEC*. This new exception now allows track lighting loads in other than dwelling units to be calculated according to the value of current limiting devices installed as required by local energy codes. These devices are installed in series with the track itself and limit the amount of current supplied to the track with a supplementary overcurrent device.

For example, if a retail store, in a state or municipality that has adopted the ASHRAE 90.1 (2001), is installing 150 feet of track, the installation would be limited to a maximum of 4500 watts. According to *NEC* 220.43(B), however, the load that must be accounted for as a demand load would be 11,250 watts. This added load could increase the size of a step-down transformer, panelboards, and service-entrance equipment. The net result of this new exception for this example results in a reduction of calculated load of 60 percent.

Change Summary

The calculation of feeder/service loads for track lighting in other than dwelling occupancies is now permitted to be based upon a the value of a current-limiting device in lieu of calculating at 150 volt-amperes per each 2 feet of track.

Comment: None

Proposal: 2-335

225.7(C)

Article 225 Outside Branch Circuits and Feeders
Part I General

277 Volts to Ground

Code Language

225.7 Lighting Equipment Installed Outdoors

(C) 277 Volts to Ground. Circuits exceeding 120 volts, nominal, between conductors and not exceeding 277 volts, nominal, to ground shall be permitted to supply luminaires for illumination of outdoor areas of industrial establishments, office buildings, schools, stores, and other commercial or public buildings.

There is no longer a prohibition to mounting luminaires supplied by 208-, 240-, or 277-volt circuits less than 900 mm (3 ft) from windows, platforms, fire escapes, and the like.

Significance of the Change

Luminaires operating at 208-, 240-, or 277-volts are now clearly permitted to be installed in any outdoor location of industrial establishments, office buildings, schools, stores, and other commercial or public buildings. The text that prohibited the installation of these luminaires less than 3 feet from windows, platforms, fire escapes, and the like has been deleted. This change is long overdue given that current luminaire construction (note that 410.6 requires that all luminaires be listed) and grounding requirements are intended to ensure that the installation does not represent a hazard. As pointed out in the substantiation for this deletion, a child or an adult could come into contact with outdoor bollard-type luminaires operating at 277 volts and yet there is no *Code* requirement to prevent such an installation.

Comment: None

Proposal: 4-25

225.27

Article 225 Outside Branch Circuits and Feeders, Part I General

Raceway Seal

Code Language

225.27 Raceway Seal. Where a raceway enters a building or structure from an underground distribution system, it shall be sealed in accordance with 300.5(G). Spare or unused raceways shall also be sealed. Sealants shall be identified for use with the cable insulation, shield, or other components.

Significance of the Change

This new requirement logically mandates that raceways containing outdoor branch circuits or feeders and spare raceways be sealed. It mirrors the existing requirement in 230.8 for raceways containing service conductors. Raceways containing outdoor branch circuits/feeders or spares for future outdoor branch circuits/feeders entering buildings or other structures are no different than services. The requirements are now the same. See 230.8.

A similar requirement exists for all underground raceways in 300.5(G), but this requirement is further qualified and applies only where "moisture may contact live parts." For example, a raceway leaving the second floor of a building and going through the first floor, then underground for some distance and back up to equipment on the third floor, would not require a seal because any water or moisture in the raceway would not contact live parts.

Another requirement, 300.7(A), addresses sealing raceways exposed to different temperatures, but it applies only where "portions of a cable, raceway, or sleeve are known to be subjected to different temperatures and where condensation is known to be a problem." This new requirement in 225.7 requires that ALL outdoor branch circuit and feeder raceways and spares entering a building or structure from an underground distribution system be sealed in accordance with 300.5(G).

Change Summary

This new section requires that all outdoor branch circuit and feeder raceways and spares be sealed to prevent condensation and the entrance of water.

Comment: 4-12

Proposal: 4-35

225.30

Article 225 Outside Branch Circuits and Feeders
Part II Buildings or Other Structures Supplied by a Feeder(s) or Branch Circuit(s)

Number of Supplies

Code Language

225.30 Number of Supplies. A building or other structure that is served by a branch circuit or feeder on the load side of a service disconnecting means shall be supplied by only one feeder or branch circuit unless permitted in 225.30(A) through (E). For the purpose of this section, a multiwire branch circuit shall be considered a single circuit.

Where a branch circuit or feeder originates in these additional buildings or other structures, only one feeder or branch circuit shall be permitted to supply power back to the original building or structure, unless permitted in 225.30(A) through (E).

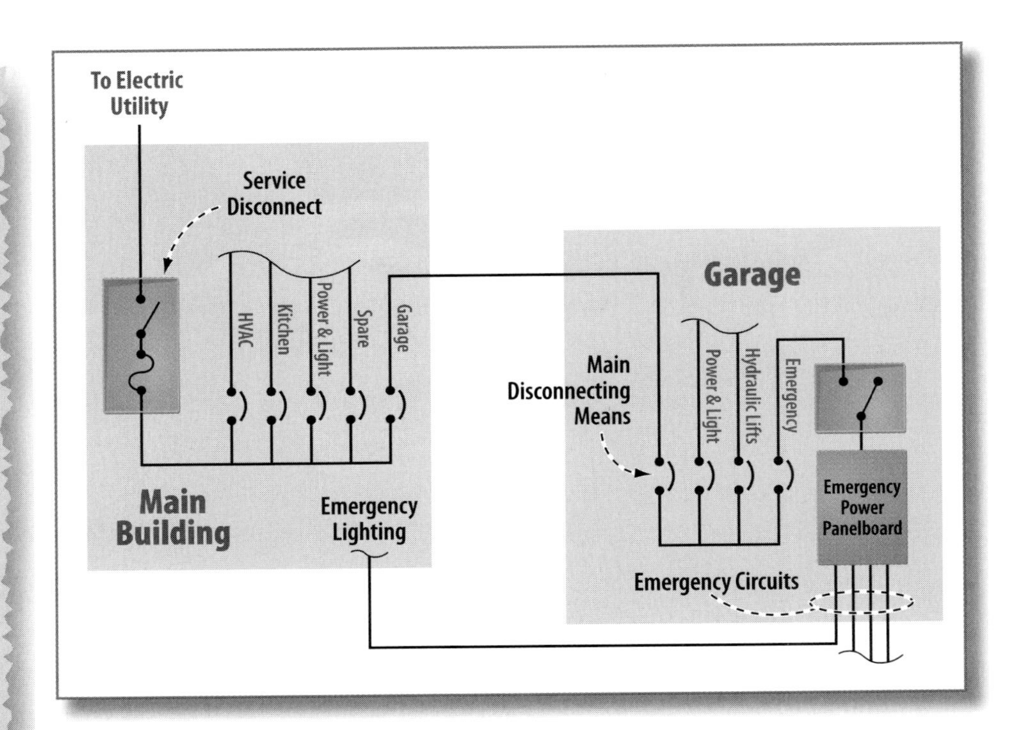

Change Summary

New text has been added to the parent text of 225.30 to address buildings and structures fed from outside feeders and branch circuits. Generally, these buildings are permitted to supply the original building or structure with only one feeder or branch circuit.

Significance of the Change

The 2008 text of 225.30 limited only the number of outside feeders or branch circuits supplying an additional building or structure where more than one building is located on the same property and under single management. The existing text, however, did not address bringing more than one feeder or branch circuit from one of these buildings or structures back to the original building. The 2008 text of 225.30 would permit an unlimited number of feeders or branch circuits to be brought back to the original building. For example, in the original building an emergency switchboard could supply a feeder to a panelboard in building No. 2 and any number of branch circuits could be fed from that emergency panelboard back to the original building.

This revision now limits (to one) the number of branch circuits or feeders in a secondary building or structure permitted to supply the original building or structure, unless permitted in 225.30(A) through (E).

Comment: 4-14

Proposal: 4-36

225.52(C) & (D)

Article 225 Outside Branch Circuits and Feeders, Part III Over 600 Volts

Locking and Indicating

Courtesy of Schneider Electric USA, Inc.

Code Language

225.52 Disconnecting Means

(C) Locking. Disconnecting means shall be capable of being locked in the open position. The provisions for locking shall remain in place with or without the lock installed.

Exception: Where an individual disconnecting means consists of fused cutouts, a suitable enclosure, capable of being locked and sized to contain all cutout fuse holders shall be installed at a convenient location to the fused cutouts.

(D) Indicating. Disconnecting means shall clearly indicate whether they are in the open "off" or closed "on" position.

Significance of the Change

These revisions are part of a task group effort to improve requirements in the *NEC* for branch circuits and feeders over 600 volts. This change is one of several in the 2011 *NEC* designed to provide the *Code* user with prescriptive requirements for high-voltage installations.

Subdivision (C) Locking has been added to require that disconnects be capable of being locked in the open position without the use of a special locking device. This will allow the installer/maintainer to effectively lock the disconnecting means in the open position with only a lock. An exception has been added for fused cutouts in an enclosure where the enclosure can be locked to prevent access to the cutouts.

New subdivision (D) Indicating mandates that all disconnects clearly indicate whether they are in the open "off" or closed "on" position.

Change Summary

Two new first-level subdivisions have been added to 225.52 to address disconnecting means over 600 volts. New 225.52(C) Locking now requires a means to lockout, and new 225.52(D) Indicating requires that a switch clearly indicate whether it is in the closed or open position.

Comment: None

Proposal: 4-21

225.52(E) & (F)

Article 225 Outside Branch Circuits and Feeders
Part III Over 600 Volts

Uniform Position and Identification

Code Language

225.52 Disconnecting Means.

(E) Uniform Position. Where disconnecting means handles are operated vertically the "up" position of the handle shall be the "on" position.

Exception: A switching device having more than one "on" position, such as a double throw switch, shall not be required to comply with this requirement.

(F) Identification. Where a building or structure has any combination of feeders, branch circuits, or services passing through it or supplying it, a permanent plaque or directory shall be installed at each feeder and branch circuit disconnect location that denotes all other services, feeders, or branch circuits supplying that building or structure or passing through that building or structure and the area served by each.

Change Summary

Two new first-level subdivisions have been added to 225.52 to address disconnecting means over 600 volts. Section 225.52(E) requires a uniform position for these disconnects, and 225.52(F) requires that a permanent plaque or directory be installed at the location of these disconnects where more than one feeder supplies or passes through a structure.

Significance of the Change

These revisions are part of a task group effort to improve requirements in the *NEC* for branch circuits and feeders over 600 volts. This change is one of several in the 2011 *NEC* designed to provide the *Code* user with prescriptive requirements for high-voltage installations.

(E) Uniform Position mandates that where handles are operated vertically, the up position of the handle be the "on" position. An exception for a double throw switch applies. In many high-voltage disconnect designs, the blades are not visible.

(F) Identification requires all disconnects over 600 volts in a building or structure that has any combination of feeders, branch circuits, or services passing through it or supplying it to identify all disconnecting means. A permanent plaque or directory is now required at each feeder and branch circuit disconnect location denoting all other sources. This new first-level subdivision mirrors the requirements in 230.2(E).

Comment: None

Proposal: 4-21

225.56

Inspections and Tests

Courtesy of Shermco Industries

Code Language

225.56 Inspections and Tests

(A) Pre-Energization and Operating Tests.

(1) Instrument Transformers.

(2) Protective Relays.

(3) Switching Circuits.

(4) Control and Signal Circuits.

(5) Metering Circuits.

(6) Acceptance Tests.

(7) Relays and Metering Utilizing Phase Differences.

(B) Test Report.

(See NEC for full text of this section.)

Change Summary

A new section is added to require pre-energization and operating tests for systems over 600 volts when first installed. Each protective, switching, and control circuit must be adjusted in accordance with the recommendations of the protective device study and tested by actual operation. A "Test Report" must be delivered to the AHJ.

Significance of the Change

Substations for outdoor branch circuits and feeders over 600 volts require the design of an overcurrent protection scheme. These protective systems consist of instrument transformers, protective relays, switching circuits, control circuits, signal circuits, metering circuits, as well as relays and metering utilizing phase differences. This type of installation requires pre-energization and operating tests to verify proper operation. The new requirement outlines the required testing for the installer and the AHJ, and an informational note refers the *Code* user to the industry standard, NETA ATS-2007, for acceptance testing.

New Section 225.56 (A) Pre-Energization and Operating Tests requires that the complete electrical system be performance tested when first installed. New Section (B) Test Report requires that the test results required by 225.56(A) be delivered to the AHJ.

Comment: 4-6

Proposal: 4-21

225.70

Article 225 Outside Branch Circuits and Feeders
Part III Over 600 Volts

Substations

Code Language

225.70 Substations

(A) Warning Signs

(1) General.

(2) Isolating Equipment.

(3) Fuse Locations.

(4) Backfeed.

(5) Metal-Enclosed and Metal-Clad Switchgear.

(See NEC for full text of this section.)

Change Summary

New requirements have been added to Article 225 to address the installation of substations on the load side of the service point.

Significance of the Change

These revisions are part of a task group effort to improve requirements in the *NEC* for branch circuits and feeders over 600 volts and installed outdoors. The text in this section is derived primarily from the *Ontario Code*, Section 36 High-Voltage Installations. Previous editions of the *NEC* provided no requirements for installers or the AHJ where a substation was installed on the load side of the service point. Additional changes to aid the *Code* user in this type of installation include a new Article 399 Outdoor, Overhead Conductors, Over 600 Volts.

This new section, 225.70, is intended to address outdoor substations for outdoor branch circuits and feeders rated over 600 volts. 225.70(A) provides signage requirements summarized as follows:

(1) *General.* General signage requirements throughout the substation.

(2) *Isolating Equipment.* Prevent opening an isolation switch under load.

(3) *Fuse Locations.* Prevent replacement of fuses while energized.

(4) *Backfeed.* Prevent injury/damage where a potential backfeed exists.

(5) *Metal Enclosed Switchgear.* Single line diagram, interlocks, isolation means, all possible sources, signs on equipment that allow access to energized parts.

Comment: None

Proposal: 4-21

Conductors Considered Outside the Building

Code Language

230.6 Conductors Considered Outside the Building. Conductors shall be considered outside of a building or other structure under any of the following conditions:

(List items (1) through (4)…No Change)

(5) Where installed in overhead service masts on the outside surface of the building traveling through the eave of that building to meet the requirements of 230.24.

Significance of the Change

Service conductors by definition must originate from a utility, and as such they are considered to be unprotected. Article 230 contains strict requirements for service conductors entering a building. Section 230.70(A)(1) requires that the service disconnecting means be installed at a readily accessible location either outside of the building or structure or inside nearest the point of entrance of the service conductors. Where a need exists to enter the structure and locate the required disconnect further, the conductors must be installed "outside of the building," in accordance with 230.6.

As soon as the outer membrane of a building or structure is penetrated, the conductors are deemed to have entered the building or structure. The intent of this change is to permit a raceway recognized in 230.43 and used as a vertical service mast to pass through an eave and be considered as outside of the building.

Change Summary

New list item (5) clarifies that where a service mast passes through the eave of a building, the raceway and conductors are considered to be outside of the building.

Comment: 4-24
Proposal: 4-64

230.24(A)

Article 230 Services
Part II Overhead Service Conductors

Clearances, Above Roofs, New Exception

Code Language

230.24 Clearances.

(A) Above Roofs. Conductors shall have a vertical clearance of not less than 2.5 m (8 ft) above the roof surface. The vertical clearance above the roof level shall be maintained for a distance of not less than 900 mm (3 ft) in all directions from the edge of the roof.

Exception No. 5: Where the voltage between conductors does not exceed 300 and the roof area is guarded or isolated, a reduction in clearance to 900 mm (3 ft) shall be permitted.

Change Summary

A new exception has been added to 230.24(A) to permit the clearance between a roof and service conductors operating at not more than 300 volts to not less than 3 feet where the roof area is guarded or isolated.

Comment: None

Proposal: 4-82

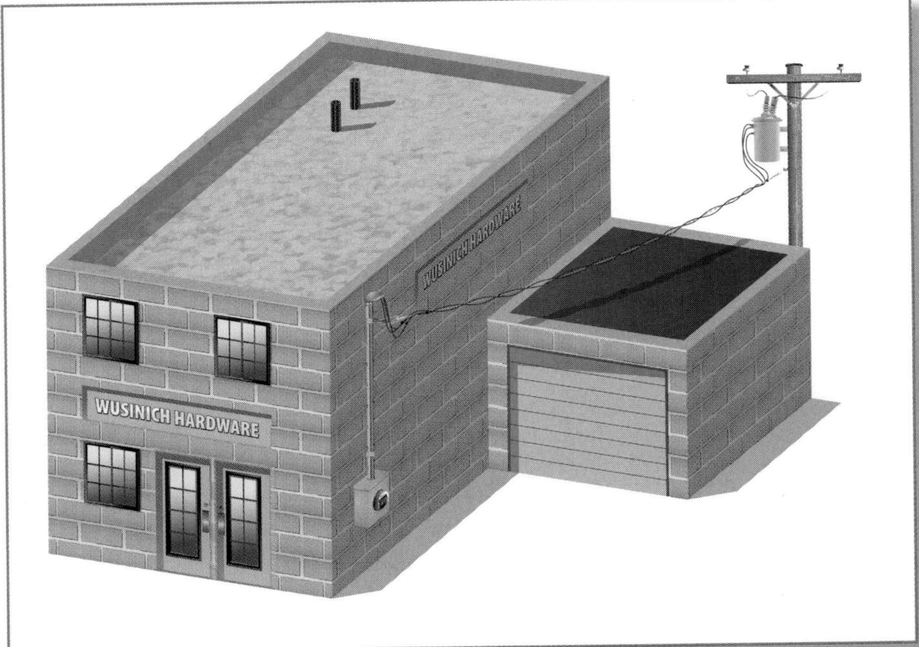

Significance of the Change

As noted in the substantiation for this proposal, this new exception is intended to correlate with allowances in the *National Electrical Safety Code* (*NESC*). This revision resolves a conflict with the present text and provides correlation between the *NEC* and the *NESC*.

The general rule in 230.24(A) requires a vertical clearance of 8 feet between the roof surface and service conductors. This exception permits the clearance between a roof and service conductors to be 3 feet where the roof area is guarded or isolated. It is important to note that this exception will apply only where both of the following limiting factors exist:

(1) The voltage between conductors does not exceed 300.

(2) The roof area is guarded or isolated.

The terms *guarded* and *isolated* are defined in Article 100 as follows:

Guarded. Covered, shielded, fenced, enclosed, or otherwise protected by means of suitable covers, casings, barriers, rails, screens, mats, or platforms to remove the likelihood of approach or contact by persons or objects to a point of danger.

Isolated (as applied to location). Not readily accessible to persons unless special means for access are used.

230.40 (Exc. No.1)

Article 230 Services
Part IV Service-Entrance Conductors

Number of Service-Entrance Conductor Sets, Exception No. 1

Code Language

230.40 Number of Service-Entrance Conductor Sets

Exception No.1 ***(Add the following to the existing text of Exception No. 1)...*** *If the number of service disconnect locations for any given classification of service does not exceed six, the requirements of 230.2(E) shall apply at each location. If the number of service disconnect locations exceeds six for any given supply classification, all service disconnect locations for all supply characteristics, together with any branch circuit or feeder supply sources, if applicable, shall be clearly described using suitable graphics or text, or both, on one or more plaques located in an approved, readily accessible location(s) on the building or structure served and as near as practicable to the point(s) of attachment or entry(ies) for each service drop or service lateral, and for each set of overhead or underground service conductors.*

Significance of the Change

As previously written this exception would permit the grouping of six of seven service disconnects for a seven-family dwelling at one point, with another set of service conductors run around the outside of the building to a remote location for the seventh disconnect. A single service could also supply a large two-story condominium with 60 units and locate service disconnects in 60 locations without marking. Because there is only one service, the marking requirements of 230.2(E) do not apply.

This additional text to the existing exception addresses a serious safety concern. In the event of a fire or other event, firemen and other emergency responders need to know how to deenergize all occupancies. This revision will require that (1) where the number of service disconnects does not exceed six, the marking requirements of 230.2(E) will apply at each location and (2) where the number of service disconnects exceeds six, all service disconnect locations for all supply characteristics, together with any branch circuit or feeder supply sources if applicable, must be clearly described on one or more plaque(s) located in a readily accessible location(s) on the building or structure served and as near as practicable to the point(s) of attachment or entry for each service.

Change Summary

New text added to 230.40 Exception No. 1 requires that where a single service supplies any number of service disconnecting means in a building with more than one occupancy, the disconnects must be marked or plaques must be located in accordance with 230.2(E).

Comment: 4-32

Proposal: 4-101

230.40 (Exc. No.4)

Article 230 Services
Part IV Service-Entrance Conductors

Number of Service-Entrance Conductor Sets, New Exception

Code Language

230.40 Number of Service-Entrance Conductor Sets. Each service drop, set of overhead service conductors, set of underground service conductors, or service lateral shall supply only one set of service-entrance conductors.

Exception No. 4: Two-family dwellings, multifamily dwellings, and multiple-occupancy buildings shall be permitted to have one set of service-entrance conductors installed to supply the circuits covered in 210.25.

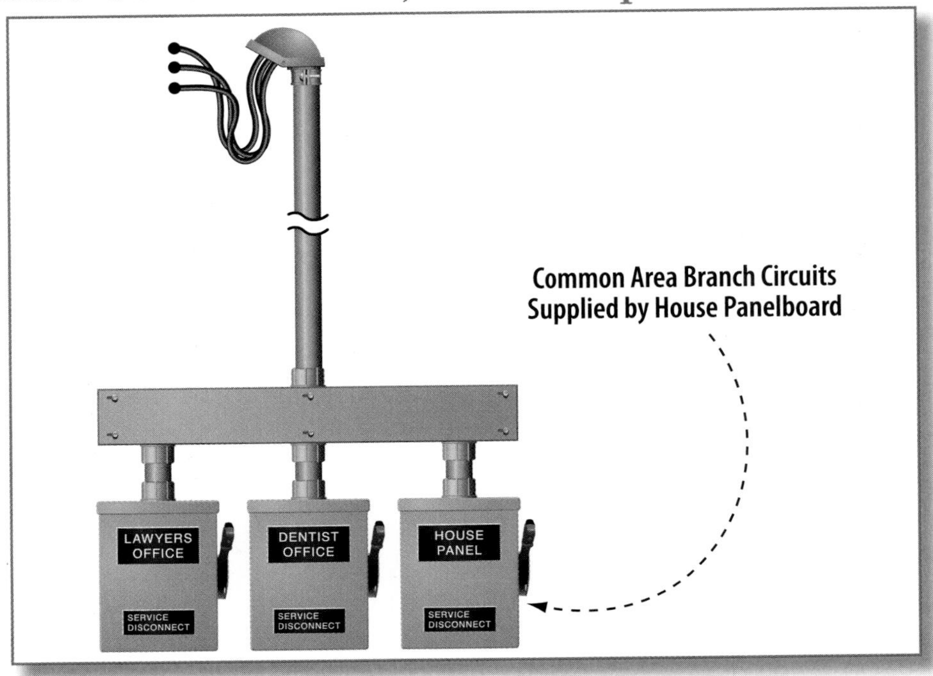

Change Summary

New text added to 230.40 Exception No. 4 now permits a single service to supply the "house panel" required by 210.25 for multiple-occupancy buildings as well as two-family dwellings and multifamily dwellings.

Significance of the Change

As previously written, this exception would permit a single service that is permitted to supply multiple occupancies (see 230.40 Exception No. 1) to also supply a "house panel," as required in 210.25 in two-family dwellings and multifamily dwellings. This permission now includes multiple-occupancy buildings such as a retail building with common areas. Section 210.25(B) requires that branch circuits installed for the purpose of lighting, central alarm, signal, communications, or other purposes for public or common areas of a two-family dwelling, a multifamily dwelling, or a multi-occupancy building not be supplied from equipment that supplies an individual dwelling unit or tenant space. The exception and 230.2 as written in the 2008 *NEC* did not permit the addition of a second service of the same voltage to an existing multiple-occupancy building. This revision logically expands the permissive text in Exception No. 4 to include multiple-occupancy buildings.

Comment: 4-35

Proposal: 4-105

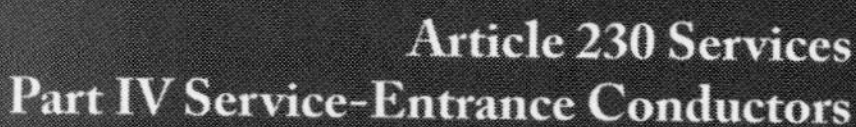

Cable Trays

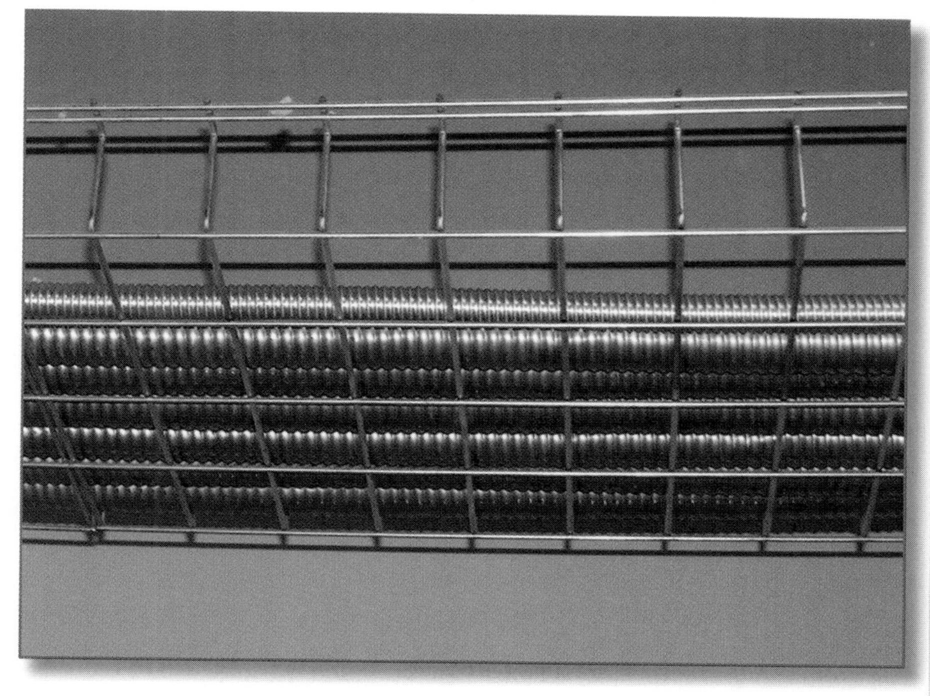

Code Language

230.44 Cable Trays. Cable tray systems shall be permitted to support service-entrance conductors. Cable trays used to support service-entrance conductors shall contain only service-entrance conductors and shall be limited to the following methods:

(1) Type SE cable
(2) Type MC cable
(3) Type MI cable
(4) Type IGS cable
(5) Single thermoplastic-insulated conductors 1/0 and larger with CT rating

Such cable trays shall be identified with permanently affixed labels with the wording "Service-Entrance Conductors." The labels shall be located so as to be visible after installation and placed so that the service-entrance conductors are able to be readily traced through the entire length of the cable tray.

Significance of the Change

This revision clearly requires all cable trays containing service conductors to be identified with permanently affixed labels. It also provides the *Code* user with a listing of permitted cable and conductor types for use as service-entrance conductors in cable tray. This change eliminates the confusion created when cable tray was used to support service conductors under previous editions of the *NEC*. For example, in the 2008 *NEC*,

— 230.43 addressed permitted wiring methods for service entrance conductors, many of which are no longer permitted.
— 230.44 generally permitted cable tray systems to support service-entrance conductors.
— 392.3 listed permitted wiring methods, cable, and conductors in cable tray.

This revision clearly permits the use of single thermoplastic-insulated conductors 1/0 and larger with CT ratings. Note that this use will be permitted only in industrial establishments in accordance with 392.3(B). The list in 230.44 addresses only permitted cables and conductors. Cable tray is not a wiring method; it is a support system. The driving force behind this revision is the permission to use single thermoplastic-insulated conductors 1/0 and larger.

Change Summary

All cable trays containing service conductors are now required to be identified with permanently affixed labels. Wiring methods that are permitted to be as installed in cable tray as service-entrance conductors are now in a user-friendly list format.

Comment: 4-39

Proposals: 4-112, 113

230.50(B)(1)

Article 230 Services
Part IV Service-Entrance Conductors

Protection Against Physical Damage

Code Language

230.50 Protection Against Physical Damage

(B) All Other Service-Entrance Conductors

(1) Service-Entrance Cables. Service-entrance cables, where subject to physical damage, shall be protected by any of the following:

(1) Rigid metal conduit
(2) Intermediate metal conduit
(3) Schedule 80 PVC conduit
(4) Electrical metallic tubing
(5) Reinforced thermosetting resin conduit (RTRC)
(6) Other approved means

Change Summary

Reinforced thermosetting resin conduit Type RTRC is now recognized in new list item (5), for the protection of service-entrance conductors where they are subject to physical damage.

Significance of the Change

This revision has occurred in many locations throughout the *NEC* where a raceway is required to be capable of withstanding physical damage. The only nonmetallic conduit recognized in previous *Code* editions for locations subject to physical damage was schedule 80 PVC. The UL White Book states, "The marking 'AG, XW, RTRC' identifies conduit suitable for use where exposed to physical damage in accordance with the *NEC*." Installers using this type of raceway in areas exposed to physical damage need to look for the marking RTRC or RTRC-AG or RTRC-XW. The suffix AG means that XW-type reinforced thermosetting resin conduit is listed for aboveground use and is suitable for use wherever IPS, ID, RTRC 40, or RTRC 80 conduit may be used. The marking "AG, XW, RTRC" identifies conduit suitable for use where exposed to physical damage in accordance with the *NEC*.

Comment: None
Proposal: 4-115

230.54

Article 230 Services
Part IV Service-Entrance Conductors

Overhead Service Locations

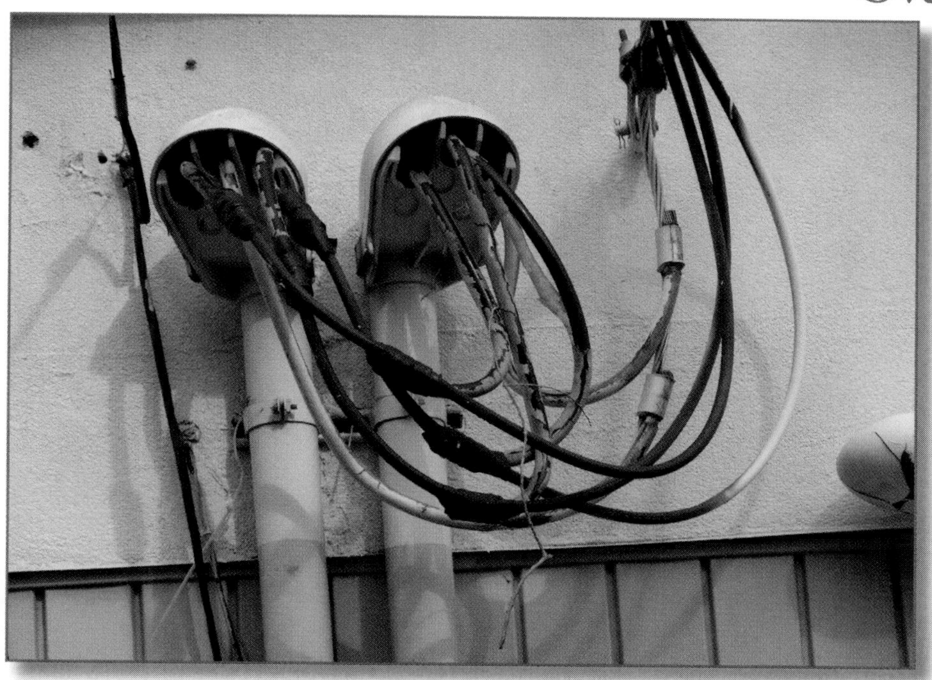

Code Language

230.54 Overhead Service Locations.

(A) Service Head. Service raceways shall be equipped with a service head at the point of connection to service drop or overhead service conductors. The service head shall be listed for use in wet locations.

(B) Service-Entrance Cable Equipped with Service Head or Gooseneck. Service-entrance cables shall be equipped with a service head. The service head shall be listed for use in wet locations.

Exception: Type SE cable shall be permitted to be formed in a gooseneck and taped with a self-sealing weather resistant thermoplastic.

Significance of the Change

Throughout 230.54, the term *service-entrance cable* is now used instead of *service cable*. They have also been revised in all of Part IV Service-Entrance Conductors. These terms are defined in Article 100 as "Service Conductors," "Service-Entrance Conductors, Overhead System," and "Service-Entrance Conductors, Underground System." This revision provides clarity through the clear separation of terms.

In first-level subdivisions 230.54(A) and (G), the term *overhead service conductors* has been added for clarity. Its usage in (A) expands coverage to both "service drop and overhead service conductors," while its usage in (G) expands coverage to both "service-entrance and overhead service conductors." The term "Service Conductors, Overhead" is new in Article 100 for the 2011 *NEC* and defines these conductors as those "overhead conductors between the service point and the first point of connection to the service-entrance conductors at the building or other structure."

The requirement in 230.54(A) and (B) mandating that the service head comply with the fitting requirement in 314.15 has been replaced with user-friendly text stating that the "service head shall be listed for use in wet locations." Sending the *Code* user to 314.15 simply required the same compliance.

Change Summary

The newly defined term *overhead service conductors* has been added to provide clarity. The requirement for service heads in wet locations no longer references 314.15; it simply requires them to be listed for use in wet locations.

Comment: None
Proposals: 4-121, 4-122

230.66

Article 230 Services
Part V Service Equipment — General

Marking

Code Language

230.66 Marking. Service equipment rated at 600 volts or less shall be marked to identify it as being suitable for use as service equipment. All service equipment shall be listed. Individual meter socket enclosures shall not be considered service equipment.

Change Summary

All service equipment is now required to be listed as well as marked to identify it as being suitable for use as service equipment.

Significance of the Change

The previous text of 230.66 required only that service equipment be marked to identify it as being suitable for use as service equipment. Anyone can manufacture or modify electrical equipment and mark it as being suitable for use as service equipment. This revision now requires that ALL service equipment is listed.

A listed piece of equipment is built and tested to recognized standards and assures the installer, maintainer, inspector, and property owner that it is acceptable for a given application. The *NEC* mandates the listing of many wiring methods, devices, and other equipment to achieve an acceptable level of safety. The term *listed* is defined in Article 100 as follows:

Listed. Equipment, materials, or services included in a list published by an organization that is acceptable to the authority having jurisdiction and concerned with evaluation of products or services, that maintains periodic inspection of production of listed equipment or materials or periodic evaluation of services, and whose listing states that either the equipment, material, or service meets appropriate designated standards or has been tested and found suitable for a specified purpose.

Comment: 4-42
Proposal: 4-126

230.72(A) (Exc.)

Article 230 Services
Part VI Service Equipment — Disconnecting Means

Grouping of Disconnects

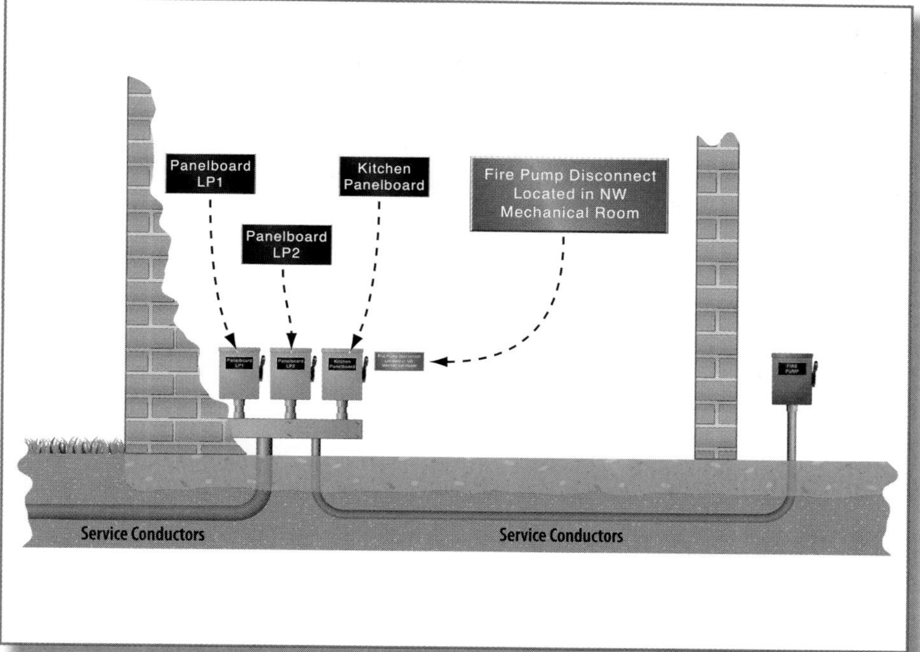

Code Language

230.72 Grouping of Disconnects.

(A) General. The two to six disconnects as permitted in 230.71 shall be grouped. Each disconnect shall be marked to indicate the load served.

Exception: One of the two to six service disconnecting means permitted in 230.71, where used only for a water pump also intended to provide fire protection, shall be permitted to be located remote from the other disconnecting means. If remotely installed in accordance with this exception, a plaque shall be posted at the location of the remaining grouped disconnects denoting its location.

Significance of the Change

The general rule of 230.72(A) is that the two to six disconnects as permitted in 230.71 be grouped and that each disconnect be marked to indicate the load served. The previous text of the exception permitted a disconnecting means only for a water pump intended to provide fire protection, to be located remote from the other disconnecting means. The marking requirement in the general rule mandated only that each disconnect be marked. An inspector, installer, maintainer, fireman, or other individual would not know there was a remote service disconnect for a fire pump.

This revision will now require, in addition to the general rule of 230.72(A), that, where a disconnecting means used only for a water pump intended to provide fire protection is located remote from the other disconnecting means, a plaque be posted at the location of the remaining grouped disconnects denoting its location.

Change Summary

A new marking requirement has been added to 230.72(A) Exception to identify the location of a service disconnect that is remotely located as permitted by this exception.

Comment: 4-49

Proposal: 4-140

230.82(9)

Article 230 Services
Part VI Service Equipment — Disconnecting Means

Equipment Connected to the Supply Side of Service Disconnect

Code Language

230.82 Equipment Connected to the Supply Side of Service Disconnect. Only the following equipment shall be permitted to be connected to the supply side of the service disconnecting means:

(9) Connections used only to supply listed communications equipment under the exclusive control of the serving electric utility, if suitable overcurrent protection and disconnecting means are provided. For installations of equipment by the serving electric utility, a disconnecting means is not required if the supply is installed as part of a meter socket, such that access can only be gained with the meter removed.

Change Summary

A new list item (9) has been added to 230.82 to permit the connection of listed communications equipment on the supply side of the service disconnecting means where it is under the exclusive control of the serving electric utility and where suitable overcurrent protection and disconnecting means are provided.

Significance of the Change

Section 230.82 provides the *Code* user with a list of equipment permitted to be connected to the supply side of the service disconnecting means. This new list item permits the use of new equipment and installations associated with Smart Grid applications and life-line (i.e., emergency calling) communications equipment powered at the premises. The new list item allows equipment to be installed without a disconnecting means where the supply is installed as part of a meter socket such that access can only be gained with the meter removed.

This revision is necessary to allow for the safe operation and installation of a new generation of metering and communications equipment. This is necessary to ensure reliable power is available for continued operation of the communications (fiber- or coax-supplied communications need a power source) equipment should an event such as a fire result in a power failure. Additionally, modern broadband systems are necessary for medical and emergency monitoring systems in addition to the traditional life-line "911" call.

The smart grid system also gives the utility the ability to monitor power and control usage via the communications equipment, including a complete power disconnect at the premises and a reconnect at a premises where power has been disconnected via a communications link.

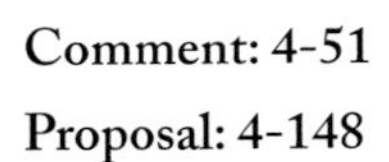

Comment: 4-51

Proposal: 4-148

230.205(A)

Article 230 Services
Part VIII Services Exceeding 600 Volts, Nominal

Location

Code Language

230.205 Disconnecting Means.

(A) Location. The service disconnecting means shall be located in accordance with 230.70.

For either overhead or underground primary distribution systems on private property, the service disconnect shall be permitted to be located in a location that is not readily accessible, if the disconnecting means can be operated by mechanical linkage from a readily accessible point, or electronically in accordance with 230.205(C), where applicable.

Change Summary

This revision mandates that a readily accessible method to operate the disconnecting means be provided. While the disconnect is not required to be readily accessible, a means to operate the device mechanically or electronically must be readily accessible.

Significance of the Change

Section 230.205 covers the requirements for all disconnecting means used for services exceeding 600 volts nominal. First-level subdivision (A) contains requirements for the location of the disconnecting means. Text was added to the 2008 *NEC* to permit a disconnecting means to be located on private property where it was not readily accessible. This would permit a pole-top switch with no linkage to the pole base. The only way that an installer/maintainer would be able to operate such a switch would be via a hot stick out of a bucket. In an emergency situation or for routine maintenance, the switch would not be capable of being opened unless a truck or other equipment were moved into place to open the switch.

This revision will now require that either (1) a mechanical linkage be installed and be readily accessible to allow persons to operate the switch from the pole base or (2) the service disconnecting means be capable of being electrically operated by a readily accessible, remote-control device in accordance with 230.205(C)..

Comment: None
Proposal: 4-169

240.4

Article 240 Overcurrent Protection
Part I General

Protection of Conductors, Informational Note

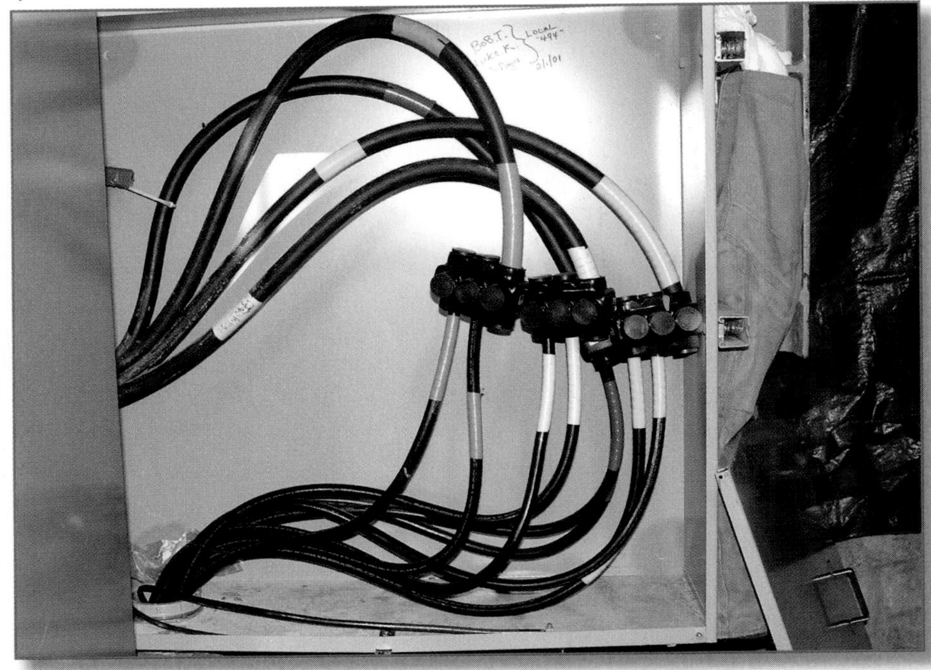

Code Language

240.4 Protection of Conductors. Conductors, other than flexible cords, flexible cables, and fixture wires, shall be protected against overcurrent in accordance with their ampacities specified in 310.15, unless otherwise permitted or required in 240.4(A) through (G).

Informational Note: See ICEA P32-382 for information on allowable short-circuit currents for insulated copper and aluminum conductors.

Change Summary

A new informational note has been added to 240.4 to reference ICEA P32-382 for information on allowable short-circuit currents for insulated copper and aluminum conductors.

Comment: 10-7

Proposal: 10-13

Significance of the Change

The new informational note references an Insulated Cable Engineers Association (ICEA) standard that utilizes a physics formula to determine how much current and time elapse before damage to conductor insulation begins. This informational note is especially useful whenever an overcurrent device is sized several times larger than the ampacity of the conductor, such as when the tap rules in 240.21 are utilized. In some cases when a tap conductor is applied within the requirements of 240.21, the overcurrent protective device can be up to 10 times the ampacity of the conductor. Product standards do not test for protection of conductors when protected at several times their ampacity. Conductors with ampacities that are many times smaller than the overcurrent device can and do vaporize under short-circuit conditions. The formulas in the standard can be used to determine whether conductor insulation will be damaged during a fault and, if so, whether the conductor size can be increased or whether a more current-limiting overcurrent protective device can be used so that the conductor insulation is not damaged under short-circuit conditions. The ICEA formulas are not new to the *NEC*. They are found in Table 240.92(B), which allows for tap conductor sizing in supervised industrial installations.

240.15(B)(1) & (2)

Article 240 Overcurrent Protection
Part I General

Circuit Breaker as Overcurrent Device

Courtesy of Eaton Corporation

Code Language

240.15 Ungrounded Conductors

(B) Circuit Breaker as Overcurrent Device. Circuit breakers shall open all ungrounded conductors of the circuit both manually and automatically unless otherwise permitted in 240.15(B)(1), (B)(2), (B)(3), and (B)(4).

(1) Multiwire Branch Circuit. Individual single-pole circuit breakers, with identified handle ties, shall be permitted as the protection for each ungrounded conductor of multiwire branch circuits that serve only single-phase line-to-neutral loads.

(2) Grounded Single-Phase ac Circuits. In grounded systems, individual single-pole circuit breakers rated 120/240 volts ac, with identified handle ties, shall be permitted as the protection for each ungrounded conductor for line-to-line connected loads for single-phase circuits.

Significance of the Change

The 2008 edition of the *NEC* revised 210.4(B) to require a means to simultaneously disconnect all ungrounded conductors of a multiwire branch circuit. That revision created serious confusion whenever *Code* users referenced 240.15(B)(1), because that section did not change to recognize the significant revision in 210.4(B). In the 2005 edition of the *NEC* the only requirement for simultaneous disconnect in 210.4(B) was for multiwire branch circuits that supplied more than one device or equipment on the same yoke.

The revised text of 240.15(B)(1) is now clearly limited to individual single-pole circuit breakers with identified handle ties. It should be noted that this second-level subdivision is limited to multiwire branch circuits that serve only single-phase line-to-neutral loads. This section is further limited by existing product standards to single-phase systems or three-phase systems where the voltage to ground does not exceed 120 volts [three-phase systems are addressed in 240.15(B)(3)]. See the product category DIVQ for Circuit Breakers, Molded Case and Circuit Breaker Enclosures in the UL *White Book*.

Change Summary

The reference to 210.4(B) has been deleted in 240.15(B)(1) because 210.4(B) now requires all multiwire branch circuits to have a simultaneous disconnect and 240.15 addresses protection requirements. The permission for protection of a multiwire circuit without handle ties has been deleted, and the requirements of 240.15(B)(2) have been clarified to apply only to single-phase AC 120/240-volt circuits.

Comment: None
Proposals: 10-30, 10-39

240.15(B)(3)

Article 240 Overcurrent Protection
Part I General

3-Phase and 2-Phase Systems and 3-Wire DC Circuits

Code Language

240.15 Ungrounded Conductors

(B) Circuit Breaker as Overcurrent Device. Circuit breakers shall open all ungrounded conductors of the circuit both manually and automatically unless otherwise permitted in 240.15(B)(1), (B)(2), (B)(3), and (B)(4).

(3) 3-Phase and 2-Phase Systems. For line-to-line loads in 4-wire, 3-phase systems or 5-wire, 2-phase systems, individual single-pole circuit breakers rated 120/240 volts ac with identified handle ties shall be permitted as the protection for each ungrounded conductor, if the systems have a grounded neutral point and the voltage to ground does not exceed 120 volts.

(4) 3-Wire Direct-Current Circuits. Individual single-pole circuit breakers rated 125/250 volts dc with identified handle ties shall be permitted as the protection for each ungrounded conductor for line-to-line connected loads for 3-wire, direct-current circuits supplied from a system with a grounded neutral where the voltage to ground does not exceed 125 V.

Change Summary

The use of an identified handle tie on individual single-pole circuit breakers in a 3-phase system is now clearly limited to single-pole circuit breakers rated 120/240 volts AC in systems with a voltage to ground that does not exceed 120 volts. A new second-level subdivision has been added to address the application of individual single-pole circuit breakers rated 125/250 volts DC with identified handle ties.

Comment: None

Proposal: 10-39

Significance of the Change

The reference to voltage limitation in 210.6 is deleted in 240.15(B)(3). The text now clearly limits the application of individual single-pole circuit breakers with identified handle ties in 2- and 3-phase systems to single-pole circuit breakers rated 120/240 volts AC where the systems have a grounded neutral point and the voltage to ground does not exceed 120 volts. For example, the use of handle ties on individual single-pole circuit breakers in a 480Y/277-volt panelboard would be prohibited. This change is driven by the product standard UL 489. These requirements are found in the UL *White Book* under the product category DIVQ for Circuit Breakers, Molded Case and Circuit Breaker Enclosures.

A new second-level subdivision 240.15(B)(4) permits the application of individual single-pole circuit breakers rated 125/250 volts DC with identified handle ties in 3-wire DC systems with a grounded neutral where the voltage to ground does not exceed 125 volts.

240.21(B)(1)

Article 240 Overcurrent Protection
Part II Location

Taps Not Over 3 m (10 ft) Long

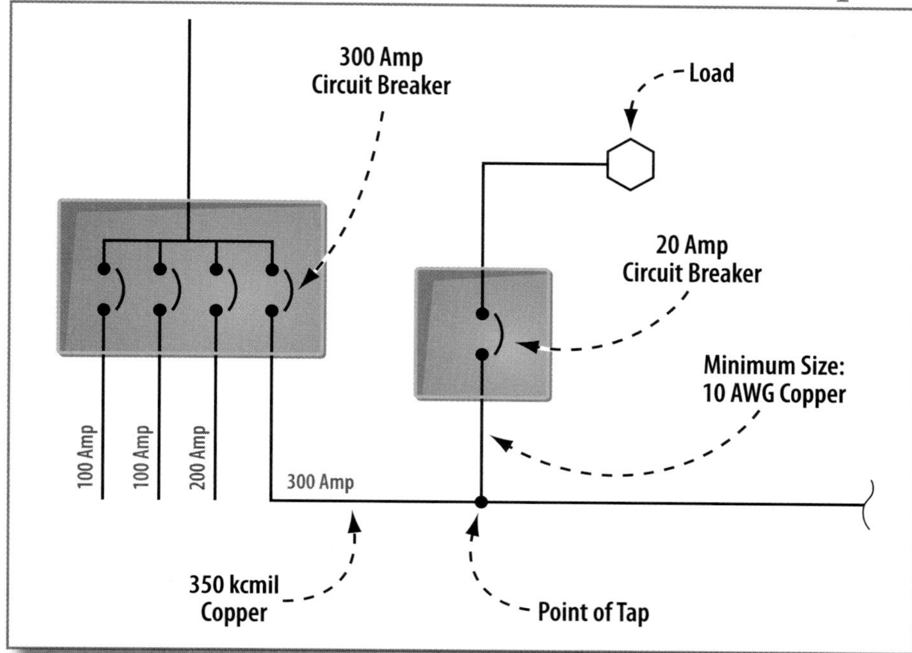

Code Language

240.21 Location in Circuit

(B) Feeder Taps

(1) Taps Not over 3 m (10 ft) Long. If the length of the tap conductors does not exceed 3 m (10 ft) and the tap conductors comply with all of the following:

[No changes in (1) through (3)]

(4) For field installations, if the tap conductors leave the enclosure or vault in which the tap is made, the ampacity of the tap conductors is not less than one-tenth of the rating of the overcurrent device protecting the feeder conductors.

Significance of the Change

Section 240.21(B)(1) provides requirements for feeder taps not over 10 feet in length. Four requirements in list format are provided in this subdivision. List item (4) governs the minimum size for tap conductors and now correlates with the text in 240.21(B)(2)(1). The result is user-friendly text for determining the minimum size conductor for a 10-foot feeder tap. For example:

Feeder protected at 400 amps

Minimum size tap conductor is required to be *not less than one-tenth of the rating of the overcurrent device protecting the feeder conductors.*

Minimum size tap conductor = 400/10 = 40 amps

Change Summary

The prescriptive text of this section to determine the minimum size of the tap conductor has been simplified.

Comment: None

Proposal: 10-48

240.24(E)

Article 240 Overcurrent Protection
Part II Location

Not Located in Bathrooms

Code Language

240.24 Location in or on Premises

(E) Not Located in Bathrooms. In dwelling units, dormitories, and guest rooms or guest suites, overcurrent devices, other than supplementary overcurrent protection, shall not be located in bathrooms.

Change Summary

The prohibition of overcurrent devices in bathrooms has been extended to bathrooms in dormitories.

Significance of the Change

A dormitory bathroom is used in a very similar manner to a dwelling unit bathroom. Thus, the prohibition against locating overcurrent devices in bathrooms is now extended to bathrooms in dormitories. This requirement now applies to bathrooms in dwelling units, dormitories, and guest rooms or guest suites. The term *bathroom* is defined in Article 100 of the *NEC* as follows:

Bathroom. An area including a basin with one or more of the following: a toilet, a urinal, a tub, a shower, a bidet, or similar plumbing fixtures.

It is not uncommon to see a panelboard installed in a bathroom in a commercial facility. In most cases a commercial facility bathroom is used only as a restroom and so it has no shower or bathing facilities.

Comment: None
Proposal: 10-65

240.87

Article 240 Overcurrent Protection
Part VII Circuit Breakers

Noninstantaneous Trip

Courtesy of Eaton Corporation

Significance of the Change

Where circuit breakers are utilized without an instantaneous trip function, documentation must be available to those authorized to design, install, operate, or inspect the installation. The documentation must provide the location of the circuit breaker(s) as well as one of several proven methods to limit energy in a fault or an approved equivalent. Large power circuit breakers used in some designs are utilized without an instantaneous trip. These installations represent the highest values of incident energy in a given facility. Three methods are listed:

(1) Zone-selective interlocking allows the upstream circuit breaker and downstream circuit breaker to communicate with each other, resulting in faster clearing time when needed.

(2) Differential relaying accomplishes the same goal by monitoring the amount of current through the upstream circuit breaker and the downstream circuit breakers.

(3) An energy-reducing maintenance switch allows the trip unit to be set on instantaneous whenever working within the flash protection boundary, and then turned back to the short-time delay mode when finished. Where an energy-reducing maintenance switch is provided, it must have a local status indicator to notify persons of the position of the switch.

Code Language

240.87 Noninstantaneous Trip. Where a circuit breaker is used without an instantaneous trip, documentation shall be available to those authorized to design, install, operate or inspect the installation as to the location of the circuit breaker(s).

Where a circuit breaker is utilized without an instantaneous trip one of the following or approved equivalent means shall be provided:

(1) Zone-selective interlocking
(2) Differential relaying
(3) Energy-reducing maintenance switching with local status indicator

Informational Note: An energy-reducing maintenance switch allows a worker to set a circuit breaker trip unit to "no intentional delay" to reduce the clearing time while the worker is working within an arc-flash boundary as defined in NFPA 70E-2009, *Standard for Electrical Safety in the Workplace*, and then to set the trip unit back to a normal setting after the potentially hazardous work is complete.

Change Summary

Section 240.87 has been added to require that where a circuit breaker is utilized without an instantaneous trip, documentation must be made available and a must means be provided to reduce the incident energy when justified energized work must be performed.

Comments: 10-41, 10-43
Proposal: 10-82

240.91

Article 240 Overcurrent Protection
Part VIII Supervised Industrial Installations

Protection of Conductors

Code Language

240.91 Protection of Conductors. Conductors shall be protected in accordance with 240.91(A) or (B).

(A) General. Conductors shall be protected in accordance with 240.4.

(B) Devices Rated over 800 Amperes. Where the overcurrent device is rated over 800 amperes, the ampacity of the conductors it protects shall be equal to or greater than 95 percent of the rating of the overcurrent device specified in 240.6 in accordance with (B)(1) and (2).

(1) The conductors are protected within recognized time vs. current limits for short-circuit currents

(2) All equipment in which the conductors terminate is listed and marked for the application

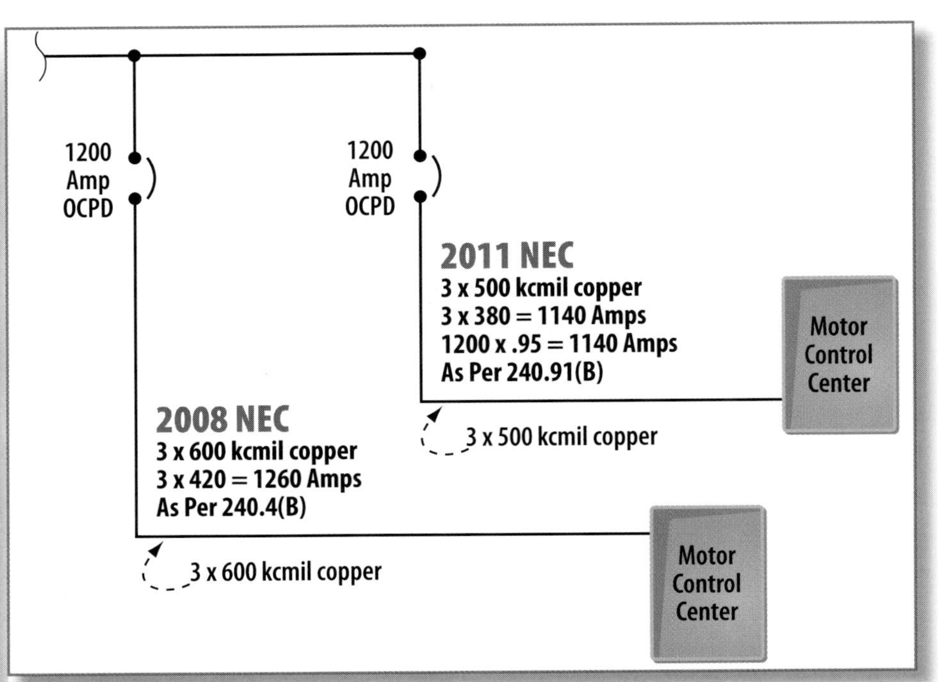

Change Summary

This new first-level subdivision 240.91(B) permits conductors in circuits rated over 800 amps to have overcurrent protection applied at values above their rated ampacity. This will apply only where the equipment is listed and marked for the application and within the limitations provided for supervised industrial installations.

Significance of the Change

New 240.91(B) now permits conductors in circuits rated over 800 amps to be protected by overcurrent protective devices rated above their ampacity. Note that this requirement is in Part VIII of Article 240 and as such is limited to supervised industrial installations. The general provisions of 240.4(B) already permit the next higher standard overcurrent device to protect a conductor(s) up to 800 amps.

This new requirement in 240.91 allows a larger overcurrent device to be applied in circuits rated over 800 amps in supervised industrial installations where both of the following apply:

(1) The conductors are protected within recognized time vs. current limits for short-circuit currents.

(2) Where all equipment in which the conductors terminate is listed and marked for the application.

The requirement that all equipment in which the conductors terminate to be "listed and marked for the application" is essential in that the additional heat created by the reduction in conductor mass could create serious problems in equipment that has not been evaluated.

Comment: 10-49

Proposal: 10-83

Definition of Supply-Side Bonding Jumper

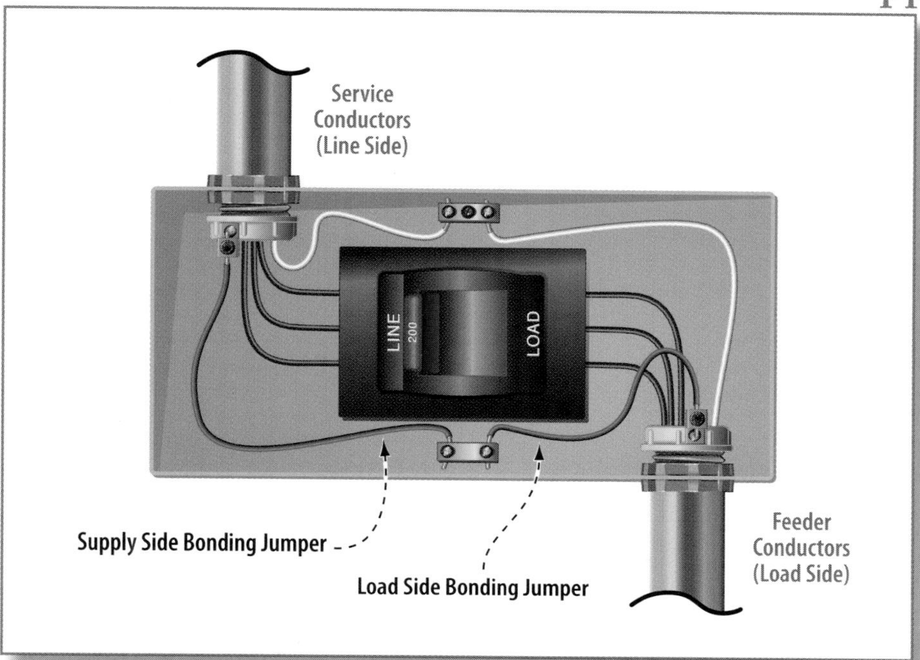

Code Language

250.2 Definitions.

Bonding Jumper, Supply-Side. A conductor installed on the supply side of a service or within a service equipment enclosure(s), or for a separately derived system, that ensures the required electrical conductivity between metal parts required to be electrically connected.

Significance of the Change

The new definition in 250.2 provides a clear differentiation between a bonding jumper installed on the supply side of an overcurrent device and one installed on the load side of an overcurrent device. The 2008 *NEC* definition of the term *bonding jumper, equipment* indicates that it is a connection between two or more portions of the equipment grounding conductor. Equipment grounding conductors are sized using 250.122 based on the rating of an overcurrent device protecting a circuit. Some equipment bonding jumpers are necessary for installation on the supply side of the service or separately derived system and would require sizing according to 250.102(C), which references Table 250.66 or the 12.5% rule, as applicable. As a result, these terms cannot be combined in one definition.

The new definition is necessary to ensure the proper identification, sizing, and installation of equipment bonding jumpers where installed within or on the supply side of service equipment or for a separately derived system. The size of the supply-side bonding jumper for separately derived systems is sized according to 250.102(C) from Table 250.66 or the 12.5% rule where the largest derived phase conductor exceeds 1100 kcmil copper or 1750 kcmil aluminum or copper-clad aluminum. The definition of the term *equipment bonding jumper* is not affected by this change.

Change Summary

A new definition that applies specifically to bonding jumpers on the supply side of services or separately derived systems has been added to 250.2.

Comments: 5-1, 5-4, 5-37

Proposal: 5-5

250.21(B) & (C)

Article 250 Grounding and Bonding
Part II System Grounding

Ground Detectors

Code Language

(B) Ground Detectors. Ground detectors shall be installed in accordance with 250.21(B)(1) and (B)(2).

(1) Ungrounded alternating current systems as permitted in 250.21(A)(1) through (A)(4) operating at not less than 120 volts and not exceeding 1000 volts shall have ground detectors installed on the system.

(2) The ground detection system sensing equipment shall be connected as close as practicable to where the system receives its supply.

(C) Marking. Ungrounded systems shall be legibly marked "Ungrounded System" at the source or first disconnecting means of the system. The marking shall be of sufficient durability to withstand the environment involved.

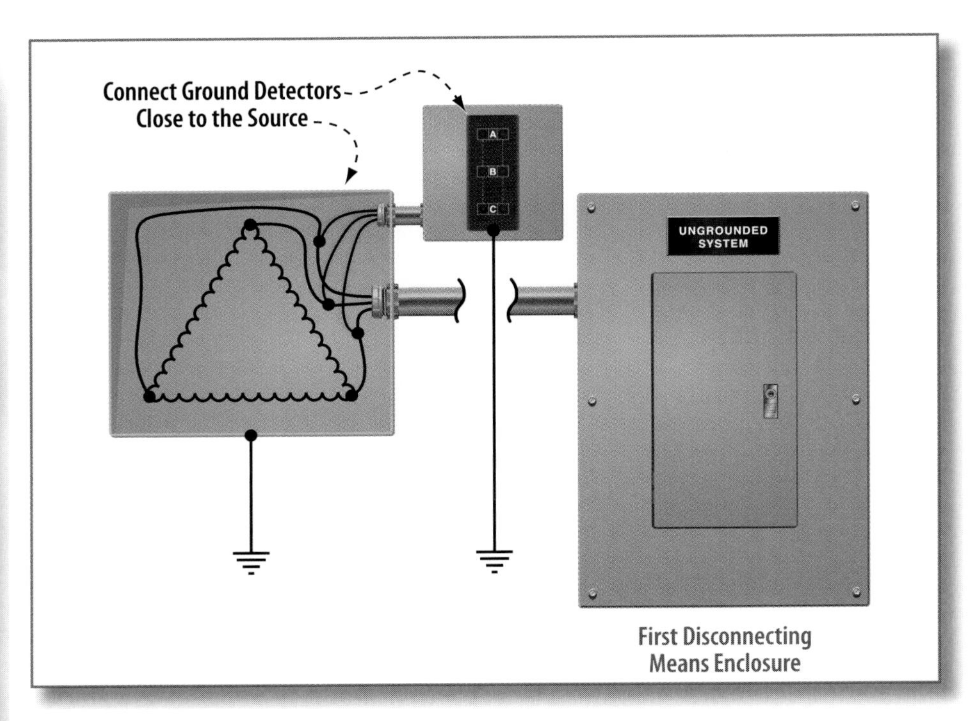

Significance of the Change

This section has been restructured to list format to meet the requirements of 3.3.2 of the *NEC Style Manual*. The revision to subdivision (B) provides direction for users regarding the functional and appropriate placement of the ground detection system sensing equipment for an ungrounded service or ungrounded separately derived system. The sensing equipment must be located as close as practicable to where the system receives its supply. The previous requirement only mentioned that ground detection be provided on the system, but it did not specify where. The problem was that for detection equipment installed at the branch circuit or feeder in a location multiple levels downstream from

Comment: 5-60

Proposals: 5-85, 5-86a

Ground Detectors (continued)

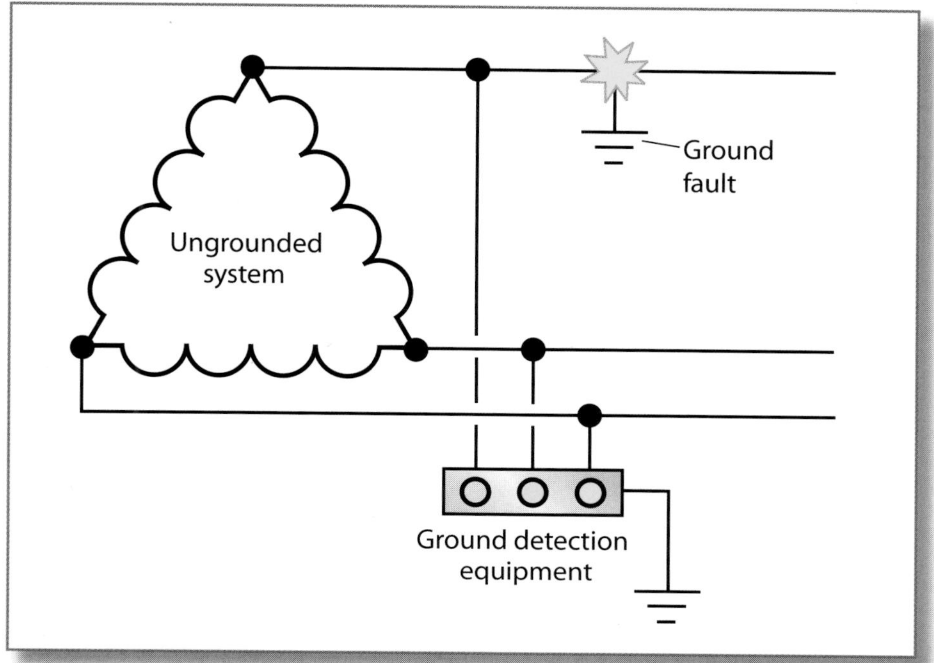

Change Summary

Section 250.21(B) has been restructured into list format and revised. A new subdivision (C) titled "Marking" has been added to this section. The marking requirement applies to the source or first disconnecting means of the system.

the system supply source or service, the disconnecting of a feeder or branch circuit to which the ground detection equipment is connected could result in an unmonitored, ungrounded system as a whole.

This revision reduces the risk of deactivation of ground detection equipment by requiring it to be connected as close to the supply source as possible. New subdivision (C) requires marking the source or first system disconnecting means with the words "Ungrounded System." This new marking requirement provides workers with a means of ready identification of an ungrounded system installation and use and, importantly, assists in distinguishing an ungrounded system from a grounded one.

250.24(C)

Article 250 Grounding and Bonding
Part II System Grounding

Grounded Conductor Brought to Service Equipment

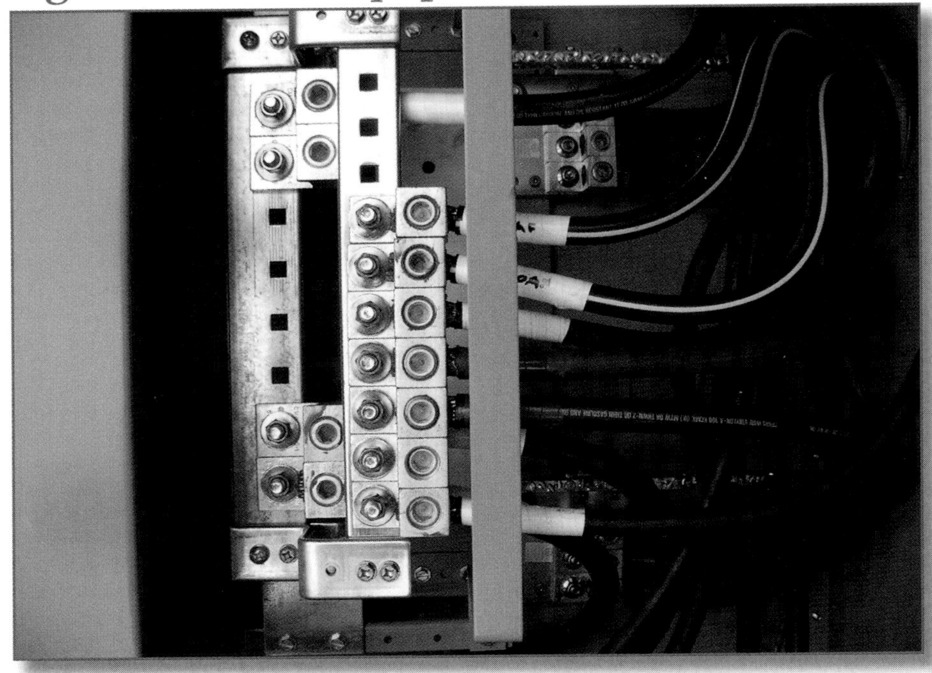

Code Language

(C) Grounded Conductor Brought to Service Equipment. Where an ac system operating at less than 1000 volts is grounded at any point, the grounded conductor(s) shall be routed with the ungrounded conductors routed to each service disconnecting means and shall be connected to each disconnecting means grounded conductor(s) terminal or bus. A main bonding jumper shall connect the grounded conductor(s) to each service disconnecting means enclosure. The grounded conductor(s) shall be installed in accordance with 250.24(C)(1) through (C)(4).

(1) Sizing for a Single Raceway. ***(Revised to address sizing for single raceways installations)***

(2) Parallel Conductors in Two or More Raceways. ***(Revised to address sizing for single raceways installations)***

Informational Note: ***(Remains unchanged)***

(3) Delta-Connected Service. ***[Last sentence of former list item (1) and located here as new list item (3)].***

(4) High Impedance. ***[Remains unchanged and renumbered as (4)]***

(Note: See NEC 250.24(C) for full text.)

Change Summary

The phrase "routed with the ungrounded conductors" has been moved from list item (1) to the driving text of subdivision (C). (C)(1) has been revised to address sizing requirements for grounded conductors where service-entrance conductors are installed in a single wireway or raceway arrangement. List item (2) has been revised to address sizing requirements for parallel arrangements where the parallel conductors are installed in two or more raceways. The last sentence of former (C)(1) has been relocated to its own list item (3), and former list item (3) addressing high-impedance grounded neutral systems has been renumbered as list item (4). The word *phase* has been removed from this section, and the word *ungrounded* has been added where it relates to service-entrance conductors. The sizing requirements for grounded conductors have been broken out into separate subdivisions to assist users in applying proper sizing rules for grounded conductors that are required to be brought to the service equipment. A clear distinction has been provided between sizing a single grounded conductor for a service and sizing each parallel grounded conductor in an installation of parallel service conductors.

Comment: 5-64, 5-65

Proposal: 5-95

Grounded Conductor Brought to Service Equipment (continued)

Significance of the Change

The changes to 250.24(C) are editorial, meant to improve clarity and usability. No technical revisions were made to current requirements. The revisions are intended to clarify which requirements apply specifically to minimum sizing requirements for grounded conductors installed in various arrangements. By locating the requirement to route the grounded conductor with the ungrounded conductors in the driving text of this subdivision, it is clear that this aspect of the installation applies to (1), (2), and (3).

The word *phase* has been removed from this section, and the word *ungrounded* has been added where it relates to the service-entrance conductor sizing requirements. The reason is because the word "phase" is undefined whereas the term "ungrounded" is clearly defined in Article 100. For services supplied by grounded systems, it is clear that the grounded service conductor must be routed with the ungrounded service conductors to the service disconnecting means and bonded to the service disconnecting means enclosure. List items (1) and (2) now provide specific direction for meeting minimum size requirements for grounded conductors at services.

Grounding Separately Derived Alternating-Current Systems

Code Language

250.30 Grounding Separately Derived Alternating-Current Systems. In addition to complying with 250.30(A) for grounded systems, or as provided in 250.30(B) for ungrounded systems, separately derived systems shall comply with 250.20, 250.21, 250.22, and 250.26.

Informational Note No. 1: An alternate ac power source, such as an on-site generator, is not a separately derived system if the grounded conductor is solidly interconnected to a service-supplied system grounded conductor. An example of such a situation is where alternate source transfer equipment does not include a switching action in the grounded conductor and allows it to remain solidly connected to the service-supplied grounded conductor when the alternate source is operational and supplying the load served.

Informational Note No. 2: See 445.13 for the minimum size of conductors that carry fault current.

(A) *(See NEC for full text.)*

(B) *(See NEC for full text.)*

(C) Outdoor Source. If the source of the separately derived system is located outside the building or structure supplied, a grounding electrode connection shall be made at the source location to one or more grounding electrodes in compliance with 250.50. In addition, the installation shall comply with 250.30(A) for grounded systems or with 250.30(B) for ungrounded systems.

Exception: The grounding electrode conductor connection for impedance grounded neutral systems shall comply with 250.36 or 250.186, as applicable.

(Note: See NEC for full text of 250.30.)

Comments: 5-69, 5-70, 5-71, 5-72, 5-73, 5-76, 5-79, 5-80, 5-81, 5-82, 5-83, 5-84, 5-85, 5-87

Proposals: 5-102, 5-101, 5-103, 5-104, 5-105, 5-106, 5-107, 5-109, 5-110, 5-111, 5-113, 5-114, 5-116, 5-117, 5-118, 5-121

Significance of the Change

Section 250.30 has been reorganized to provide a more logical order for the rules applicable to separately derived systems. The requirements for the grounded conductor have been moved to 250.30(A)(3) to follow the requirements for the equipment bonding jumper, because the grounded conductor should be the next consideration when there is a grounded conductor and the system bonding jumper is not located at the source. The term *supply-side bonding jumper* now replaces the term *equipment bonding jumper* in 250.30(A)(2). This term is also now defined in 250.2 and is required to be installed in accordance with 250.102(C).

The next considerations are typically the grounding electrode, so the requirements have been moved to 250.30(A)(4) and text has been added to identify that a separately derived system is clearly required to have a grounding electrode. Exception No. 2 to (1) and (2) is new and addresses separately derived systems that originate in listed equipment suitable for use as service equipment.

This exception allows the grounding electrode used for the service or feeder equipment to be used as the grounding electrode for the separately derived system as long as the required grounding electrode conductor is large enough. This exception was previously located in

Grounding Separately Derived Alternating-Current Systems (cont'd)

Change Summary

Section 250.30(A) and (B) were revised to include both editorial and technical revisions. Section 250.30(C) is new and covers outdoor sources. Two new informational notes have been added following the driving text in 250.30. Informational Note No. 1 is the relocated text from 250.20(D).

250.30(A)(3) Exception No. 2.

Although the rest of the items in 250.30(A)(5) through (8) have been renumbered accordingly, the requirements have not been changed significantly.

A new 250.30(C) has been added to address situations where a building or other structure has been supplied by a separately derived system, for example, where the transformer or source is located outside the building and the system is customer owned. This new section provides guidance not previously included, and the requirements parallel the efforts in recent *NEC* cycles toward not using the grounded conductor to ground or bond equipment or to connect to a grounding electrode. The new (B)(1) and (B)(2) are consistent with the language in 250.35 and in a companion proposal submitted for 250.32. Additionally, the revised text adds consistency to all three sections.

The following technical revisions were also incorporated into this section as a result of action on other proposals. The system bonding jumper is required to not extend beyond the enclosure where it originates. The connection of a grounding electrode conductor tap to a common grounding electrode conductor is required to be made with a connector listed as grounding and bonding equipment instead of just a listed connector providing consistency with that connection required for services.

The new subdivision (C) addresses outdoor sources and requires a grounding electrode connection at the source location. The term *equipment bonding jumper* has been changed in 250.30(A)(2) to *supply-side bonding jumper*.

250.32(B)(1)

Article 250 Grounding and Bonding
Part II System Grounding

Grounded Systems

Code Language

(B) Grounded Systems.

(1) Supplied by a Feeder or Branch Circuit. ***(Text unchanged except for minor editorial revisions)***

Exception: For installations made in compliance with previous editions of this Code that permitted such connection, the grounded conductor run with the supply to the building or structure shall be permitted to serve as the ground-fault return path if all of the following requirements continue to be met:

(1) An equipment grounding conductor is not run with the supply to the building or structure.

(2) There are no continuous metallic paths bonded to the grounding system in each building or structure involved.

(3) Ground-fault protection of equipment has not been installed on the supply side of the feeder(s).

If the grounded conductor is used for grounding in accordance with the provision of this exception, the size of the grounded conductor shall not be smaller than the larger of either of the following:

(1) That required by 220.61

(2) That required by 250.122

(2) Supplied by Separately Derived System.

(a) With Overcurrent Protection. If overcurrent protection is provided where the conductors originate, the installation shall comply with (B)(1).

(b) Without Overcurrent Protection. If overcurrent protection is not provided where the conductors originate, the installation shall comply with 250.30(A). If installed, the supply-side bonding jumper shall be connected to the building or structure disconnecting means and to the grounding electrode(s).

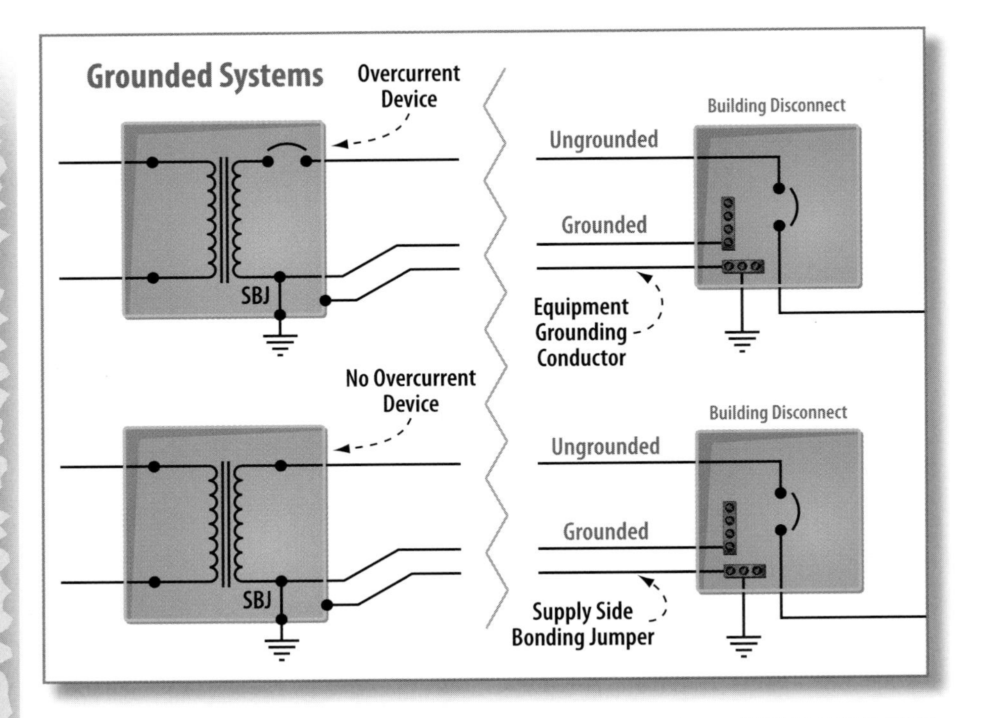

Change Summary

Section 250.32(B) has been restructured to a list format in accordance with 3.3.2 of the *NEC Style Manual*. The existing text of (B) and the (revised) exception have been renumbered as list item (1), and a new list item (2) has been added to address separate buildings or structures supplied by separately derived systems. List item (2) includes third-level subdivisions (a) and (b) that address feeders from separately derived systems that do not have overcurrent protection and feeders from separately derived systems that are provided with overcurrent protection. The difference relates to sizing rules for either equipment grounding conductor or a supply-side bonding jumper.

Grounded Systems (continued)

Significance of the Change

The exception to (B)(1) has been revised to clarify that existing installations can continue the use of the grounded conductor for grounding as in the past if all the conditions of (1), (2), and (3) of the exception continue to be met. This allowance is similar to that for using grounded conductors for range and dryer circuits, as provided in 250.140. This revision also provides users with clear direction on installing equipment grounding conductors or supply-side bonding jumpers from separately derived systems installed in outdoor locations and to supply separate buildings or structures.

Where the separately derived system is installed outside and overcurrent protection is provided for the feeder at the point at which the feeder conductors originate, an equipment grounding conductor sized according to 250.122 must be installed with the feeder conductors routed to the building or structure. Where the separately derived system is installed outside and overcurrent protection is not installed at the point where the feeder originates, a supply-side bonding jumper in accordance with 250.30(A)(2) must be installed with the feeder conductors to the disconnecting means at the building or structure supplied by the feeder. The supply-side bonding jumper must be connected to the disconnecting means enclosure and to the required grounding electrode. This revision also correlates with companion revisions in 250.30(C) that now address separately derived systems installed in outdoor locations.

Comments: 5-91, 5-93
Proposal: 5-126

250.32(C)

Article 250 Grounding and Bonding
Part II System Grounding

Ungrounded Systems

Code Language

(C) Ungrounded Systems.

(1) Supplied by a Feeder or Branch Circuit. An equipment grounding conductor, as described in 250.118, shall be installed with the supply conductors and be connected to the building or structure disconnecting means and to the grounding electrode(s). The grounding electrode(s) shall also be connected to the building or structure disconnecting means.

(2) Supplied by a Separately Derived System.

(a) With Overcurrent Protection. If overcurrent protection is provided where the conductors originate, the installation shall comply with (C)(1).

(b) Without Overcurrent Protection. If overcurrent protection is not provided where the conductors originate, the installation shall comply with 250.30(B). If installed, the supply-side bonding jumper shall be connected to the building or structure disconnecting means and to the grounding electrode(s).

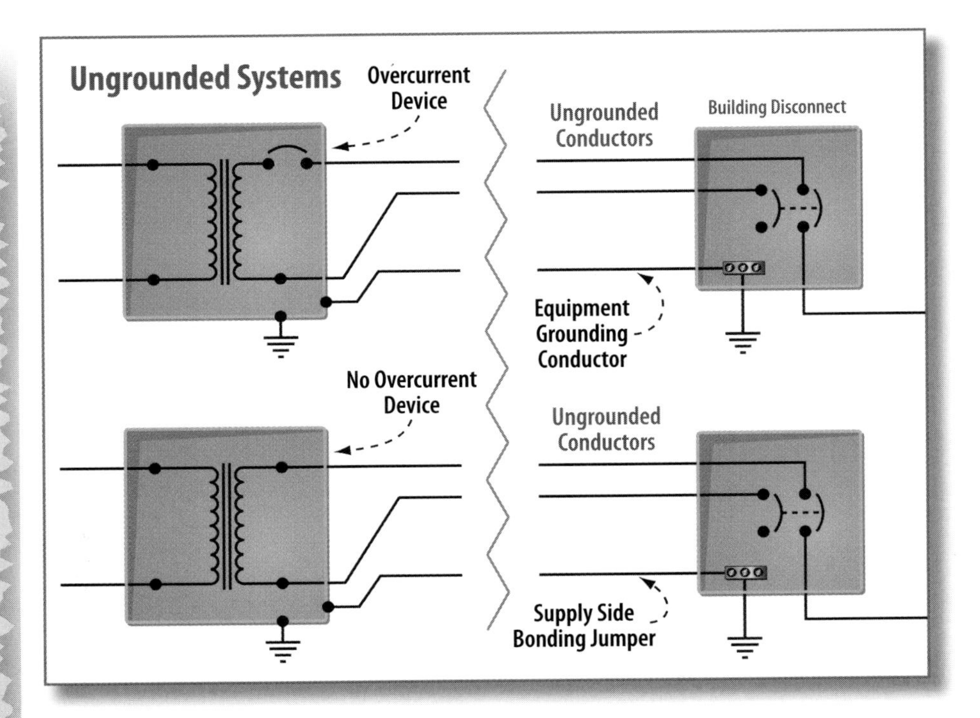

Change Summary

Section 250.32(C) has been revised and expanded to address buildings or structures supplied by feeders or branch circuits and those supplied by ungrounded separately derived systems. The section has also been revised into a list structure in accordance with the *NEC Style Manual*. List item (2) includes feeders from an ungrounded separately derived system where no overcurrent protection is provided at the point the feeders receive their supply and feeders that have no overcurrent protection at their point of supply but instead at the building disconnect. List item (2) includes third-level subdivisions (a) and (b) that address feeders from ungrounded separately derived systems that do not have overcurrent protection and feeders from ungrounded separately derived systems that are provided with overcurrent protection. The difference relates to sizing rules for either equipment grounding conductor or a supply-side bonding jumper.

Ungrounded Systems (continued)

Courtesy of Cogburn Brothers, Inc.

Significance of the Change

This section previously provided a requirement only to connect a building or structure disconnecting means to a grounding electrode. Other than the requirements in 215.6, no specific *Code* requirement addressed the installation of equipment grounding conductors with the feeders supplying a separate building or structure when the supply system is ungrounded. This revision corrects this deficiency and aligns with the requirements provided in 250.4(B)(4) addressing effective fault current paths to facilitate overcurrent device operation for ungrounded systems. It is fairly well understood that a first phase-to-ground fault condition in an ungrounded system activates only the required ground fault detection systems, not overcurrent devices. A second fault on another ungrounded phase conductor from the same ungrounded system would cause the feeder overcurrent device to operate.

This can happen only when an effective path for fault current is provided. Section 250.4(B)(4) is clear that the earth is not permitted as an effective path for fault current. This revision adds a requirement to install an equipment grounding conductor or supply-side bonding jumper with the feeder conductors supplying a second building or structure, which is consistent with the same requirement for grounded systems as provided in 250.32(B)(1) and (B)(2). The equipment grounding conductor can be one of the types described in 250.118 and the supply-side bonding jumper (wire-type) has to be installed and sized according to 250.102(C).

Comment: 5-93

Proposal: 5-129

250.52(A)(1)

Article 250 Grounding and Bonding
Part III Grounding Electrode System
and Grounding Electrode Conductor

Metal Underground Water Pipe

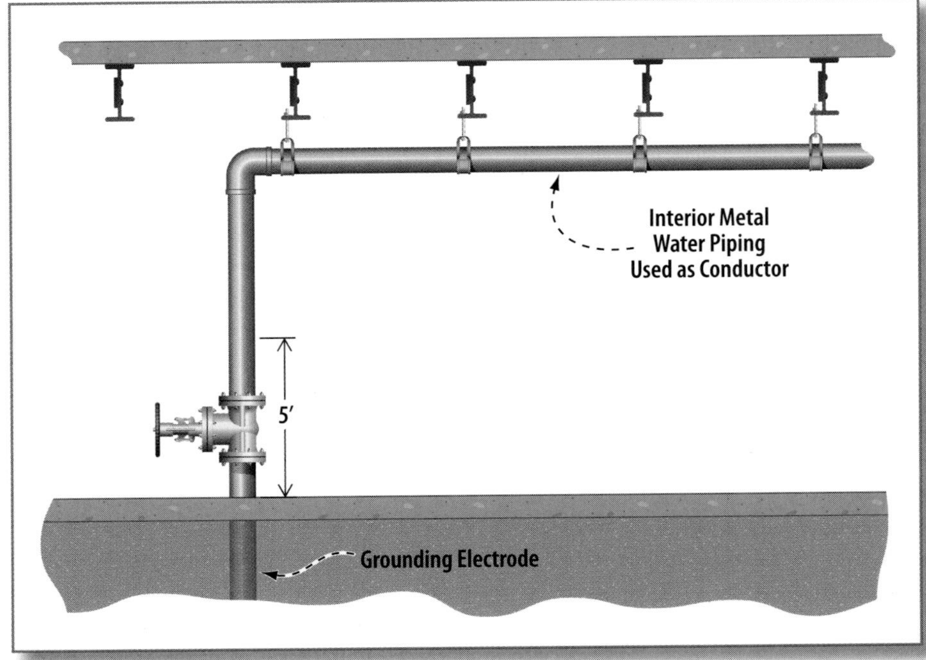

Code Language

(1) Metal Underground Water Pipe. A metal underground water pipe in direct contact with the earth for 3.0 m (10 ft) or more (including any metal well casing bonded to the pipe) and electrically continuous (or made electrically continuous by bonding around insulating joints or insulating pipe) to the points of connection of the grounding electrode conductor and the bonding conductor(s) or jumper(s) if installed.

Change Summary

The last sentence and the exception to this section have been relocated to a new 250.68(C). The description remaining in this section is now consistent with the definition of the term *grounding electrode*.

Comment: None

Proposal: 5-146

Significance of the Change

This revision results in the description of a water pipe grounding electrode in 250.52(A)(1) aligning with the definition of a grounding electrode in Article 100. Section 250.52(A) provides the details and descriptions of various grounding electrodes recognized for use. Section 250.52(A)(1) describes condition(s) under which a metal water pipe is recognized as a grounding electrode. The relocated portions of the previous *Code* text related to functions of the water pipe other than that making a direct connection to the earth. Metallic water pipes located above the earth may function as a conductive path or similarly to a grounding electrode conductor and cannot or should not be considered a part of a grounding electrode.

The text covering the portion of the water piping above the earth, previously included in this section, has been relocated to a new section, 250.68(C), that addresses conductive paths and grounding electrode conductors that connect to defined grounding electrodes.

The result of this revision is the description of a water pipe electrode in 250.52(A)(1) is now consistent with how grounding electrodes are defined in Article 100, thus adding clarity and enhancing usability. The requirements for a water pipe electrode to be in contact with the earth for at least 10 feet have not been altered by this revision.

250.52(A)(2)

Article 250 Grounding and Bonding
Part III Grounding Electrode System and Grounding Electrode Conductor

Metal Frame of the Building or Structure

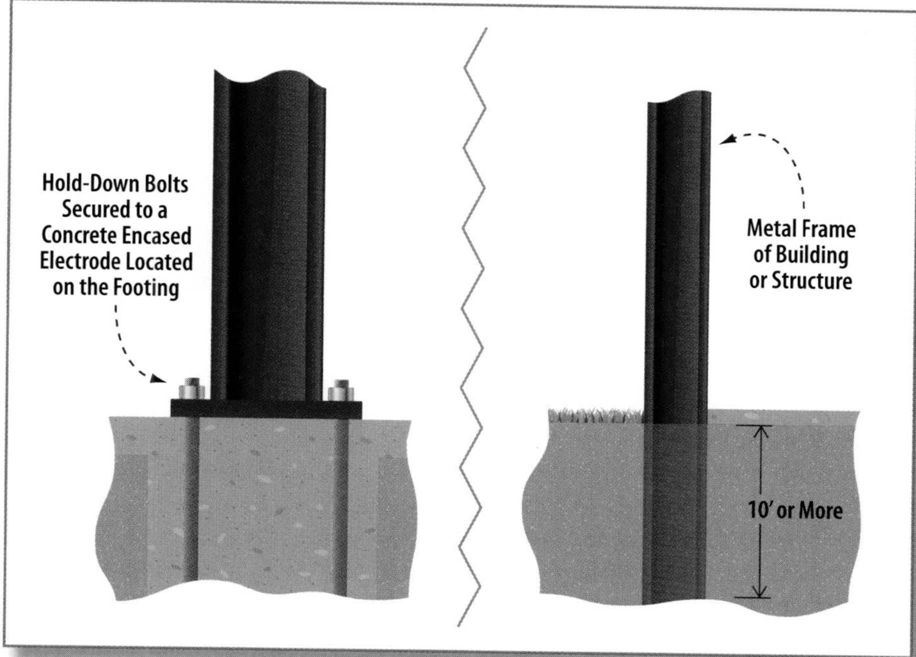

Code Language

250.52(A)(2) Metal Frame of the Building or Structure. The metal frame of the building or structure that is connected to the earth by one or more of the following methods:

(1) At least one structural metal member that is in direct contact with the earth for 3.0 m (10 ft) or more, with or without concrete encasement.

(2) Hold-down bolts securing the structural steel column are connected to a concrete-encased electrode that complies with 250.52(A)(3) and is located in the support footing or foundation. The hold-down bolts shall be connected to the concrete-encased electrode by welding, exothermic welding, the usual steel tie wires, or other approved means.

Significance of the Change

The revisions to 250.52(A)(2) establish more consistency between the definition of the term *grounding electrode* and the description of the term *structural metal building frame electrode* provided in this section. As a result, list item (1) remains as it did in the 2008 *NEC*, and list item (2) has been revised to clarify that building metal frames that qualify as grounding electrodes are those that are effectively connected to a concrete-encased grounding electrode through steel anchor bolts. This section specifies three acceptable methods of connecting the building frame anchors to the steel reinforcing members in the footing or foundation of the structure.

The revision provides users with clear criteria that constructed building metal frames must meet in order to use building metal frame as a grounding electrode. Metal building frames can continue to serve as effective conductive paths to the grounding electrode system. For this reason, new text covering metal building frames that function as conductive paths to the electrode system has been placed in 250.68(C). Grounding electrode conductors will continue to be connected to grounding electrodes through the metal building frame that provides an effective and electrically continuous path as it functions presently.

Change Summary

Section 250.52(A)(2) has been revised, and list items (3) and (4) have been removed and partially relocated to 250.68(C).

Comment: None

Proposal: 5-150

Related Proposals: 5-149, 5-151, 5-152

250.52(A)(3)

Article 250 Grounding and Bonding
Part III Grounding Electrode System and Grounding Electrode Conductor

Concrete-Encased Electrode

Code Language

250.52 Grounding Electrodes.

(A) Electrodes Permitted for Grounding

(3) Concrete-Encased Electrode. A concrete-encased electrode shall consist of at least 6.0 m (20 ft) of either (1) or (2):

(1) One or more bare or zinc galvanized or other electrically conductive coated steel reinforcing bars or rods of not less than 13 mm (½ in.) in diameter, installed in one continuous 6.0 m (20 ft) length, or if in multiple pieces connected together by the usual steel tie wires, exothermic welding, welding, or other effective means to create a 6.0 m (20 ft) or greater length; or

(... Continued on next page...)

Change Summary

Section 250.52(A)(3) has been revised for clarity and restructured into a list format. A new informational note has been added.

Courtesy of Cogburn Brothers, Inc.

Comments: 5-103, 5-104

Proposal: 5-158

Concrete-Encased Electrode (continued)

Courtesy of Cogburn Brothers, Inc.

Code Language

(... Continued from previous page...)

(2) Bare copper conductor not smaller than 4 AWG

Metallic components shall be encased by at least 50 mm (2 in.) of concrete and shall be located horizontally within that portion of a concrete foundation or footing that is in direct contact with the earth or within vertical foundations or structural components or members that are in direct contact with the earth. If multiple concrete-encased electrodes are present at a building or structure, it shall be permissible to bond only one into the grounding electrode system.

Informational Note: Concrete installed with insulation, vapor barriers, films or similar items separating the concrete from the earth is not considered to be in "direct contact" with the earth.

Significance of the Change

Section 250.52(A)(3) provides the details and description of what constitutes a concrete-encased electrode. Revision of this section results in two list items that describe the creation of a concrete-encased electrode by reinforcing steel of a footing or by installing a copper wire. Each of these methods provides a conductive means that establishes a connection to the concrete. Concrete-encased grounding electrodes are the combination of concrete and the conductive component that connects to the concrete such as steel or an electrical conductor. The second paragraph clarifies that concrete-encased electrodes can be established in both a horizontal or vertical orientation as long as a minimum of 20 feet of the electrode are in direct contact with the earth. If multiple (separate) concrete-encased electrodes are present at a building or structure, it is permitted to use only one in the grounding electrode system, but the use of one is required.

The new informational note addresses conditions that disqualify concrete-encased electrodes from use in the grounding electrode system, because the isolation created between the concrete and the earth obstructs or impedes the direct connection between the two.

250.53(A)(2)

Article 250 Grounding and Bonding
Part III Grounding Electrode System and Grounding Electrode Conductor

Rod, Pipe, and Plate Electrodes

Code Language

(A) Rod, Pipe, and Plate Electrodes. Rod, pipe, and plate electrodes shall meet the requirements of 250.53(A)(1) through (A)(3).

(1) Below Permanent Moisture Level. If practicable, rod, pipe, and plate electrodes shall be embedded below permanent moisture level. Rod, pipe, and plate electrodes shall be free from nonconductive coatings such as paint or enamel.

(2) Supplemental Electrode Required. A single rod, pipe, or plate electrode shall be supplemented by an additional electrode of a type specified in 250.52(A)(2) through (A)(8). The supplemental electrode shall be permitted to be bonded to one of the following:

(1) Rod, pipe, or plate electrode
(2) Grounding electrode conductor
(3) Grounded service-entrance conductor
(4) Nonflexible grounded service raceway
(5) Any grounded service enclosure

Exception: If a single rod, pipe, or plate grounding electrode has a resistance to earth of 25 ohms or less, the supplemental electrode shall not be required.

(3) Supplemental Electrode. If multiple rod, pipe, or plate electrodes are installed to meet the requirements of this section, they shall not be less than 1.8 m (6 ft) apart.

Informational Note: The paralleling efficiency of rods is increased by spacing them twice the length of the longest rod.

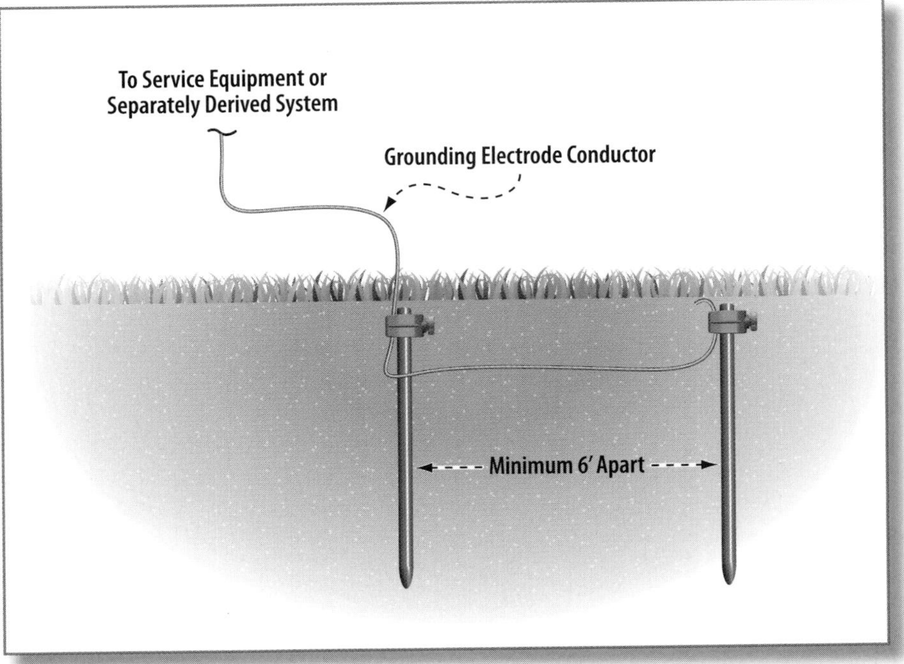

Significance of the Change

CMP-5 acted favorably to the concepts of deleting the 25 ohm requirement for a single rod, pipe, or plate electrode. The justification for this revision far outweighed the need to keep a requirement that can fall in and out of *NEC* compliance owing to seasonal conditions or changes in earth electrode contact resistance resulting from natural environmental differences over time. As a result, 250.56 has been deleted and a new (2) and an exception have been added to 250.53(A). Section 250.53(A)(2) is new and requires that a single rod, pipe, or plate electrode be supplemented by an additional electrode of any type provided in (1) through (5). The 25 ohm resistance requirement for a single rod, pipe, or plate electrode is now addressed only as an alternative by exception to a new requirement to install two electrodes as a first choice.

Installers choosing a single rod, pipe, or plate electrode are now required to install an additional electrode during the initial installation. This new requirement is essentially how workers are currently deal-

Rod, Pipe, and Plate Electrodes (continued)

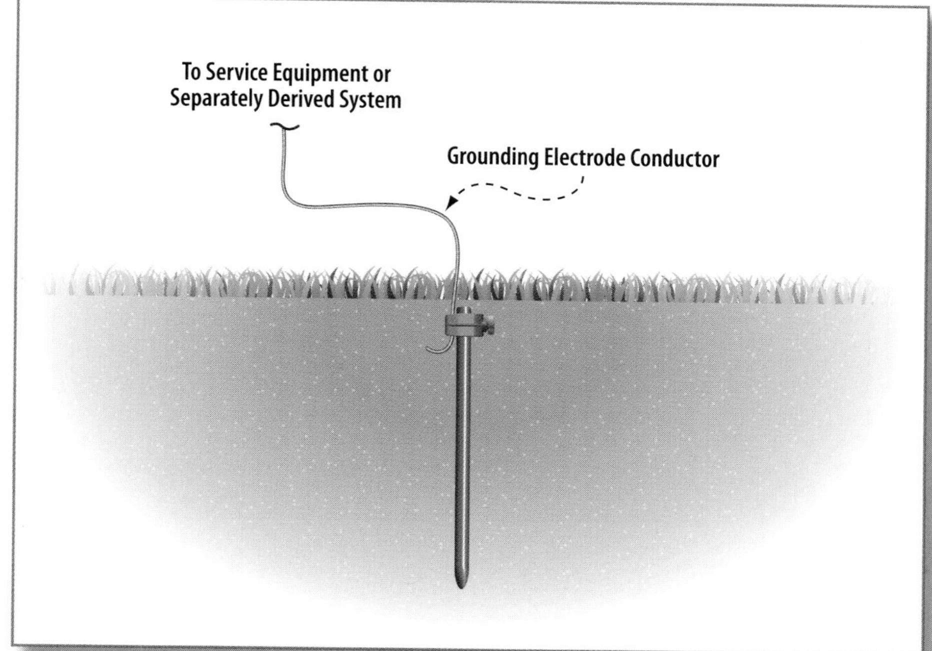

Change Summary

Section 250.56 has been deleted. A new list item (2) and exception have been added to 250.53(A). New list item (3) includes the last sentence of former 250.56 addressing minimum spacing requirements between installed electrodes. The existing informational note to 250.56 has been relocated to follow (3) of this section.

ing with the requirements in former 250.56. The common practice is to install two ground rods so that a test to prove the 25 ohm requirement for a single electrode is no longer required. The *Code* rules now parallel what the installers are doing as a normal practice. The revisions to this section are an incremental approach to removing a long-standing contact resistance requirement that has little or no value in relation to what constitutes an effective ground-fault current path, since the earth is prohibited from this use.

It should be understood that the *NEC* provides minimum requirements for safe electrical installations. Many more elaborate and detailed electrical designs may specify lower electrode contact resistant values by engineering specification. These requirements are over and above the *NEC* minimums and become requirements by authority of the project design and engineering team. The revision to this section is an effort to align the *Code* with actual practices in the field relative to installations of single electrodes of the rod, pipe, or plate types.

Comment: 5-112

Proposal: 5-169a

Related Proposals: 5-174, 176a

Grounding Electrode Conductor Installation

Code Language

250.64 Grounding Electrode Conductor Installation.

(A) Aluminum or Copper-Clad Aluminum Conductors. *(Editorial)*

(B) Securing and Protection Against Physical Damage. *(Technical and Editorial)*

(C) Continuous. *(Editorial)*

(D) Service with Multiple Disconnecting Means Enclosures. *(Editorial)*

(E) Enclosures for Grounding Electrode Conductors. *(Technical)*

(F) Installation to Electrodes. *(Editorial)*

(See the NEC for full text revisions.)

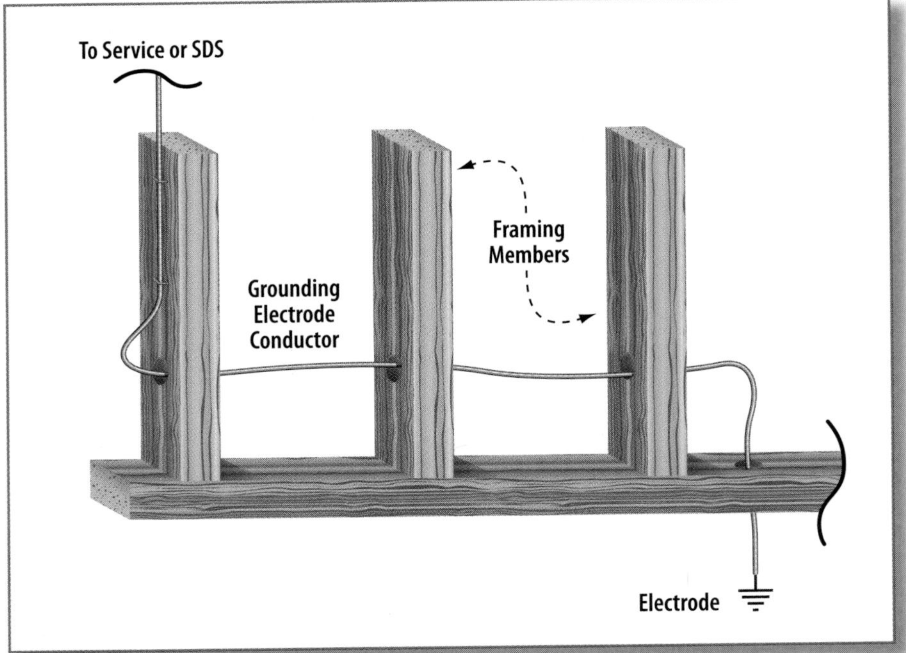

Change Summary

Both editorial and technical revisions have been incorporated in 250.64. The acronyms RMC, IMC, EMT, PVC, and RTRC have replaced the spelled out wiring methods in this section. A new sentence added to 250.64(B) reads as follows: "Grounding electrode conductors shall be permitted to be installed on or through framing members." The words "methods in compliance with 250.92(B) for installations at service equipment locations and with 250.92(B)(2) through (B)(4) for other than service equipment locations" have been added to the fourth sentence of subdivision (E).

Grounding Electrode Conductor Installation (continued)

Significance of the Change

Section 250.64 has been revised to include new technical provisions and to clarify current requirements. Many of the revisions to this section are editorial and improve clarity and usability of the requirements. Section 250.64(B) now recognizes that it is permitted to route grounding electrode conductors through or on framing members. Without this provision being specifically stated, some enforcement authorities have been prohibiting such installations. It is recognized that physical protection for grounding electrode conductors can be achieved by the method in which they are installed. Installing the grounding electrode conductor through horizontal or vertical framing members often provides satisfactory degrees of protection against physical damage.

Section 250.64(E) addresses specific bonding rules that apply to grounding electrode conductors installed in ferrous metal enclosures for protection against possible damage from magnetic field stresses. Revision of subdivision (E) clarifies the locations where additional bonding of the grounding electrode conductor and enclosures is required to be in accordance with 250.92(B) at service equipment. Bonding methods in 250.92(B)(2) through (B)(4) apply to metal enclosures for grounding electrode conductors installed in other than service equipment locations.

Comment: 5-123

Proposals: 5-194, 5-195, 5-196, 5-199

250.68(C)

Article 250 Grounding and Bonding
Part III Grounding Electrode System and Grounding Electrode Conductor

Metallic Water Pipe and Structural Metal

Code Language

(C) Metallic Water Pipe and Structural Metal. Grounding electrode conductors and bonding jumpers shall be permitted to be connected at the following locations and be used to extend the connection to an electrode(s):

(1) Interior metal water piping located not more than 1.52 m (5 ft) from the point of entrance to the building shall be permitted to be used as a conductor to interconnect electrodes that are part of the grounding electrode system.

*Exception: [**Same as the exception following 250.52(A)(1) that appeared in the 2008 NEC (no changes and see NEC for full text of the exception).**]*

(2) The structural metal frame of a building that is directly connected to a grounding electrode as specified in 250.52(A)(2) or 250.68(C)(2)(a), (b), or (c) shall be permitted as a bonding conductor to interconnect electrodes that are part of the grounding electrode system, or as a grounding electrode conductor.

a. By connecting the structural metal frame to the reinforcing bars of a concrete-encased electrode as provided in 250.52(A)(3) or ground ring as provided in 250.52(A)(4)

b. By bonding the structural metal frame to one or more of the grounding electrodes, as specified in 250.52(A)(5) or (A)(7) that comply with 250.53(A)(2)

c. By other approved means of establishing a connection to earth

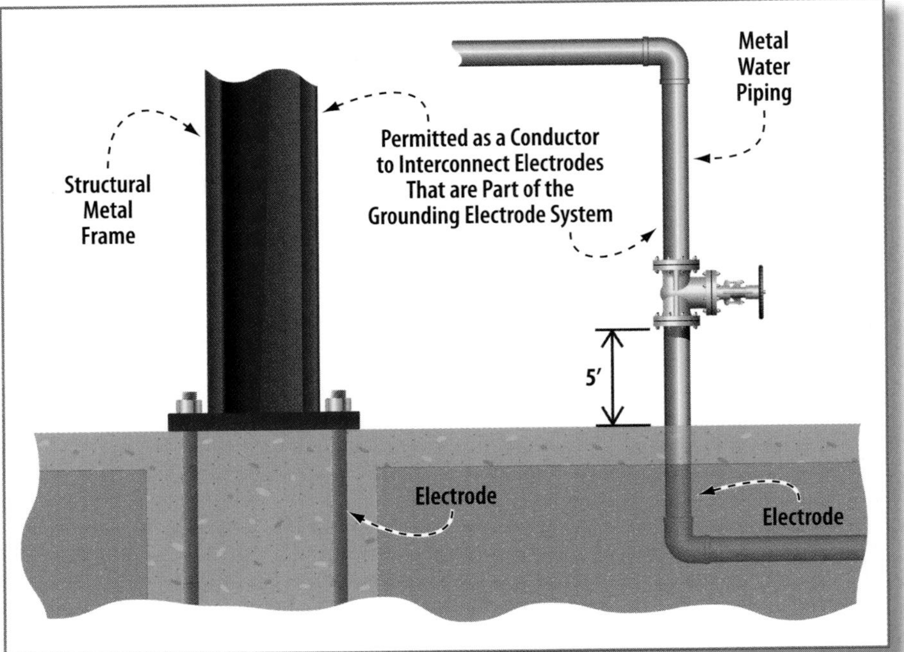

Metallic Water Pipe and Structural Metal (continued)

Change Summary

A new subdivision (C) titled "Metallic Water Pipe and Structural Metal" has been added to 250.68. This new subdivision includes text that was removed from 250.52(A)(1) and (A)(2). No new technical requirements are included as part of this revision.

Significance of the Change

The text from 250.52(A)(1) and (A)(2) has been relocated here because items that are not in the earth should not be considered to be grounding electrodes but can be used as a conductor to create a conductive path to the electrode system. This change continues the effort to align *Code* rules with how grounding and bonding words and terms are defined in the *NEC*. Because metal water piping systems and interconnected structural metal building frames perform grounding electrode conductor functions, they have been included in the Part III requirements for grounding electrode conductors.

This new subdivision recognizes metal water piping and interconnected metal building frames as conductors that provide effective path to grounding electrodes, but they are not grounding electrodes by definition. This revision provides the allowances to connect grounding electrode conductors and bonding jumpers of a grounding electrode system to metal water piping and structural metal building frames that provide a path to ground (the earth).

Comment: 5-131

Proposals: 5-212, 5-170

Method of Bonding at the Service

Code Language

(B) Method of Bonding at the Service. Bonding jumpers meeting the requirements of this article shall be used around impaired connections, such as reducing washers or oversized, concentric, or eccentric knockouts. Standard locknuts or bushings shall not be the only means for the bonding required by this section but shall be permitted to be installed to make a mechanical connection of the raceway(s).

Electrical continuity at service equipment, service raceways, and service conductor enclosures shall be ensured by one of the following methods:

(1) Bonding equipment to the grounded service conductor in a manner provided in 250.8

(2) Connections utilizing threaded couplings or threaded hubs on enclosures if made up wrenchtight

(3) Threadless couplings and connectors if made up tight for metal raceways and metal-clad cables

(4) Other listed devices, such as bonding-type locknuts, bushings, or bushings with bonding jumpers

Change Summary

The driving text of 250.92(B) has been revised by the addition of a new first sentence. Additionally, the word *oversized* has been added to the first sentence of 250.92(B).

Significance of the Change

Section 250.92(B) provides requirements for more robust bonding methods applicable to service equipment, service raceways, and service conductor enclosures. The reasoning behind the revision is to ensure effective bonding for metallic enclosures containing system conductors that are typically not equipped with the same overcurrent protection as feeders and branch circuits installed on the load side of the service disconnecting means. This revision clarifies that the strengthened bonding requirements of this section apply to installations of reducing washers or when encountering oversized knockouts in addition to concentric or eccentric knockouts that present possible compromises in the bonding integrity between raceways and enclosures where they are encountered.

This revision also establishes consistency between the exception to 250.97 where oversized, concentric, and eccentric knockouts are addressed. The substantiation provided with the proposal identified the need for including reducing washers and oversized knockouts in this bonding rule for consistency and performance concerns.

Comment: 5-138

Proposals: 5-223, 5-225

250.94

Article 250 Grounding and Bonding
Part V Bonding

Bonding for Other Systems

Courtesy of ERICO International Corporation

Code Language

250.94 Bonding for Other Systems. An intersystem bonding termination for connecting intersystem bonding conductors required for other systems shall be provided external to enclosures at the service equipment or metering equipment enclosure and at the disconnecting means for any additional buildings or structures. The intersystem bonding termination shall comply with the following:

(1) Be accessible for connection and inspection.

(2) Consist of a set of terminals ... ***(See NEC for full text.)***

(3) Not interfere with opening the enclosure ... ***(See NEC for full text.)***

(4) At the service equipment ... ***(See NEC for full text.)***

(5) At the disconnecting means ... ***(See NEC for full text.)***

(6) The terminals shall be listed as grounding and bonding equipment.

Significance of the Change

Section 250.94 requires that an intersystem bonding termination be provided for connecting required bonding conductors from limited energy systems covered in Articles 770 and Chapter 8. This section has been reorganized into a list of acceptable installations of an intersystem bonding termination device. The exception continues to address alternative connection locations for existing installations that do not have an intersystem bonding termination provided. The revisions to the informational notes clarify that Articles 770 and 830 also include intersystem grounding and bonding requirements for conductive optical fiber cables and network-powered broadband communications systems.

Change Summary

Section 250.94 has been restructured into a list format to improve clarity and usability. The revisions to this section do not affect the existing exception or the existing Informational Note No.1. Informational Note No. 2 has been revised to include references to grounding and bonding rules in Articles 770 and 830.

Comments: 5-142, 5-143

Proposal: 5-227

250.104(C)

Article 250 Grounding and Bonding
Part V Bonding

Structural Metal

Code Language

(C) Structural Metal. Exposed structural metal that is interconnected to form a metal building frame and is not intentionally grounded or bonded and is likely to become energized shall be bonded to the service equipment enclosure; the grounded conductor at the service; the disconnecting means for buildings or structures supplied by a feeder or branch circuit; the grounding electrode conductor, if of sufficient size; or to one or more grounding electrodes used. The bonding jumper(s) shall be sized in accordance with Table 250.66 and installed in accordance with 250.64(A), (B), and (E). The points of attachment of the bonding jumper(s) shall be accessible unless installed in compliance with 250.68(A), Exception No. 2.

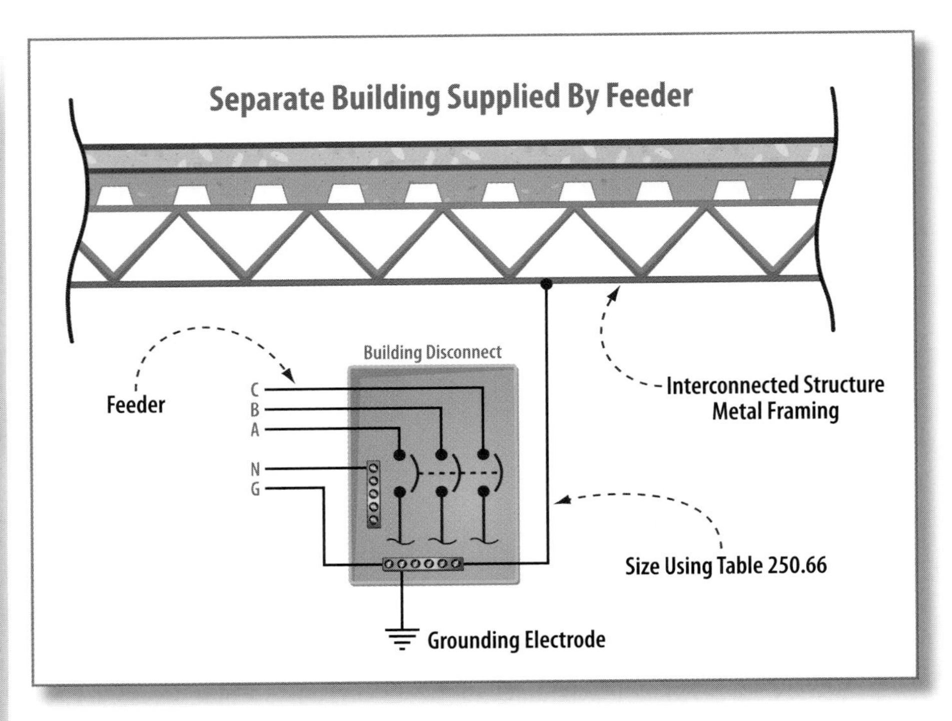

Change Summary

The words "the disconnecting means for buildings or structures supplied by a feeder or branch circuit" have been added to the first sentence of 250.104(C). The words "unless installed in compliance with 250.68(A) Exception No. 2" have been added to the last sentence.

Significance of the Change

The structural metal framing bonding requirements in 250.102(C) apply to buildings supplied by services or by feeders or branch circuits. Most often, electrical inspectors enforce bonding of structural metal if the building or structure has exposed interconnected structural metal framing members and the building or structure is supplied with a service, feeder, or branch circuit. This revision provides a consistent structural metal framing bonding rule that applies whether the building is supplied by a service or other than a service without having to apply the required judgment about all the possibilities presented in the concepts of determining what would be likely to become energized. Referencing 250.68 Exception No. 2 permits the connection of bonding conductors or jumpers to be fireproofed and not be accessible if appropriate.

Comment: None
Proposals: 5-253, 5-254

250.119 (Exc.)

Article 250 Grounding and Bonding
Part VI Equipment Grounding
and Equipment Grounding Conductors

Identification of Equipment Grounding Conductors, Exception

Code Language

Exception: Power-limited Class 2 or Class 3 cables, power-limited fire alarm cables or communications cables, containing only circuits operating at less than 50 volts where connected to equipment not required to be grounded in accordance with 250.112(I) shall be permitted to use a conductor with green insulation or green with one or more yellow stripes for other than equipment grounding purposes.

Significance of the Change

The significant revision in this exception is its expansion to apply to other limited energy cables, including and beyond those containing just Class 2 or Class 3 power-limited circuits. A conductor of these cable assemblies that is green or green with one or more yellow stripes can be used for other than equipment grounding purposes. This revision makes the exception current with actual practices in the field relative to manufacturing, installation, and use of these limited energy cables.

The exception has also been revised to clarify its application only if these cables are used with equipment that is not required to be grounded. Where the equipment is required to be grounded, the equipment grounding conductor identification requirements in the rule apply. The intent of the exception is to allow specific wiring methods, connected to specific equipment, to be exempt from the rule in 250.119 regarding the use of a conductor with green insulation. This exception lists the wiring methods, while 250.112(I) addresses the equipment grounding requirements or exemptions as applicable.

Change Summary

The exception to 250.119 has been revised and expanded by adding the words "power-limited fire alarm cables or communications cables"; the words "where connected to equipment not required to be grounded in accordance with 250.112(I)" after the word "volts"; and the words "or green with one or more yellow stripes" after the words "green insulation."

Comment: 5-175
Proposal: 5-275

Use of Equipment Grounding Conductors

Code Language

250.121 Use of Equipment Grounding Conductors.
An equipment grounding conductor shall not be used as a grounding electrode conductor.

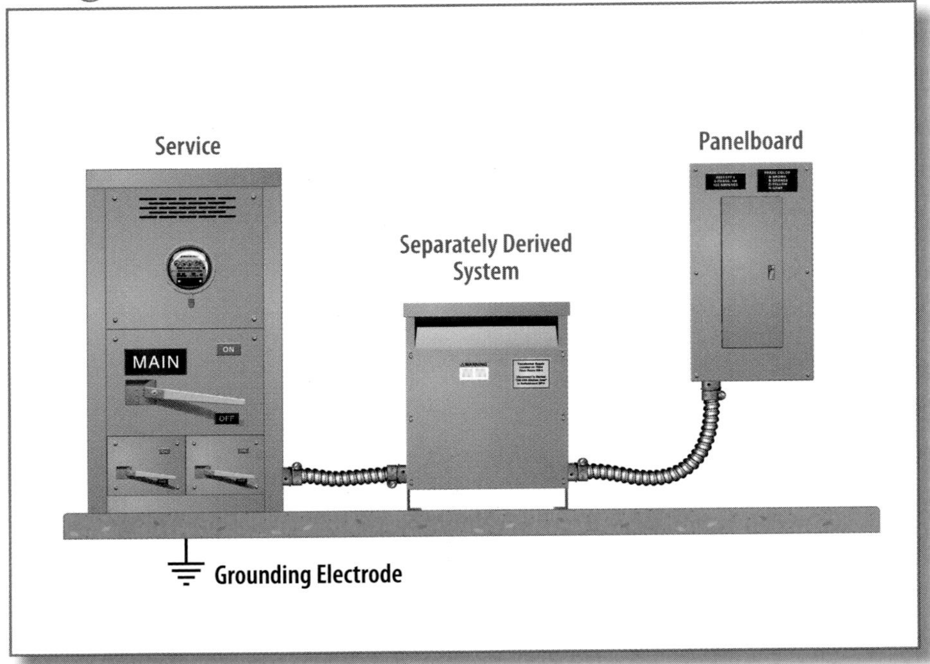

Change Summary

A new section, 250.121 Use of Equipment Grounding Conductors, has been added to Part VI of Article 250.

Comments: 5-162, 5-163

Proposal: 5-259

Significance of the Change

This new rule prohibits equipment grounding conductors from being used simultaneously as grounding electrode conductors. The *NEC* was silent on this issue in previous editions, so this practice was never restricted. Problems were encountered in attaining compliance with all applicable rules pertaining to both grounding electrode conductors and equipment grounding conductors, which is complicated and difficult in most instances.

The other problem addressed by this new rule is the probability of current introduced on the equipment grounding conductor during normal operation. Equipment grounding conductors do not normally carry current, whereas a grounding electrode conductor may normally carry current since it is often in parallel with the neutral conductor. Additionally, grounding electrode conductors can carry varying amounts of current during normal operation, and equipment grounding conductors should not. Equipment grounding conductors should carry current only during ground-fault conditions when serving as an effective ground-fault current path to facilitate overcurrent device operation. This new section makes clear that grounding electrode conductors and equipment grounding conductors serve different purposes in the electrical safety system, are sized differently, and have different installation requirements.

250.122(F)

Article 250 Grounding and Bonding
Part VI Equipment Grounding and Equipment Grounding Conductors

Conductors in Parallel

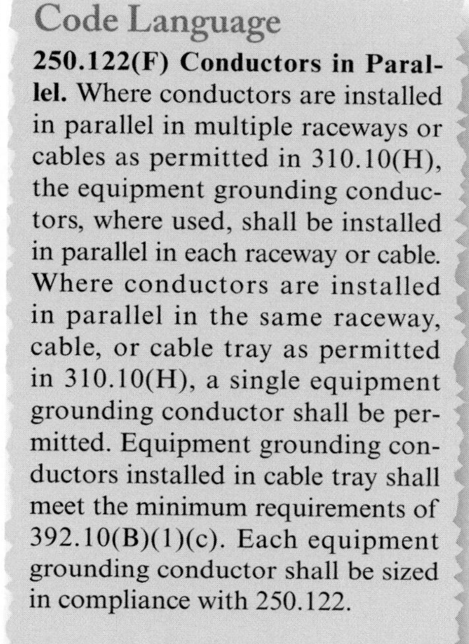

Code Language

250.122(F) Conductors in Parallel. Where conductors are installed in parallel in multiple raceways or cables as permitted in 310.10(H), the equipment grounding conductors, where used, shall be installed in parallel in each raceway or cable. Where conductors are installed in parallel in the same raceway, cable, or cable tray as permitted in 310.10(H), a single equipment grounding conductor shall be permitted. Equipment grounding conductors installed in cable tray shall meet the minimum requirements of 392.10(B)(1)(c). Each equipment grounding conductor shall be sized in compliance with 250.122.

Significance of the Change

This section has been revised to clarify the equipment grounding conductor installation requirements for parallel conductors installed in cable tray arrangements. When ungrounded conductors are installed in parallel in the same raceway, cable, or cable tray, only one equipment grounding conductor is required for the entire parallel set. The *Code* did not specifically address parallel installations in a single raceway or cable tray. Without such language, the requirement was often misinterpreted to mean that an equipment grounding conductor must be installed for each conductor set installed in the same cable tray, which would be unnecessary.

The ungrounded conductors are installed in parallel to create a combined current-carrying capacity. The single equipment grounding conductor is required to be sized in compliance with 250.122 and would be at sufficient size to serve as the grounding means for equipment and to also serve as the ground-fault current path for the installation. The revised text should add clarity on how an equipment grounding conductor should be installed in a cable tray installation containing conductors installed in parallel. The new reference to 392.10(B)(1)(c) is appropriate, given that specific requirements apply when metallic cable tray serves as the equipment grounding conductor for the circuit(s) contained in the tray.

Change Summary

A new second sentence has been added 250.122(F). The last sentence of this section has been revised to refer to all of 250.122 rather than just the table.

Comment: 5-184
Proposal: 5-295

Grounding of Equipment

Code Language

250.190 Grounding of Equipment.

(A) Equipment Grounding. All non-current-carrying metal parts of fixed, portable, and mobile equipment and associated fences, housings, enclosures, and supporting structures shall be grounded.

Exception: Where isolated from ground and located such that any person in contact with ground cannot contact such metal parts when the equipment is energized, shall not be required to be grounded.

Informational Note: See 250.110, Exception No. 2, for pole-mounted distribution apparatus.

(B) Grounding Electrode Conductor. If a grounding electrode conductor connects non-current-carrying metal parts to ground, the grounding electrode conductor shall be sized in accordance with Table 250.66, based on the size of the largest ungrounded service, feeder, or branch circuit conductors supplying the equipment. The grounding electrode conductor shall not be smaller than 6 AWG copper or 4 AWG aluminum.

(... Continued on next page ...)

Change Summary

Section 250.190 has been revised and expanded to include the concepts in Proposals 5-313 and 5-314. It has been reorganized into a list format to meet the *NEC Style Manual* requirements. The existing exception and informational note have been retained.

Comments: 5-192, 5-193

Proposals: 5-313, 5-314

Courtesy of Cogburn Brothers, Inc.

Significance of the Change

This revision expands this rule to include requirements for equipment grounding conductor use, sizing, and connections. These changes continue the process of including medium- and high-voltage requirements in the *NEC* to apply to many installations that previously were covered by the *NESC*. Connections to and minimum sizing of grounding electrode conductors are clarified in new subdivision (B). The changes in subdivision (C) provide users with information that clarifies when shielding on cables can serve as the required equipment grounding conductor for the circuit. Where the shield of cables is insufficient to serve as the equipment grounding conductor, a separate equipment grounding conductor must be installed and sized using Table 250.122 based on the rating of the fuse size or the setting of the overcurrent protective relay.

The revisions to this section have not impacted the original requirement for sizing equipment grounding conductors not less than 6 AWG copper or 4 AWG aluminum. The substantiation in Proposal 5-314 indicated that there have been instances of medium-voltage distribution systems 24940Y/14400 being installed using shield MV cable and the ribbon (tape shield) being used as the required equip-

Grounding of Equipment (continued)

Courtesy of Cogburn Brothers, Inc.

Code Language

(... Continued from previous page ...)

(C) Equipment Grounding Conductor. Equipment grounding conductors shall comply with 250.190(C)(1) through (C)(3).

(1) General. Equipment grounding conductors that are not an integral part of a cable assembly shall not be smaller than 6 AWG copper or 4 AWG aluminum.

(2) Shielded Cables. The metallic insulation shield encircling the current-carrying conductors shall be permitted to be used as an equipment grounding conductor, if it is rated for clearing time of ground fault current protective device operation without damaging the metallic shield. The metallic tape insulation shield and drain wire insulation shield shall not be used as an equipment grounding conductor for solidly grounded systems.

(3) Sizing. Equipment grounding conductors shall be sized in accordance with Table 250.122 based on the current rating of the fuse or the overcurrent setting of the protective relay.

Informational Note: The overcurrent rating for a circuit breaker is the combination of the current transformer ratio and the current pickup setting of the protective relay.

ment grounding conductor. The cross-sectional area of the copper ribbon tape shields typically only equates to an 8 AWG copper conductor, which could be too small to meet the requirements of 250.190 in some instances. The new text in this section clarifies that an equipment grounding conductor is required to be installed along with the medium-voltage feeder conductors if the shielding (ribbon tape or concentric) is insufficient to serve as an equipment grounding conductor and an effective ground-fault current path.

250.191

Article 250 Grounding and Bonding
Part X Grounding of Systems and Circuits Over 1 kV

Grounding System at AC Substations

Code Language

250.191 Grounding System at Alternating-Current Substations. For ac substations, the grounding system shall be in accordance with Part III of Article 250.

Informational Note: For further information on outdoor ac substation grounding, see ANSI/IEEE 80-2000, *IEEE Guide for Safety in AC Substation Grounding.*

I-Stock Photo, Courtesy of NECA

Change Summary

A new section, 250.191 Grounding System at Alternating-Current Substations, and an associated informational note have been added to Part X of Article 250.

Significance of the Change

Previous editions of the *NEC* did not effectively address grounding systems commonly installed and used in AC high-voltage substations. The *Code* addressed these requirements only generally, based on a substantiation that concerns about addressing step and touch potential were not *NEC* requirements. However, IEEE 80 goes into much more detail on how to design a proper grounding system for AC substations. It is important to recognize that many substations are not under the scope of the *NESC* or a serving utility and that *NEC* rules are typically applicable.

This revision is a good first step toward expanding the requirements for substation grounding electrode systems and provides an appropriate reference to Part III of Article 250. An extremely important informational note has been added to provide users with a reference to the appropriate IEEE guidelines for installing substation grounding systems. This change reflects the continuing effort to include appropriate medium- and high-voltage requirements in the *Code* that apply to installations falling under the *NEC* scope.

Comment: None
Proposal: 5-315

Listing

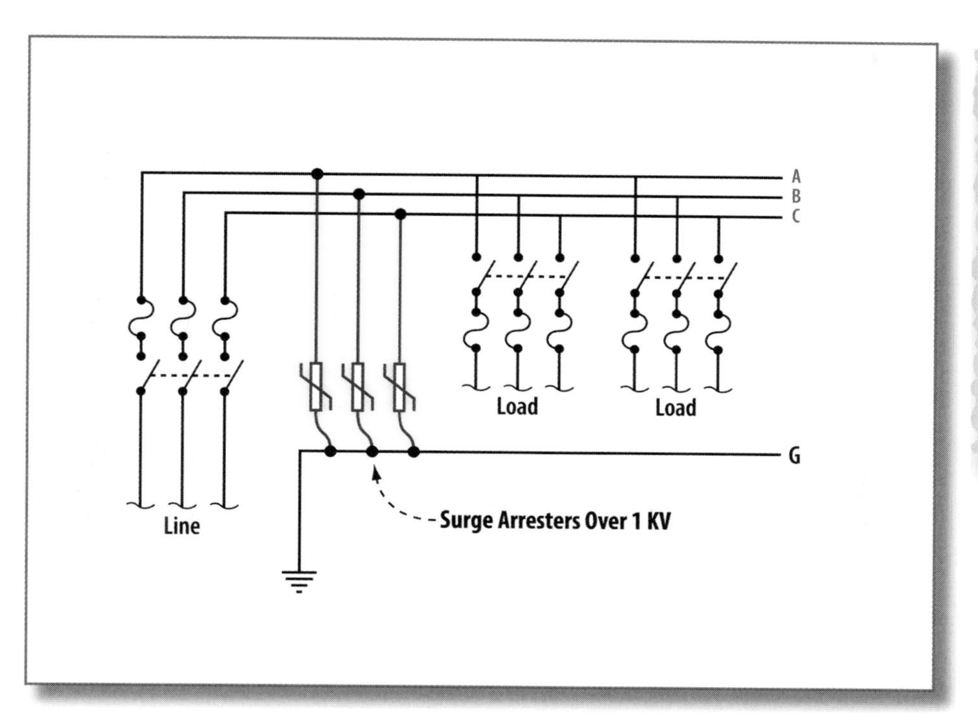

Code Language

280.5 Listing. A surge arrester shall be a listed device. ***(Deleted Text)***

Significance of the Change

Section 280.5 was new to the 2008 *NEC*, but action on Proposal 5-316 has deleted the listing requirement for surge arresters rated for use in systems over 1000 volts. Seldom are requirements for listed equipment removed from the *NEC*, given that product evaluation and certification by qualified testing laboratories provide evidence that a product meets minimum standards for safety and use. Inspection authorities also use product certification and listing as a basis for approvals. This requirement was deleted based on substantiation that indicated there are currently no listed surge arresters rated for 1000 volts and higher. However, there are standards to which these devices can be evaluated. Category VZQK in the *UL Guide Information for Electrical Equipment* provides additional information. This category, entitled "Surge Arresters 1000 Volts and Higher," currently includes no manufacturers that list to this category.

Deleting this requirement does not eliminate the possibility of inspectors requiring some form of product evaluation or other evidence of meeting minimum safety standards in their approval process. Many surge arrester installations at this voltage level were not previously covered by *NEC* but now are, and listed and labeled equipment may be a requirement for jurisdictions having approving authority.

Change Summary

Section 280.5 has been deleted from Article 280.

Comments: 5-195, 5-201

Proposals: 5-316, 5-317

Type 3 SPDs

Code Language

285.25 Type 3 SPDs. Type 3 SPDs (TVSSs) shall be permitted to be installed on the load side of branch-circuit overcurrent protection up to the equipment served. If included in the manufacturer's instructions, the Type 3 SPD connection shall be a minimum 10 m (30 ft) of conductor distance from the service or separately derived system disconnect.

Change Summary

Section 285.25 has been revised by adding the words "if included in the manufacturer's instructions" and removing the word "anywhere" from this requirement.

Significance of the Change

Article 285 was extensively revised during the 2008 *NEC* development process, at the very same time UL 1449 was being revised. Because UL 1449 was not finalized at the time of Code-Making Panel 5 discussions for the 2008 *NEC*, no allowances were made for this exception in the 2008 *Code*. The current revision corrects this deficiency and aligns this requirement and exceptions with those of the product standard for surge protective devices (SPDs). This section has been revised to specify a minimum distance of 30 feet for the conductors of a Type 3 SPD if so instructed by the manufacturer.

Note that if an SPD is provided with a cautionary marking indicating specific installation requirements, installers are required to comply with all those installation instructions included in the listing and labeling of electrical equipment. The revisions to this requirement are intended to ensure that SPD equipment is installed and used within the limitations of its ratings and its evaluation for use. Other benefits from these new requirements ensure that SPD products are correctly installed on systems and that the maximum short-circuit current ratings of the device are not exceeded. The revision emphasizes the importance of following manufacturer's installation guidelines and instructions.

Comments: 5-204, 5-205

Proposal: 5-324

The *NEC* Process

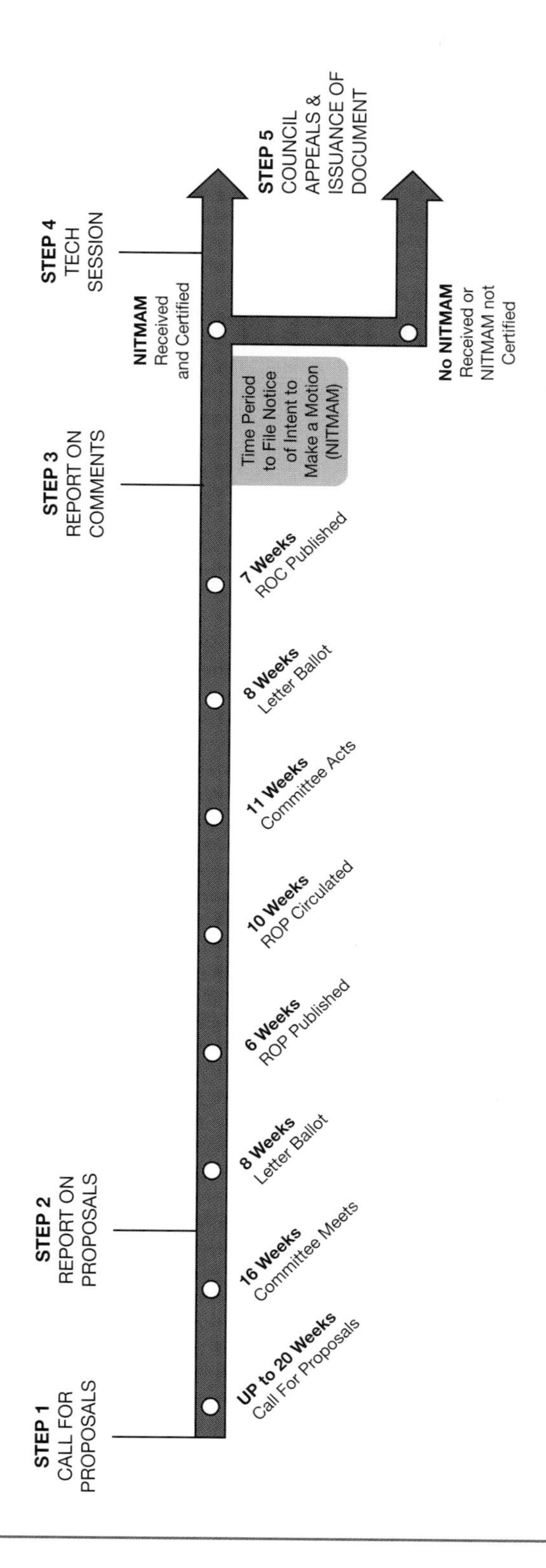

Chapter 3

Articles 300-399
Wiring Methods and Materials

Significant Changes
TO THE NEC® 2011

300.4(E)

Article 300 Wiring Methods
Part I General Requirements

Cables, Raceways, or Boxes Installed In or Under Roof Decking

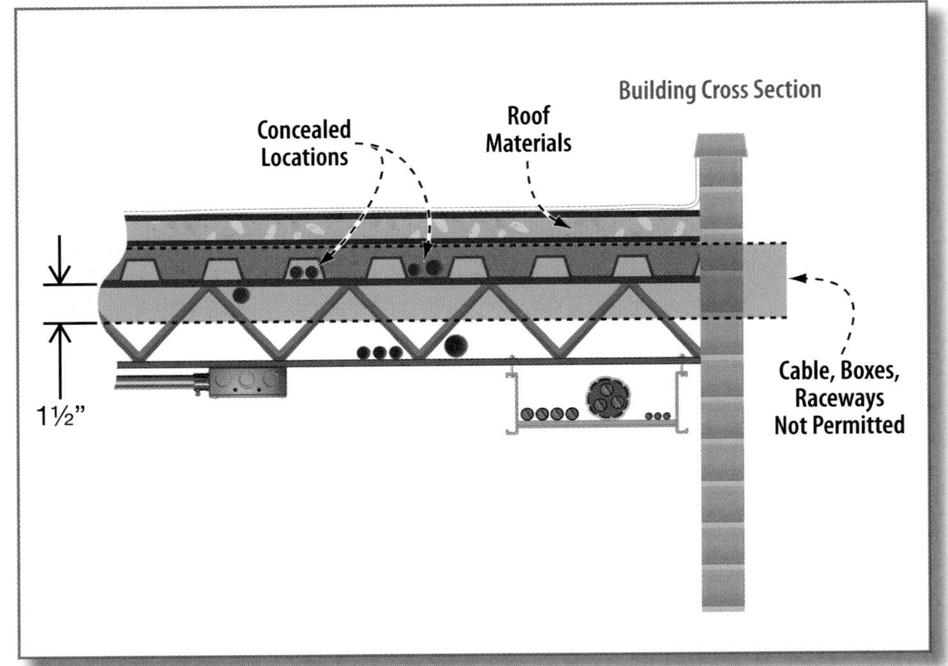

Code Language

(E) Cables, Raceways, or Boxes Installed In or Under Roof Decking. A cable, raceway, or box, installed in exposed or concealed locations under metal-corrugated sheet roof decking, shall be installed and supported so there is not less than 38 mm (1 ½ in.) measured from the lowest surface of the roof decking to the top of the cable, raceway, or box. A cable, raceway, or box shall not be installed in concealed locations in metal-corrugated, sheet decking-type roof.

(The existing exception and informational note remain unchanged.)

Change Summary

The phrase "measured from the lowest" has been incorporated into the first sentence of this section. Boxes have been added to the requirement, and a new second sentence addresses concealed locations in metal-corrugated sheet metal roof decking.

Comment: None

Proposals: 3-34, 3-39

Significance of the Change

This rule was new in the 2008 *NEC* and only addressed protection for cables and raceways installed under roof decking. This current revision provides protection for boxes and makes clear that the minimum distance from the (lowest) bottom of the decking material to the top of the wiring method or box is not permitted to be less than 1 ½ inches. The previous language required the measurement to be from the "nearest" surface of the decking material.

This section has also been revised to restrict wiring from being installed in concealed spaces above the roof decking between the insulation and roofing material. Wiring installed in such locations is subject to physical damage during roofing and reroofing operations. Substantiation provided with the proposal indicated that junction and pull boxes are exposed to the same damage as are cables and raceways when installed on the underside of metal roof decking. These revisions prohibit electrical wiring from being installed in concealed locations above the roof decking and from within 1 ½ inches from the lowest portion of the roof decking material. The existing exception continues to exempt rigid metal conduit and intermediate metal conduit from these installation restrictions.

Table 300.5

Article 300 Wiring Methods
Part I General Requirements

Minimum Cover Requirements, 0 to 600 Volts, Nominal...

Note: Reproduction of new row 3 in Table 300.5 (in part)

Location of Wiring Method or Circuit	Column 1	Column 4	Column 5
Under a Building	0 0 (in a raceway or Type MC or Type MI cable identified for direct burial)	0 0 (in a raceway or Type MC or Type MI cable identified for direct burial)	0 0 (in a raceway or Type MC or Type MI cable identified for direct burial)

A new row has been added to Table 300.5 and addresses burial depth requirements for raceway and Type MC or Type MI cables identified for direct burial.

Code Language

New text added to Table 300.5 in row 3 (Under a building) in columns 1, 4, and 5 as follows:

0 mm and 0 in.
(in a raceway or Type MC or Type MI cable identified for direct burial)

See reproduction of Table 300.5 (in part).

Significance of the Change

Some types of MC cable and some types of MI cable are rated for direct burial and for encasement in concrete. Substantiation in the proposal indicated that it was not clear that the "Under a Building" rule in Table 300.5 required that the cable or conductors be in a raceway. For example, Type MC cable that is listed and identified for direct burial is certainly suitable for installation in a crawl space under a building, for direct burial, as well as encased in concrete, although there was no mention of these conditions within the table. This change is a companion revision to the new exceptions included in 300.5(C). The new wording in columns 1, 4, and 5 of Table 300.5 clarifies that Type MC and Type MI cables can be installed in the earth under a building as allowed by the new exceptions to 300.5(C) and by specific provisions within each respective wiring method article. The cables used in these applications must be suitable for installation in direct burial with the earth. There is no depth requirement for Type MC and Type MI cables installed in the earth underneath a building.

Change Summary

The text in Columns 1, 4, and 5 of Table 300.5 has been revised to include the wording "(in a raceway or Type MC or Type MI cables identified for direct burial)."

Comment: None

Proposal: 3-49

300.5(C)

Article 300 Wiring Methods
Part I General Requirements

Underground Cables Under Buildings

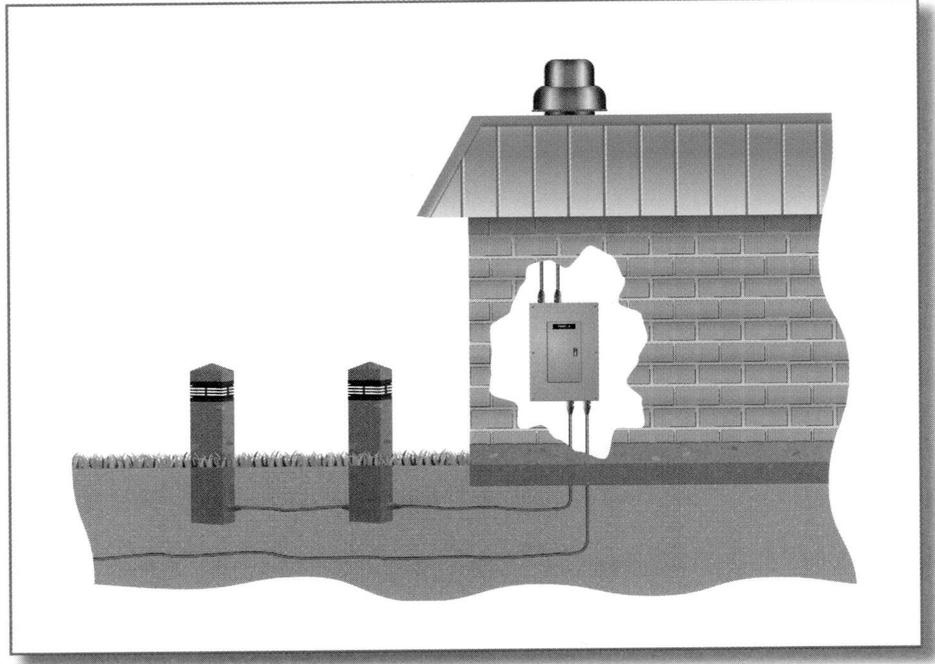

Code Language

300.5(C) Underground Cables Under Buildings. Underground cable installed under a building shall be in a raceway.

Exception No. 1: Type MI cable shall be permitted under a building without installation in a raceway where embedded in concrete, fill, or other masonry in accordance with 332.10(6) or in underground runs where suitably protected against physical damage and corrosive conditions in accordance with 332.10(10).

Exception No. 2: Type MC cable listed for direct burial or concrete encasement shall be permitted under a building without installation in a raceway in accordance with 330.10(A)(5) and in wet locations in accordance with 330.10(11).

Change Summary

Two new exceptions have been added to 300.5(C). These exceptions correlate with the additional information in Table 300.5 that now provides the minimum burial depths for MC cables and MI cable where installed underneath a building.

Significance of the Change

Substantiation with the proposals indicated that Type MC cable and Type MI cable are available for use where listed for direct earth burial and concrete encasement. These cable assemblies do not require protection in a raceway under a building for suitable installation. CMP-3 acted favorably to the addition of MI cable as a new Exception No.1, given that certain conditions based on requirements in 332.10 apply before MI cable can be installed under a building without an additional raceway requirement. The revision expands the uses of Types MI and MC cables that are suitable for use in direct burial or concrete encasement applications. Permitting these types of cable to be installed underneath buildings or structures without being contained in a raceway means that they are not easily removed or replaced without the removal of concrete and earth beneath the building. The important feature of this new allowance (use) is that the product used in these applications are suitable for the use. The MC cable has to be listed for this use and the MI cable should be the copper-sheathed type, as opposed to the steel-alloy type.

Comment: 3-17
Proposals: 3-48, 3-52

300.5(I) (Exc. No. 1)

Article 300 Wiring Methods
Part I General Requirements

Conductors of the Same Circuit, Exception No. 1

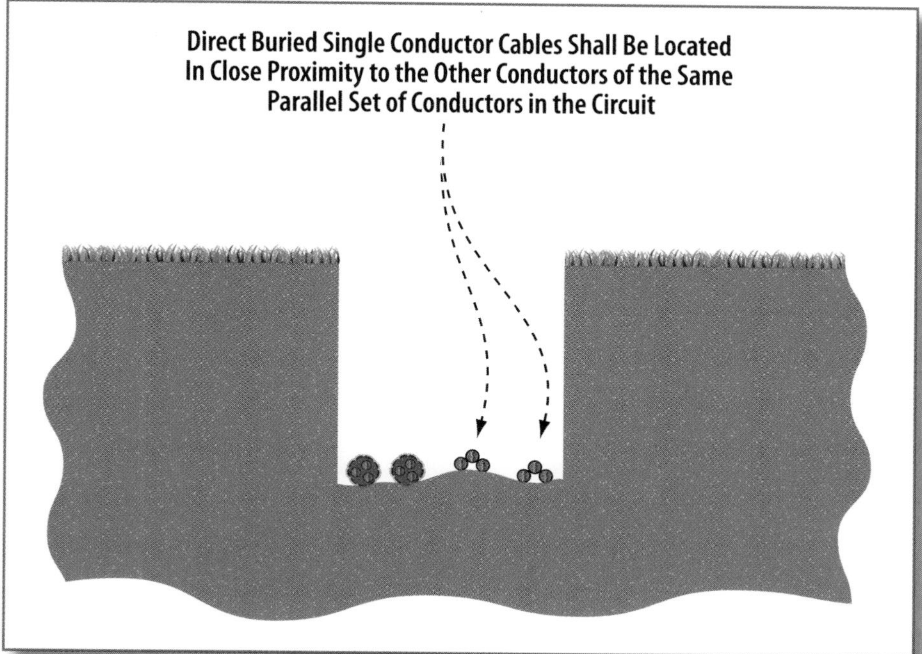

Code Language

Exception No. 1: Conductors shall be permitted to be installed in parallel in raceways, multiconductor cables, or direct-buried single conductor cables. Each raceway or multiconductor cable shall contain all conductors of the same circuit including equipment grounding conductors. Each direct-buried single conductor cable shall be located in close proximity in the trench to the other single conductor cables in the same parallel set of conductors in the circuit, including equipment grounding conductors.

Significance of the Change

This exception was revised to include multiconductor underground cables (USE multiple conductor cables) as well as single underground conductor cables (single USE conductor cables), as provided in Article 338. The text of the 2008 *NEC* required that for single-conductor cables run in parallel, all of the conductors must run together, because the exception only covered cables or raceways that are capable of containing all of the circuit conductors, which left out provisions for single conductors. Since derating requirements apply to all conductors run in the same trench, the derating penalties for doing this are severe. The revised wording corrects this inadvertent omission of single conductor direct-buried cables from this exception. The revised last sentence clarifies that single conductors of a direct-burial installation are required to be placed close together for single conductor feeders and branch circuits as well as those installed in parallel arrangements.

Change Summary

Exception No. 1 has been revised to include the wording "direct-buried single conductor cables." The revision to Exception No. 1 clarifies that the conductors of a parallel circuit must be installed in close proximity to each other.

Comment: None

Proposal: 3-61

300.11(A)(2)

Article 300 Wiring Methods
Part I General Requirements

Non–Fire-Rated Assemblies

Code Language

(2) Non–Fire-Rated Assemblies. Wiring located within the cavity of a non–fire-rated floor–ceiling or roof–ceiling assembly shall not be secured to, or supported by, the ceiling assembly, including the ceiling support wires. An independent means of secure support shall be provided and shall be permitted to be attached to the assembly. Where independent support wires are used, they shall be distinguishable by color, tagging, or other effective means.

(No changes to the exception)

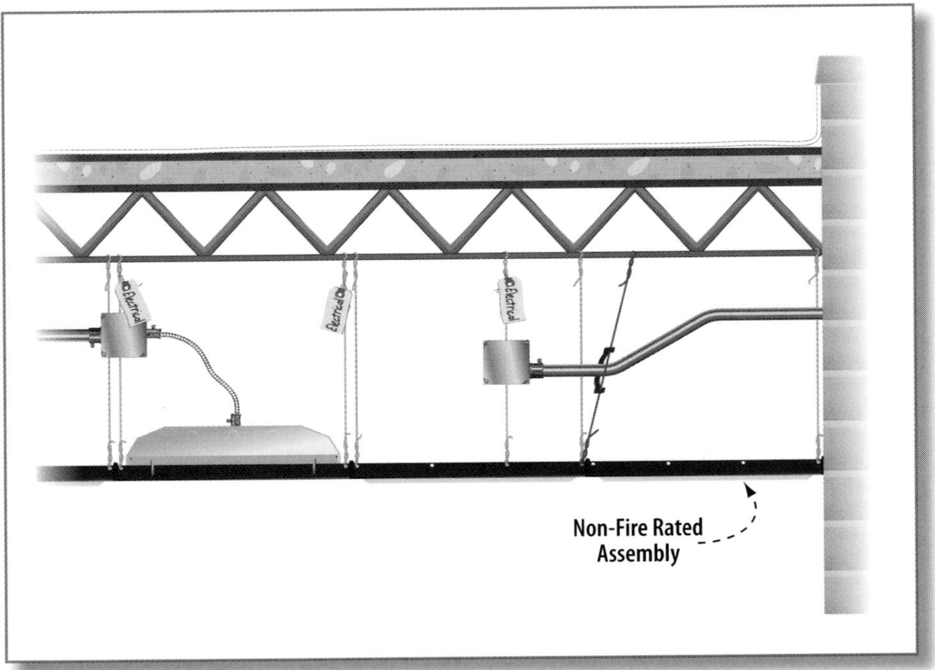

Change Summary

This section has been revised by adding a new last sentence as follows: Where independent support wires are used, they shall be distinguishable by color, tagging, or other effective means. This revision aligns Section 300.11(A)(2) with similar requirements contained in Section 300.11(A)(1).

Significance of the Change

The purpose of this revision was to include an identification requirement for support wires that are installed in accordance with approved wiring methods and are located above a non-fire-rated floor-ceiling or roof-ceiling assembly. This change aligns this identification requirement with an identical requirement in 300.11(A)(1) for identification of additional support wires for wiring methods installed above a fire-rated floor-ceiling or roof-ceiling assemblies. The revision improves consistency for identical installations covered in different *Code* sections. The benefits of identifying support wires that are installed in addition to those for suspended ceiling support are ease of recognizing those wires that are for electrical wiring and distinguishing them from those that have a structural support function for the building construction. As provided in 300.11(A)(1), independent identification of support wire can be by color, tagging, or other effective means. Substantiation in the proposal also indicated that inspectors would be able to readily differentiate between construction support wires that serve different purposes.

Comment: None
Proposal: 3-73

Spread of Fire or Products of Combustion

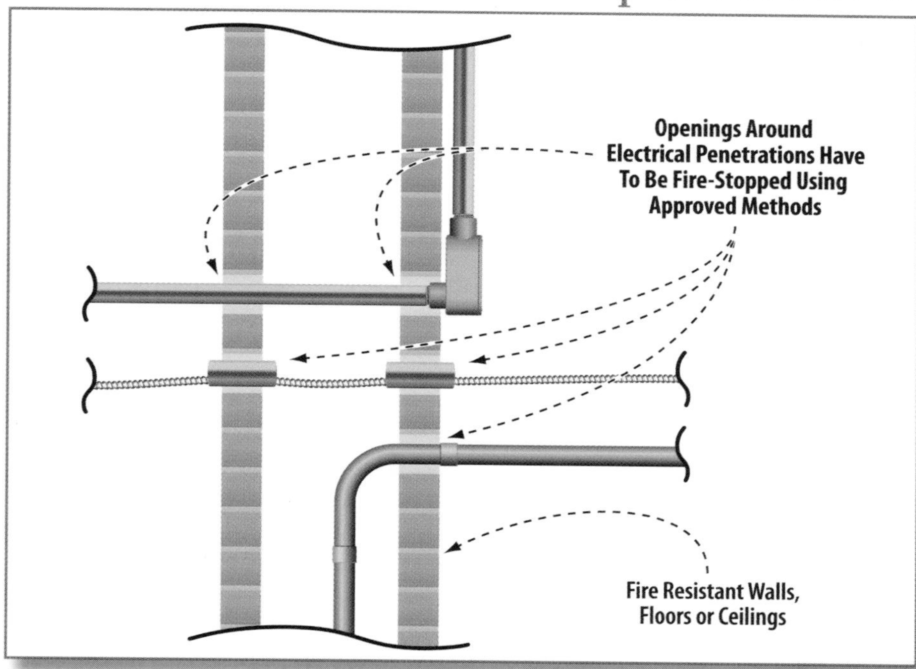

Code Language

300.21 Spread of Fire or Products of Combustion. Electrical installations in hollow spaces, vertical shafts, and ventilation or air-handling ducts shall be made so that the possible spread of fire or products of combustion will not be substantially increased. Openings around electrical penetrations into or through fire-resistant-rated walls, partitions, floors, or ceilings shall be fire-stopped using approved methods to maintain the fire resistance rating.

(The Informational Note remains unchanged.)

Significance of the Change

Section 300.21 provides a requirement to maintain fire resistance ratings around electrical penetrations of fire-rated walls, ceilings, partitions, and floors. This revision provides additional clarification as to which openings and penetrations the fire-stopping requirement applies. The previous text only addressed conditions where electrical wiring passed through fire-resistant-rated walls, partitions, floors, or ceilings. In some instances, however, a penetration is made into a chamber that has a fire resistance rating but the wiring does not pass through, such as in a wall or other construction cavity or in a plenum or other space used for environmental air. Adding the term "into or" makes it clear that not only through-penetrations, but also membrane penetrations, into fire-resistive construction must be provided with fire-stopping to maintain the fire-resistive rating.

Change Summary

The words "into or" have been added in the second sentence before the word "through." This revision allows this rule to apply to not only electrical penetrations through a fire-rated floor, wall, ceiling, or other partition, it also applies to electrical penetrations into a building component that must maintain a fire rating.

Comment: None

Proposal: 3-92

Wiring in Ducts Not Used for Air Handling...

Courtesy of Chris Bayer

Code Language

300.22 Wiring in Ducts Not Used for Air Handling, Fabricated Ducts for Environmental Air, and Other Spaces For Environmental Air (Plenums). The provisions of this section shall apply to the installation and uses of electrical wiring and equipment in ducts used for dust, loose stock, or vapor removal; ducts specifically fabricated for environmental air; and other spaces used for environmental air (plenums).

(Informational Note remains unchanged.)

(A) Ducts for Dust, Loose Stock, or Vapor Removal. ***(No changes to this section)***

(B) Ducts Specifically Fabricated for Environmental Air. ***(See NEC for full text.)***

(C) Other Spaces Used for Environmental Air (Plenums). This section shall apply to spaces not specifically fabricated for environmental air-handling purposes but used for air-handling purposes as a plenum. This section shall not apply to habitable rooms or areas of buildings, the prime purpose of which is not air handling.

(Informational Note No. 1 remains unchanged.)

Informational Note No. 2: The phrase "Other Spaces Used for Environmental Air (Plenum)" as used in this section correlates with the use of the term "plenum" in NFPA 90A-2009, the *Standard for the Installation of Air-Conditioning and Ventilating Systems*, and other mechanical codes where the plenum is used for return air purposes, as well as some other air-handling spaces.

(1) Wiring Methods. ***(No changes to this section)***

(... Continued on next page ...)

Change Summary

The title of 300.22 was changed to Wiring in Ducts Not Used for Air Handling, Fabricated Ducts for Environmental Air, and Other Spaces for Environmental Air (Plenums). Subdivision titles have been revised, in (B) to handle fabrication, and in (C) to cover spaces not specifically fabricated for air-handling purposes. A new list item (2), Cable Tray Systems, has been added to 300.22(C). Former 300.22(C)(2), titled Equipment, has been renumbered as list item (3) and slightly revised. A new Informational Note No. 2 has been added following 300.22(C).

Comment: 3-38

Proposals: 3-94, 3-97

Wiring in Ducts Not Used for Air Handling... (continued)

Significance of the Change

The revisions to this section incorporate the language "space used as a plenum for environmental air" to more clearly describe a space such as that above a ceiling or below a raised floor that is not built solely to move air. Revision of the subdivision titles aligns 300.22 with other relevant codes and standards, including NFPA 90A, the *International Mechanical Code*, and the *Uniform Mechanical Code*. Additionally, revisions of certain terms were made in accordance with directives to maintain consistency and the status quo and to better align terms in this section of the *NEC* with those used in NFPA 90A.

This new language in 300.22(C)(2) clarifies the use of metallic cable tray systems in spaces intended for environmental air-handling applications (plenums). Provisions for cable tray installations are now included in 300.22(C). Section 300.22(C)(2)(a) clearly permits metal cable tray systems to support the wiring methods specifically addressed in 300.22(C)(1). Section 300.22(C)(2)(b) makes it clear that solid bottom metal cable tray systems with solid covers are permitted to be used with wiring methods and cables not already covered in (C)(1) and installed in accordance with 392.10(A) and (B). Cable tray is not a raceway, but rather a support system for wiring methods. Former list item (2) covering equipment was renumbered as (3), and the word *equipment* was added to the first sentence to clarify that "equipment with nonmetallic enclosures," not just "nonmetallic enclosures," is covered by the provisions in this rule and that both types must be listed for use in air-handling spaces.

New Informational Note No. 2 explains the meaning of the phrase "other space used for environmental air (plenums)" as used in this section and correlates this usage with that in NFPA 90A, the *Standard for the Installation of Air-Conditioning and Ventilating Systems*, 2009, and in other mechanical codes governing plenums used for return-air purposes, in addition to other air-handling spaces.

Code Language

(... Continued from previous page ...)

(2) Cable Tray Systems. The provisions in (a) or (b) shall apply to the use of metallic cable tray systems in other spaces used for environmental air (plenums), where accessible, as follows:

(a) Metal Cable Tray Systems. Metal cable tray systems shall be permitted to support the wiring methods in 300.22(C)(1).

(b) Solid Side and Bottom Metal Cable Tray Systems. Solid side and bottom metal cable tray systems with solid metal covers shall be permitted to enclose wiring methods and cables, not already covered in 300.22(C)(1), in accordance with 392.10(A) and (B).

(3) Equipment. Electrical equipment with a metal enclosure, or electrical equipment with a nonmetallic enclosure listed for use within an air-handling space and having adequate fire-resistant and low-smoke-producing characteristics, and associated wiring material suitable for the ambient temperature shall be permitted to be installed in such other space unless prohibited elsewhere in this *Code*.

Informational Note: One method of defining adequate fire-resistant and low-smoke producing characteristics for electrical equipment with a nonmetallic enclosure is in ANSI/UL 2043-2008, *Fire Test for Heat and Visible Smoke Release for Discrete Products and Their Accessories Installed in Air-Handling Spaces.*

Conductors for General Wiring

Code Language

Article 310 Conductors for General Wiring

Part I General

310.2. Definitions

Electrical Ducts. Electrical conduits, or other raceways round in cross section, that are suitable for use underground or embedded in concrete.

Thermal Resistivity. As used in this *Code*, the heat transfer capability through a substance by conduction. It is the reciprocal of thermal conductivity and is designated Rho and expressed in the units °C-cm/W.

Part II Installation

310.10. Uses Permitted. The conductors described in 310.104 shall be permitted for use in any of the wiring methods covered in Chapter 3 and as specified in their respective tables or as permitted elsewhere in this *Code*.

Part III Construction Specifications

(See NEC text for full article renumbering and technical revisions accepted from separate proposals.)

Reorganization of Article 310 Conductors for General Wiring

Part I General

310.2 Definitions.

Electrical Ducts. Electrical conduits, or other raceways round in cross section, that are suitable for use underground or embedded in concrete.

Thermal Resistivity. As used in this *Code*, the heat transfer capability through a substance by conduction. It is the reciprocal of thermal conductivity and is designated Rho and expressed in the units °C-cm/W.

Part II Installation (See *NEC* for full text.)

310.10 Uses Permitted. The conductors described in 310.104 shall be permitted for use in any of the wiring methods covered in Chapter 3 and as specified in their respective tables or as permitted elsewhere in this *Code*.

Part III Construction Specifications (See *NEC* for full text.)

Note: See NEC text for full article renumbering and arrangement.

Change Summary

Article 310 and its contained tables were renumbered and reorganized. This article is now subdivided into three parts titled "General," "Installation," and "Construction Specifications." New section 310.2 simply relocates the definitions for the terms *Electrical Ducts* and *Thermal Resistivity* from 310.60(A) to this position to conform to the *NEC Style Manual.* Various revisions have also been incorporated into the article using the new numbering sequence. Separate technical revisions have also been incorporated throughout Article 310 in addition to the renumbering and reorganization.

Comment: 6-4
Proposal: 6-8

Conductors for General Wiring (continued)

Courtesy of Cogburn Brothers, Inc.

Significance of Change

Article 310 was completely reorganized and renumbered to provide a more logical and sequential order, and the tables within have been renumbered according to that new layout and numbering sequence. These revisions bring Article 310 into conformance with the *NEC Style Manual* and provide more consistent correlation with other Chapter 3 articles. The restructuring of Article 310 also incorporates an expanded numbering range to allow for future revisions and growth of the *NEC*. In addition, the article has been segmented into three parts. Part I covers general requirements, Part II covers conductor installation, and Part III covers construction specifications for conductors. The rules and contents of tables generally remain as in the 2008 *NEC*. Proposal 6-8 (Log No. 2395) did not include any technical revisions to the current requirements in Article 310, although separate changes were accepted that resulted in various revisions to some of the existing rules.

Section 310.6 was renumbered as 310.10(H) and revised to include conductors rated over 2000 V, since this section now addresses all conductor types within Article 310. As part of this revision, some existing rules had to be divided into separate sections to comply with 2.3.1 of the *NEC Style Manual*, which states that "tables and figures shall be referenced in the text and shall be designated by the number of the *NEC* rule in which they are referenced." Without this division, confusion could exist over the table or figure designations to the *NEC* rules. This reorganization is an overall improvement in the arrangement of the article and enhances clarity and usability.

310.10(E)

Article 310 Conductors for General Wiring
Part II Installation

Shielding

Code Language

(E) Shielding. Non-shielded, ozone-resistant insulated conductors with a maximum phase-to-phase voltage of 5000 volts shall be permitted in Type MC metal clad cables in industrial establishments where the conditions of maintenance and supervision ensure that only qualified persons service the installation. For other establishments, solid dielectric insulated conductors operated above 2000 volts in permanent installations shall have ozone-resistant insulation and shall be shielded. All metallic insulation shields shall be connected to a grounding electrode conductor, a grounding busbar, an equipment grounding conductor or a grounding electrode.

Informational Note: The primary purposes of shielding are to confine the voltage stresses to the insulation, dissipate insulation leakage current, drain off the capacitive charging current, and carry ground fault current to facilitate operation of ground fault protective devices in the event of an electrical cable fault.

Exception No. 2: Nonshielded insulated conductors listed by a qualified testing laboratory shall be permitted for use up to 5000 volts to replace existing nonshielded conductors, on existing equipment in industrial establishments only, under the following conditions:

(a) Where the condition of maintenance and supervision ensures that only qualified personnel install and service the installation.

(b) Conductors shall have insulation resistant to electric discharge and surface tracking, or the insulated conductor(s) shall be covered with a material resistant to ozone, electric discharge, and surface tracking.

(c) Where used in wet locations, the insulated conductor(s) shall have an overall nonmetallic jacket or a continuous metallic sheath.

(d) Insulation and jacket thicknesses shall be in accordance with Table 310.13(D).

Informational Note: Relocation or replacement of equipment may not comply with the term *existing* as related to this exception.

Courtesy of Cogburn Brothers, Inc.

Significance of the Change:

Section 310.6 was renumbered as 310.10(E) as a result of action on Proposal 6-8 to reorganize Article 310 into more logical order. In addition to the renumbering, 310.10(E) was revised to include two technical changes. The first sentence was revised to clarify that non-shielded, ozone-resistant insulated conductors having a maximum phase-to-phase voltage of 5000 volts are permitted in Type MC metal clad cables in industrial establishments under controlled conditions that include qualified persons servicing the installation. The words "equipment grounding conductor" were added to the second sentence of the requirement. This revision clarifies that the shielding of medium- and high-voltage cables can be connected to an equipment grounding conductor in addition to a grounding electrode, grounding electrode conductor, or grounding busbar within equipment. The second revision required a new Exception No. 2, to allow listed non-shielded cables up to 5000 volts to be used as replacements on existing equipment under restrictive conditions and only in industrial establishments. Exception Nos. 1 and 3 to 310.10(E) remain unchanged.

Shielding (continued)

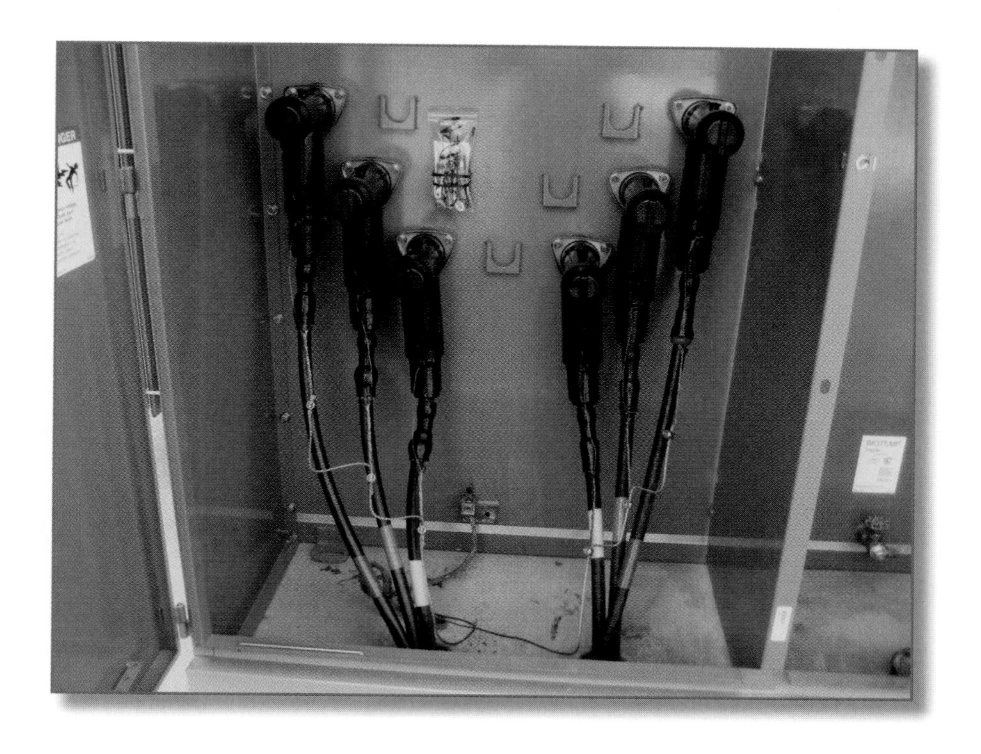

Change Summary

The first sentence of 310.10(E) [Former 310.6] has been revised, and the second sentence now includes the words "equipment grounding conductor." A new Exception No. 2 and two new informational notes have been added to this rule. Existing Exception No. 2 has been renumbered as Exception No. 3 and remains unchanged.

The conditions of acceptability under the exception include the installation being serviced by qualified persons under conditions of maintenance and supervision. The other conditions are as follows:

1. The conductors have insulation resistance to electric discharge and surface tracking, or the conductors are covered with an insulation resistant to ozone, surface tracking, and electric discharge.
2. For conductors in wet locations, the insulated conductor has an overall nonmetallic jacket or a continuous metallic sheath.
3. The insulation and jacket thickness meet the requirements in Table 310.13(D).

The new informational note following 310.10(E) describes the purpose of cable shielding and how it functions. The new informational note following the new exception provides users with the limits of the exception and emphasizes that the provisions of the new exception may apply to existing cable replacements on equipment as long as the conditions within the exception are met. This informational note emphasizes the involvement and decision making that may have to include the approving authority for acceptance.

Comment: 6-7, 6-9

Proposals: 6-21a, 6-21b, 6-24

310.10(H)

Article 310 Conductors for General Wiring
Part II Installation

Conductors in Parallel

Code Language

310.10(H) Conductors in Parallel

(1) General. Aluminum, copper-clad aluminum, or copper conductors, for each phase, polarity, neutral, or grounded circuit shall be permitted to be connected in parallel (electrically joined at both ends) only in sizes 1/0 AWG and larger where installed in accordance with 310.10(H)(2) through (H)(6).

Exception No. 1 ***(remains unchanged)***

Exception No. 2: Under engineering supervision, 2 AWG and 1 AWG grounded neutral conductors shall be permitted to be run in parallel for existing installations.

Change Summary

Section 310.4, covering requirements for parallel conductors, has been renumbered as 310.10(H). As revised, the general rule recognizes only conductors in sizes 1/0 and larger for parallel arrangements. Exception No. 2 has been revised to reference grounded neutral conductors in sizes 1 AWG and 2 AWG only.

Significance of the Change

This revision provides users with additional clarity about conductors that qualify for parallel installations and those that do not. As revised, the general restriction on conductors installed in parallel arrangements applies to sizes 1/0 AWG and larger, other than as allowed by the exceptions. Conductors in sizes 1/0 AWG and larger are permitted to be installed in parallel arrangements as long as they meet all applicable requirements in 310.10(H)(2) through (H)(6). The difference in numbering results from the reorganization and renumbering of Article 310, per action on Proposal 6-8.

Exception No. 1 has not been affected by these changes; however, Exception No. 2 has been revised to clarify that both 1 AWG and 2 AWG sizes for grounded neutral conductors are permitted under the conditions of the exception. Previously, Exception No. 2 would restrict 1 AWG sizes from being installed in parallel applications. As revised, only sizes 1 and 2 AWG are recognized under this exception, which expands the allowance only to cover 1 AWG not previously mentioned. The revisions to 310.10(H)(1) [previously 310.4] result in improved clarity and usability without any technical changes in requirements.

Comment: None

Proposals: 6-16, 6-19

310.10(H)(2) & (6)

Article 310 Conductors for General Wiring
Part II Installation

Conductors in Parallel

Code Language

310.10(H) Conductors in Parallel.

(2) Conductor Characteristics. The paralleled conductors in each phase, polarity, neutral, grounded circuit conductor, equipment grounding conductor, or equipment bonding jumper shall comply with all of the following:

(1) Be the same length
(2) Consist of the same conductor material
(3) Be the same size in circular mil area
(4) Have the same insulation type
(5) Be terminated in the same manner

(6) Equipment Bonding Jumpers. Where parallel equipment bonding jumpers are installed in raceways, they shall be sized and installed in accordance with 250.102.

Significance of the Change

Former 310.4(B) has been renumbered as 310.10(H)(2), and new list item (6) covers sizing requirements for equipment bonding jumpers installed with conductors in parallel. This revision expands requirements for parallel conductors to include provisions for equipment bonding jumpers installed with other conductors installed in parallel arrangements. The conductor that joins the metallic parts of the transformer and the switchboard, panelboard, or other equipment it supplies is actually an equipment bonding jumper as covered by 250.30(A)(2), not an equipment grounding conductor.

Section 250.102(C) provides the sizing requirements for the supply-side equipment bonding jumper in this case and clarifies that the supply-side equipment bonding jumper (wire type) must be installed in the parallel conduits and be sized based on the circular mil (cm) area of the conductors in the conduit. New list item (6) provides the reference to 250.102(C) clarifying how to size a supply-side equipment bonding jumper when it is a wire type. The change enhances usability and correlates with the sizing requirements for equipment bonding jumpers for separately derived systems as provided in 250.30(A)(2).

Change Summary

The words "equipment bonding jumper" have been added to 310.10(H)(2) [former 310.4(B)]. A new list item (6) covering equipment bonding jumpers has been added to this section.

Comment: None

Proposal: 6-17

310.15(B)

Article 310 Conductors for General Wiring
Part II Installation

Tables

Code Language

310.15(B) Tables. Ampacities for conductors rated 0 to 2000 volts shall be as specified in the Allowable Ampacity Table 310.15(B)(16) through Table 310.15(B)(19), and Ampacity Table 310.15(B)(20) and Table 310.15(B)(21) as modified by 310.15(B)(1) through (B)(7).

The temperature correction and adjustment factors shall be permitted to be applied to the ampacity for the temperature rating of the conductor, if the corrected and adjusted ampacity does not exceed the ampacity for the temperature rating of the termination in accordance with the provisions of 110.14(C). ***(Informational note remains unchanged.)***

Change Summary

The first sentence of 310.15(B) has been revised to reflect the changes to table numbers resulting from the acceptance of Proposal 6-52 and Comment 6-21. A new second sentence has been added to refer users to 110.14(C).

Courtesy of Cogburn Brothers, Inc.

Significance of the Change

The tables in Article 310 have been renumbered. The second sentence is new and provides a significant correlation to 110.14(C), in language intended to provide further information to the user. This revision clarifies that the ampacity for the temperature rating of a conductor may be used for application of the temperature correction and adjustment factors, but the lower of the corrected and adjusted ampacity or the ampacity for the temperature rating of the conductor must be used. The actual temperature rating of the conductor can be used as the starting point for ampacity adjustment or conductor correction factors as long as the resultant ampacity does not exceed the temperature limitations of the termination (60 or 75 degrees C).

While this process may have been familiar to many *Code* users, the revision provides the specific wording needed to clarify what otherwise often has to be explained as something that must be read into the requirements of the two sections working in cooperation with each other. The tie to 110.14 and the additional text help to promote more consistent application of ampacity adjustment and correction factors.

Comment: 6-21
Proposals: 6-51, 6-52

310.15(B)(2)

& Ambient Tables

Article 310 Conductors for General Wiring
Part II Installation

Tables

$$I' = I\sqrt{\frac{T_c - T_a'}{T_c - T_a}}$$

where:

I' = ampacity corrected for ambient temperature
I = ampacity shown in the tables
T_c = temperature rating of conductor (°C)
T_a' = new ambient temperature (°C)
T_a = ambient temperature used in the table (°C)

(3) Adjustment Factors.

Table 310.15(B)(2)(a) Ambient Temperature Correction Factors Based on 30°C (86 °F)

Table 310.15(B)(2)(b) Ambient Temperature Correction Factors Based on 40°C (104 °F)

Note: Reproduction of equation in Section 310.15(B)(2) [only 310.15(B)(2) table titles shown]

Code Language

(Code Text and Tables (in part): Note: See 2011 NEC for full text and tables)

(2) Ambient Temperature Correction Factors. Ampacities for ambient temperatures other than those shown in the ampacity tables shall be corrected in accordance with Table 310.15(B)(2)(a), Table 310.15(B)(2)(b), or shall be permitted to be calculated using the following equation:

(See Equation and Legend, Left. See NEC Section 310.15(B)(2) for complete equation and legend)

Table 310.15(B)(2)(a) Ambient Temperature Correction Factors Based on 30°C (86°F)

Table 310.15(B)(2)(b) Ambient Temperature Correction Factors Based on 40°C (104°F)

Change Summary

Sections 310.15(B)(2) through (6) have been renumbered as 310.15(B)(3) through (7). Section 310.15(B)(2) was expanded to included a new equation and two new tables, which are the relocated ambient temperature correction factor values that previously were located at the bottom of conductor ampacity Tables 310.16, 310.17, 310.18, 310.19, and 310.20.

Significance of the Change

This revision harmonizes the ampacity correction factors for various ambient temperatures between the *NEC* and the Canadian Electrical Code (CEC), Part I; consolidates the ampacity correction and adjustment factors into a single section [310.15(B)(2)]; and relocates and consolidates seven ambient temperature correction tables into two tables [Table 310.15(B)(2)(a) and 310.15(B)(2)(b)]. The new equation is the same as permitted in 310.60(C)(4) to calculate the ampacity correction factors for various ambient temperatures based on the conductor temperature ratings in the tables. This equation also appears in 3.4.1 of IEEE STD 835, *IEEE Standard Power Cable Ampacity Tables*.

Because the *NEC* is used internationally, the lower and higher ambient temperatures were added to provide the appropriate ampacity correction factors for colder and warmer climates. Some of the ambient temperature ranges were revised to 5°C increments to harmonize with the proposal submitted for the CEC, to correlate with existing *NEC* Tables 310.20 and B310.3, and to provide consistency with the 30°C table. This revision resulted from the work of the NFPA/CSA NEC/CEC Ampacity Harmonization Task Group. The new equation provides an additional method to calculate ampacity correction for ambient temperature and improves usability by consolidating all ambient correction factor values into two tables in a revised 310.15(B)(2).

Comment: None

Proposal: 6-53

Table 310.15(B)(16)

Article 310 Conductors for General Wiring
Part II Installation

Allowable Ampacities of Insulated Conductors...

Code Language

Reproduction of *NEC* Table 310.15(B)(16) [in part]

	Copper			Aluminum or Copper-Clad Aluminum		
Size AWG or kcmil	60°C (140°F)	75°C (167°F)	90°C (194°F)	60°C (140°F)	75°C (167°F)	90°C (194°F)
14	15			15		
12	20					
8				35		
6						55
3			115			
1			145			
300				195		260
600	350					
700				315		425
800						445
1500	525					
2000	555					

Revised ampacity values provided in Table 310.15(B)(16)

Allowable Ampacities of Insulated Conductors... (continued)

Change Summary

Ampacity values were changed in Table 310.15(B)(16) as follows:

60°C Column(Copper): 14 AWG = 15 A; 12 AWG = 20 A; 600 kcmil = 350 A; 1500 kcmil = 525 A; 2000 kcmil = 555 A

75°C Column (Copper): No changes

90°C Column (Copper): 3 AWG = 115 A; 1 AWG = 145 A

60°C Column (Aluminum): 12 AWG = 15 A; 8 AWG = 35 A; 300 kcmil = 195 A; 700 kcmil = 315 A

75°C Column (Aluminum): No changes

90°C Column (Aluminum): 6 AWG = 55 A; 300 kcmil = 260 A; 700 kcmil = 425 A; 800 kcmil = 445 A

Significance of the Change

The revision of various ampacity values in Table 310.15(B)(16) [formerly Table 310.16] resulted from work of the NFPA/CSA NEC/CEC Ampacity Harmonization Task Group. These revisions align the values of *NEC* Table 310.15(B)(16) with those of CEC Part I, Tables 2 and 4. The ampacity values in the 75°C columns of both the CEC and *NEC* are identical, so no changes were made to these values. The values in the 60°C and 90°C columns were calculated using defined equations from the following standards:

— IEEE Std 835-1994, *IEEE Standard Power Cable Ampacity Tables*, Section 3.4.2, Adjustment for Change in Maximum Conductor Temperature or Temperature Due to Dielectric Loss.

— *AIEE-IPCEA Power Cable Ampacities*, 1962 (AIEE Pub. No. S-135-1 / IPCEA Pub. No. P-46-426) Volume 1 – *Copper Conductors, Adjustment for Change in Parameters*, page III.

The revised value in each case was a difference of 5 amperes, either higher or lower. Proposals have been submitted to the CEC Part I to correct inconsistencies resulting from the work on this proposal. The revision requires users to exercise additional care when dealing with correction factors for conductors where the ampacity values were lowered, especially for sizes 14, 12, 8, and 6 AWG.

Comment: 6-65

Proposals: 6-99, 6-100

312.8

Article 312 Cabinets, Cutout Boxes, and Meter Socket Enclosures
Part I Installation

Switch and Overcurrent Device Enclosures...

Code Language

312.8 Switch and Overcurrent Device Enclosures with Splices, Taps, and Feed-Through Conductors. The wiring space of enclosures for switches or overcurrent devices shall be permitted for conductors feeding through, spliced, or tapping off to other enclosures, switches, or overcurrent devices where all of the following conditions are met:

(1) The total of all conductors installed at any cross section of the wiring space does not exceed 40 percent of the cross-sectional area of that space.

(2) The total area of all conductors, splices, and taps installed at any cross section of the wiring space does not exceed 75 percent of the cross-sectional area of that space.

(3) A warning label is applied to the enclosure that identifies the closest disconnecting means for any feed-through conductors.

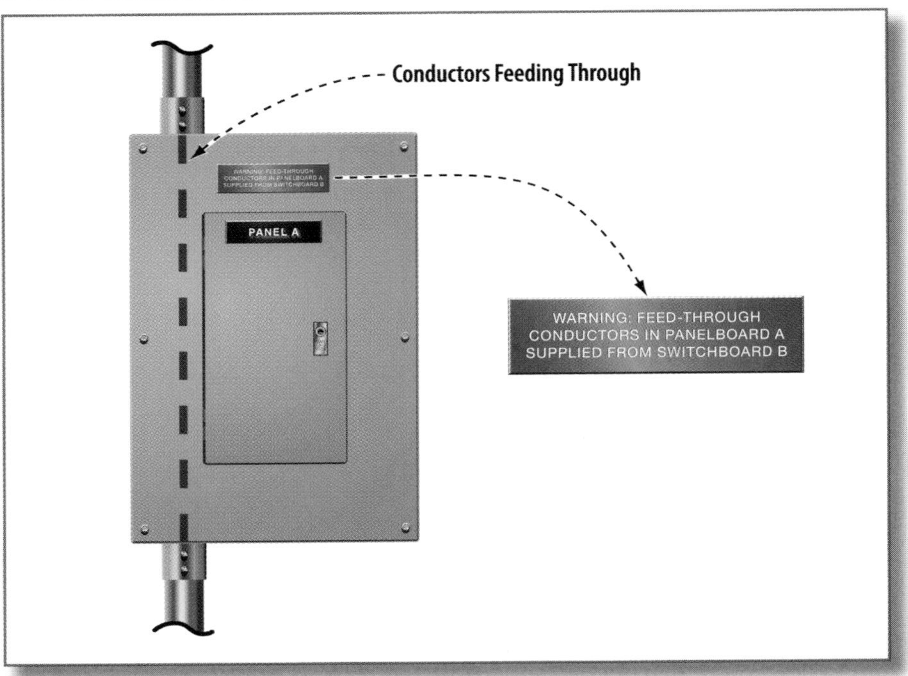

Change Summary

The title of 312.8 was revised to read as follows: Switch and Overcurrent Device Enclosures with Splices, Taps, and Feed-Through Conductors. This section was restructured to list format in accordance with the *NEC Style Manual*, and a new sentence was added as list item (3).

Significance of the Change

The revisions to this section improve clarity and usability and change the driving language from restrictive to permissive text with conditions. The section is now presented in list format, with list items (1) and (2) addressing fill capacity of cross-sectional areas of enclosures previously covered by this rule. New list item (3) incorporates the concepts introduced by Proposal 9-33. The result is a marking requirement in item (3) that identifies the closest disconnecting means for conductors that pass through an enclosure. This revision is an improvement in worker safety because it draws attention to the possibilities of the presence of energized conductors in the enclosure even when the supply circuit is de-energized, because the feed-through conductors could be supplied from another source. The warning label provides workers with direction to the disconnecting means for the conductors feeding through the equipment so they can be de-energized as may be necessary so that certain work tasks can be performed in a safe manner.

Comment: None

Proposals: 9-32,9-33, 9-34

314.16(C)(3)

Article 314 Outlet, Device, Pull, and Junction Boxes; Conduit Bodies; Fittings; and Handhole Enclosures
Part II Installation

Short Radius Conduit Bodies

Code Language

314.16(C) Conduit Bodies

(3) Short Radius Conduit Bodies. Conduit bodies such as capped elbows and service-entrance elbows that enclose conductors 6 AWG or smaller, and are only intended to enable the installation of the raceway and the contained conductors, shall not contain splices, taps, or devices and shall be of sufficient size to provide free space for all conductors enclosed in the conduit body.

Significance of the Change

This revision is a relocation of the text in 314.5 to 314.16(C)(2), with 314.5 being deleted as a result. All the information related to splices, taps, or devices should be located in the same section for usability purposes. Part II of Article 314 provides the installation requirements for outlet, device, pull, and junction boxes; conduit bodies; fittings; and handholes. Section 314.16(C) already includes requirements for conduit bodies and is located in Part II of the article dealing with installation requirements. No technical changes resulted from the acceptance of this proposal.

Change Summary

A new list item (3) titled Short-Radius Conduit Bodies has been added to 314.16(C). The new (3) includes the requirements relocated from former Section 314.5.

Comment: 9-24

Proposal: 9-48

314.27(A)(1) & (2)

Article 314 Outlet, Device, Pull, and Junction Boxes; Conduit Bodies; Fittings; and Handhole Enclosures
Part II Installation

Outlet Boxes

Code Language

314.27 Outlet Boxes.

(A) Boxes at Luminaire or Lampholder Outlets. Outlet boxes or fittings designed for the support of luminaires and lampholders, and installed as required by 314.23, shall be permitted to support a luminaire or lampholder.

(1) Wall Outlets. Boxes used at luminaire or lampholder outlets in a wall shall be marked on the interior of the box to indicate the maximum weight of the luminaire that is permitted to be supported by the box in the wall, if other than 23 kg (50 lb).

*Exception **(unchanged except for adding the words "or the lampholder")***

(2) Ceiling Outlets. At every outlet used exclusively for lighting, the box shall be designed or installed so that a luminaire may be attached. Boxes shall be required to support a luminaire or lampholder weighing a minimum of 23 kg (50 lb). A luminaire that weighs more than 23 kg (50 lb) shall be supported independently of the outlet box, unless the outlet box is listed and marked for the maximum weight to be supported.

Change Summary

Section 314.27(A) has been restructured into a list format. Subdivision (B) has been incorporated into the text of 314.27(A)(1) and (2). Subdivisions (C), (D), and (E) have been reidentified as (C) and (D) accordingly. The words "or lampholders" have been incorporated into this section and exception.

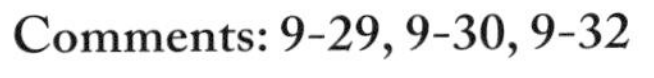

Comments: 9-29, 9-30, 9-32

Proposal: 9-77

Significance of the Change

This revision reorganizes 314.27(A) into list format and removes redundant text in subdivision (B) by incorporating its concepts into (A). New list item (1) addresses boxes for luminaire and lampholder wall outlets. New list item (2) addresses luminaire and lampholder outlets installed in a ceiling. The existing exception to 314.27(A) has been revised to include the words "or lampholder" and follows 314.27(A)(1) since it applies to wall outlet boxes.

Action by CMP-9 also reinserted language in list item (1) addressing the marking that was included in the 2008 cycle as part of Proposal 9-56 but was omitted in error from the first printing of the 2008 *NEC*. If the marking were on the outside of the box, it would be impossible to judge on a final inspection if a wall finish is applied. Luminaire or lampholder selections are frequently made after the rough inspection is performed. The revision clarifies that the marking addressing use suitability is required on the inside of the outlet box.

314.28(E)

Article 314 Outlet, Device, Pull, and Junction Boxes; Conduit Bodies; Fittings; and Handhole Enclosures
Part II Installation

Power Distribution Blocks

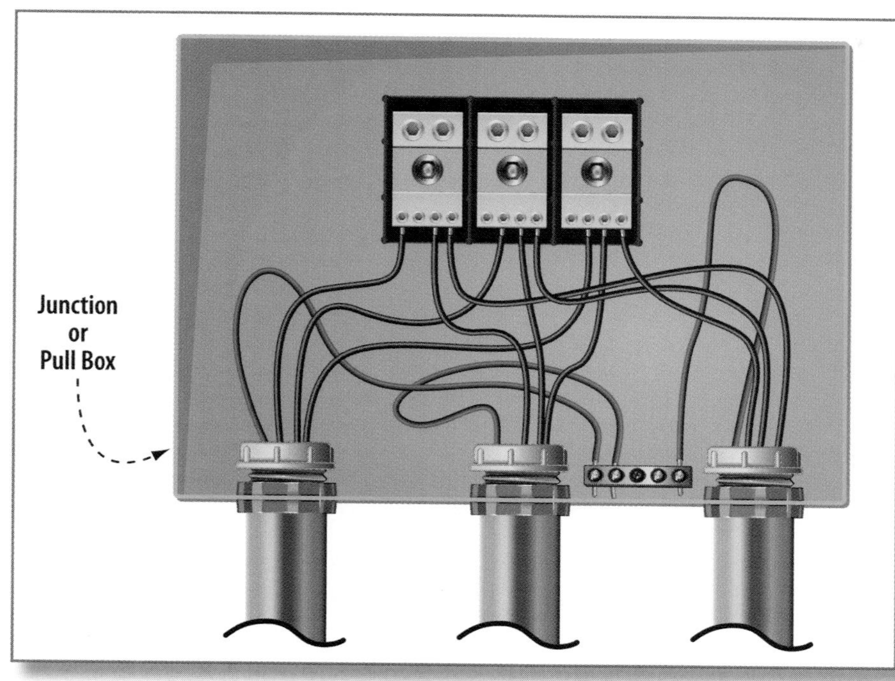

Code Language

(E) Power Distribution Blocks. Power distribution blocks shall be permitted in pull and junction boxes over 1650 cm^3 (100 in.^{3}) for connections of conductors where installed in boxes and where the installation complies with (1) through (5).

Exception: Equipment grounding terminal bars shall be permitted in smaller enclosures.

(1) Installation. Power distribution blocks installed in boxes shall be listed.

(2) Size. In addition to the overall size requirement in the first sentence of 314.28(A)(2), the power distribution block shall be installed in a box with dimensions not smaller than specified in the installation instructions of the power distribution block.

(3) Wire Bending Space. Wire bending space at the terminals of power distribution blocks shall comply with 312.6.

(4) Live Parts. Power distribution blocks shall not have uninsulated live parts exposed within a box, whether or not the box cover is installed.

(5) Through Conductors. Where the pull or junction boxes are used for conductors that do not terminate on the power distribution block(s), the through conductors shall be arranged so the power distribution block terminals are unobstructed following installation.

Significance of the Change

This new section incorporates provisions for power distribution blocks in Article 314. Power distribution blocks were previously included for metal wireways as indicated in 376.56(B). This new provision aligns Article 314 with Article 376 relative to installing power distribution blocks. The option of landing conductors to terminals in boxes is often utilized within larger boxes in addition to wireways, as a convenient method of splicing conductors or complying with a feeder tap rule such as in 240.21(B). This change expands the use of listed power distribution blocks to other enclosures but retains the same restrictions as those for wireways. New list item (5) requires that any conductors passing through the box without terminating on the block have to be arranged in a manner that does not obstruct the terminal blocks in the completed installation. This revision is consistent with current practices that include the use of boxes covered by Article 314 to install power distribution blocks.

Change Summary

A new subdivision (E) titled Power Distribution Blocks has been added to 314.28.

Comments: 9-37, 9-38

Proposal: 9-87

314.70 & 314.71

Article 314 Outlet, Device, Pull, and Junction Boxes; Conduit Bodies; Fittings; and Handhole Enclosures
Part IV Pull and Junction Boxes, Conduit Bodies...

Pull and Junction Boxes, Conduit Bodies, and Handhole Enclosures...

Code Language

IV. Pull and Junction Boxes, Conduit Bodies, and Handhole Enclosures for Use on Systems over 600 Volts, Nominal

314.70 General.

(A) Pull and Junction Boxes. Where pull and junction boxes are used on systems over 600 volts, the installation shall comply with the provisions of Part IV and with the following general provisions of this article:

(1) Part I, 314.2; 314.3; and 314.4

(2) Part II, 314.15; 314.17; 314.20; 314.23(A), (B), or (G); 314.28(B); and 314.29

(3) Part III, 314.40(A) and (C); and 314.41

(B) Conduit Bodies. Where conduit bodies are used on systems over 600 volts, the installation shall comply with the provisions of Part IV and with the following general provisions of this article:

(1) Part I, 314.4

(2) Part II, 314.15; 314.17; 314.23(A), (E), or (G); and 314.29

(3) Part III, 314.40(A); and 314.41

(... Continued on next page ...)

Change Summary

The title of Part IV of Article 314 has been revised as follows: Pull and Junction Boxes, Conduit Bodies, and Handhole Enclosures for Use on Systems over 600 Volts, Nominal. Two new subdivisions have been added to 314.70(B) to address conduit bodies and (C) to address handhole enclosures. Section 314.71 has been revised to include conduit bodies in the sizing requirements.

Pull and Junction Boxes, Conduit Bodies, and Handhole Enclosures...

Code Language

(... Continued from previous page ...)

(C) Handhole Enclosures. Where handhole enclosures are used on systems over 600 volts, the installation shall comply with the provisions of Part IV and with the following general provisions of this article:

(1) Part I, 314.3; and 314.4

(2) Part II, 314.15; 314.17; 314.23(G); 314.28(B); 314.29; and 314.30

314.71 Size of Pull and Junction Boxes, Conduit Bodies, and Handhole Enclosures. Pull and junction boxes and handhole enclosures shall provide adequate space and dimensions for the installation of conductors, and they shall comply with the specific requirements of this section. Conduit bodies shall be permitted if they meet the dimensional requirements for boxes.

(No other changes in this section)

Significance of the Change

The title and text of 314.70 have been revised to incorporate conduit bodies into the requirements of Part IV of Article 314. The applicable requirements for conduit bodies in Parts I and II of the article differ from those for conduit bodies used for installations over 600 volts and thus need to be listed separately. The revisions to this section incorporate conduit bodies under the applicable rules. Where conduit bodies are used in installations over 600 volts, they are required to meet the applicable requirements contained in Part IV of Article 314. Panel action also incorporated a new subdivision (C) to include handhole enclosures in the requirements of Part IV, which was inadvertently overlooked when handholes were added to Article 314. The result is requirements that address conduit bodies and handholes used in over 600-volt installations where none previously existed.

Comment: None

Proposal: 9-92

328.14

Article 328 Medium Voltage Cable: Type MV
Part: II Installation

Installation

Code Language

328.14 Installation. Type MV cable shall be installed, terminated and tested by qualified persons.

Informational Note: IEEE 576-2000 *Recommended Practice for Installation, Termination, and Testing of Insulated Power Cables as Used in Industrial and Commercial Applications* includes installation information and testing criteria for MV cable.

Change Summary

A new section has been added to Article 328 to require that Type MV cable be installed, terminated, and tested only by qualified persons.

Significance of the Change

Proper installation of any raceway with conductors or cable assembly requires a certain amount of expertise and experience. The installation of MV cable is very often misunderstood, in that installers are often under the misconception that these cables can be installed like any other conductor or cable. Manufacturers of Type MV cable provide various recommendations on cable-pulling tensions, sidewall bearing pressures, and bending radius for the cable that must be followed when medium-voltage cables are installed, to ensure the cable is not damaged during the installation. This new requirement mandates that only qualified persons install Type MV cable.

An informational note has been added to reference IEEE 576, which is the recommended practice for installing, terminating, and splicing medium-voltage power cables in commercial and industrial locations.

Type MV cable is always tested before it is energized to detect any damage the cable may have sustained during the installation process. Qualified installers are trained to install raceways with adequate bending radius, size splice and pull boxes, pull the cable without creating damage, and terminate the cable.

Comments: 7-14, 7-15

Proposal: 7-36

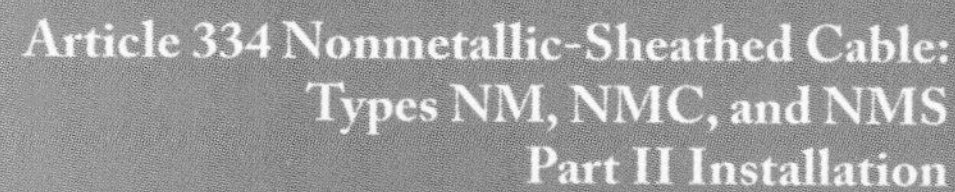

Uses Permitted

Code Language

334.10 Uses Permitted. Type NM, Type NMC, and Type NMS cables shall be permitted to be used in the following:

(1) One- and two-family dwellings and their attached or detached garages, and their storage buildings.

Significance of the Change

The text of the 2008 *NEC* in 334.10(1) did not specifically permit the use of Type NM cable in accessory structures of a dwelling unit. "Dwelling Unit" is defined in Article 100 as follows:

Dwelling Unit. A single unit, providing complete and independent living facilities for one or more persons, including permanent provisions for living, sleeping, cooking, and sanitation.

While 334.10(1) provided a blanket permission for all dwelling units, a structure that is detached would not have been considered part of a dwelling unit, and 334.10(3) would therefore have applied and necessitated a 15-minute finish rating. This revision clearly permits the use of Type NM cable in all attached or detached garages and storage buildings. Other accessory structures of a dwelling unit, which may include but not be limited to bungalows, playhouses, and workshops, do not fall under 334.10(1); therefore, the remainder of Article 334 must be applied. For example, a bungalow used for sleeping quarters that does not meet the definition of a dwelling unit would not be considered as a garage or a storage building. Furthermore, if the bungalow includes a dropped or suspended ceiling, the use of Type NM cable would be prohibited by 334.12(A)(2).

Change Summary

Type NM cable is now clearly permitted for use in all attached or detached garages, and in storage buildings associated with a dwelling unit.

Comment: 7-26

Proposal: 7-77

338.10(B)(2)

Article 338 Service-Entrance Cable: Types SE and USE
Part II Installation

Use of Uninsulated Conductor

Code Language

338.10 Uses Permitted

(B) Branch Circuits or Feeders.

(2) Use of Uninsulated Conductor. Type SE service entrance cable shall be permitted for use where the insulated conductors are used for circuit wiring and the uninsulated conductor is used only for equipment grounding purposes.

Exception: In existing installations, uninsulated conductors shall be permitted as a grounded conductor in accordance with 250.32 and 250.140, where the uninsulated grounded conductor of the cable originates in service equipment, and with 225.30 through 225.40.

Change Summary

The term *uninsulated conductor* is now used to describe the permitted use of the concentric conductors in Type SE cable where it is used as a branch circuit or feeder.

Comment: None

Proposals: 7-131, 7-132

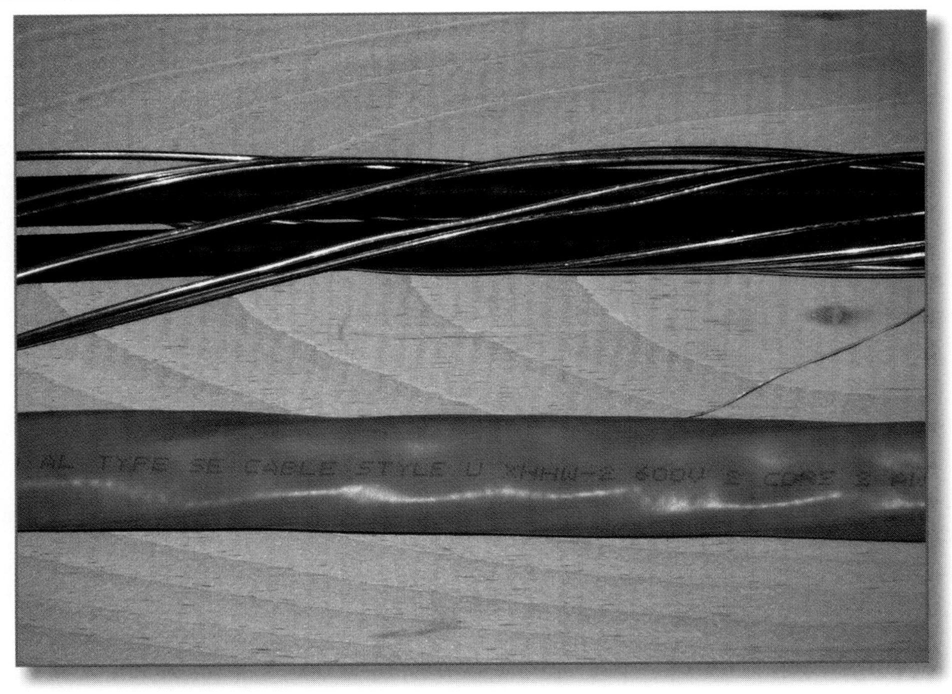

Significance of the Change

The title of 338.10(B)(2) in the 2008 *NEC*, Grounded Conductor Not Insulated, was misleading and confusing. When an SE cable is used as service-entrance conductors for a dwelling unit, the concentric conductors in the SE cable are the "grounded conductor." However, this section details requirements for the use of SE cable as a branch circuit or feeder; thus the new title, "Use of Uninsulated Conductor," is appropriate. The general rule for the use of uninsulated concentric conductor in an SE cable installed as a branch circuit or feeder is that it be used as an equipment grounding conductor. The exception has been revised to clearly inform the *Code* user that the uninsulated concentric conductor in an SE cable may be used for the grounded conductor in EXISTING installations only.

Note that in most cases the uninsulated concentric conductor in an SE cable is smaller in size than the insulated conductors. The UL category for Type SE cable is "SERVICE CABLE TXKT" or "SERVICE-ENTRANCE CABLE TYLZ." Based upon tests which have been made involving the maximum heating that can be produced, an uninsulated conductor in a service cable assembly is considered to have the same current-carrying capacity as the insulated conductors even though it may be smaller in size.

338.10(B)(4)

Article 338 Service-Entrance Cable: Types SE and USE
Part II Installation

Installation Methods for Branch Circuits and Feeders

Code Language

338.10 Uses Permitted

(B) Branch Circuits or Feeders.

(4) Installation Methods for Branch Circuits and Feeders.

(a) *Interior Installations.* In addition to the provisions of this article, Type SE service-entrance cable used for interior wiring shall comply with the installation requirements of Part II of Article 334, excluding 334.80.

Where installed in thermal insulation, the ampacity shall be in accordance with the 60°C (140°F) conductor temperature rating. The maximum conductor temperature rating shall be permitted to be used for ampacity adjustment and correction purposes, if the final derated ampacity does not exceed that for a 60°C (140°F) rated conductor.

Informational Note No.1: See 310.15(A)(3) for temperature limitation of conductors.

Informational Note No. 2: For the installation of main power feeder conductors in dwelling units refer to 310.15(B)(7).

Significance of the Change

Where Type SE cable is used as a branch circuit or feeder on the interior of a building or structure, 338.10(B)(4)(a) applies. Section 338.10(B)(4) provides installation methods for the use of Type SE in both interior and exterior installations. Third-level subdivision 338.10(B)(4)(a) for interior installations has been revised to exclude the requirements of 334.80. This change is significant, in that the general use of SE cable indoors (see below for installations in thermal insulation) is no longer limited to a final ampacity in accordance with the 60°C rating of the conductor unless installed in thermal insulation. This means that an SE cable rated at 75°C can have overcurrent protection provided at the 75°C rating. All Type NM cables are required by 334.112 to be rated at 90°C, and 334.80 requires all Type NM cables to have a final derated ampacity at not more than the 60°C rating of the conductor.

A second paragraph has been added to limit conductor ampacity when Type SE is installed in thermal insulation. This new provision requires the final ampacity of Type SE installed in thermal insulation to be at the 60°C rating of the conductor. A new informational note informs the *Code* user of the permissive requirements of 310.15(B)(7) for the installation of main power feeder conductors in dwelling units.

Change Summary

The use of Type SE cable is no longer subject to the ampacity requirements of 334.80. The use of Type SE cable in thermal insulation requires a final derated ampacity of 60°C.

Comment: 7-48

Proposal: 7-133

3XX.30(C)

Articles 342 (IMC), 344 (RMC), 352 (PVC), 355 (RTRC) and 358 (EMT)
Part II Installation

Unsupported Raceways

Code Language

3XX.30~~(C) Unsupported Raceways.~~ ~~Where oversized, concentric or eccentric knockouts are not encountered, Type IMC shall be permitted to be unsupported where the raceway is not more than 450 mm (18 in.) and remains in unbroken lengths (without coupling). Such raceways shall terminate in an outlet box, junction box, device box, cabinet, or other termination at each end of the raceway.~~

Change Summary

The permissive text allowing unsupported raceways of lengths up 18 inches where oversized concentric or eccentric knockouts were not encountered has been deleted.

Comment: None

Proposals: 8-24a, 35, 78, 105, 125

Significance of the Change

The 2008 *NEC* added to five sections a new second-level subdivision containing permissive text allowing unsupported raceways of lengths up 18 inches where oversized concentric or eccentric knockouts were not encountered. This text was incorporated into the 3XX.30 sections of the following raceway articles that address "Securing and Supporting": 342 (IMC), 344 (RMC), 352 (PVC), 355 (RTRC), and 358 (EMT). This 2008 revision was intended to permit the installation of short lengths of raceway (up to 18 inches) without support. The technical committee has reversed its position and now requires that short nipples be installed with support. This text has been completely deleted in all five of these articles.

The existing requirements in these articles will now require support of these short sections of raceway as follows:

342.30(A) Securely Fastened. IMC shall be secured in accordance with one of the following:

(1) IMC shall be securely fastened within 900 mm (3 ft) of each outlet box, junction box, device box, cabinet, conduit body, or other conduit termination.

An 18-inch or smaller nipple will now require a means of securement.

348.30(A) & 350.30(A)

Articles 348 Flexible Metal Conduit: Type FMC
and 350 Liquidtight Flexible Metal Conduit: Type LFMC
Part II Installation

Securely Fastened

Code Language

348.30(A) and 350.30(A) Securely Fastened. [FMC, or LFMC] shall be securely fastened in place by an approved means within 300 mm (12 in.) of each box, cabinet, conduit body, or other conduit termination and shall be supported and secured at intervals not to exceed 1.4 m (4 ½ ft).

(Exception No. 1 was revised editorially only.)

Exception No. 2: Where flexibility is necessary after installation, lengths from the last point where the raceway is securely fastened shall not exceed the following:

(1) 900 mm (3 ft) for metric designators 16 through 35 (trade sizes ½ through 1¼)

(2) 1200 mm (4 ft) for metric designators 41 through 53 (trade sizes 1½ through 2)

(3) 1500 mm (5 ft) for metric designators 63 (trade size 2½) and larger

Significance of the Change

The general rule for securing flexible metal conduit (FMC) and liquidtight flexible metal conduit (LFMC) is that they be securely fastened in place by an approved means within 12 inches of each box, cabinet, conduit body, or other conduit termination and that they be supported and secured at intervals not to exceed 4 ½ feet. Exception No. 2 permits the 12-inch distance to be increased where flexibility is necessary after installation. The 2008 text of the *NEC* seemed to be unclear on this issue, and many *Code* users interpreted this exception to permit no means of securing the conduit where the total length of the conduit did not exceed the lengths provided in Exception No. 2. For example, where a 3-foot length of 1-inch FMC or LFMC is installed between a condulet and a motor, the 2008 *NEC* would seem to permit the conduit installation without support. This was not the intent of the technical committee.

This revision clarifies that the length of unsupported conduit may be measured only from the last point it was securely fastened in place. Therefore, the preceding example of a 3-foot length of 1-inch FMC or LFMC would now require that the conduit be secured within 12 inches of the condulet. From that point, a length of up to 3 feet of unsupported conduit would be permitted to the motor.

Change Summary

New qualifying text has been added to the exception that permits support in excess of 12 inches of conduit termination where flexibility is required after installation. This new text requires that the distance permitted be measured from the last point of support.

Comment: None

Proposals: 8-47, 8-61

348.60 & 350.60

Articles 348 Flexible Metal Conduit: Type FMC
and 350 Liquidtight Flexible Metal Conduit: Type LFMC
Part II Installation

Grounding and Bonding

Code Language

348.60 & 350.60 Grounding and Bonding. If used to connect equipment where flexibility is necessary to minimize the transmission of vibration from equipment or to provide flexibility for equipment that requires movement after installation, an equipment grounding conductor shall be installed.
(Remainder of text unchanged)

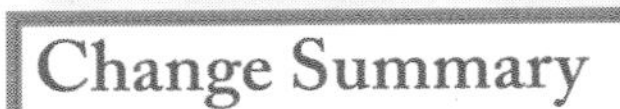

Change Summary

The requirement for an equipment grounding conductor in flexible metal conduit or liquidtight flexible metal conduit where flexibility is necessary has been clarified.

Comments: 8-51, 8-29
Proposals: 8-51, 8-65

Significance of the Change

This revised text clarifies that an equipment grounding conductor is required in short lengths of FMC and LFMC, subject to vibration or where flexibility is required. The intent of this requirement is to ensure the continuity of the ground fault path where movement or vibration of the flexible metal conduits could compromise it. This revised text now clarifies that, where FMC or LFMC is used for one of the following reasons, an equipment grounding conductor is required:

(1) Flexibility is necessary to minimize the transmission of vibration from equipment, for example, as a connection to a motor or transformer.
(2) Flexibility for equipment that requires movement after installation, for example, an industrial control panel mounted to a hinged panel to allow access to equipment.

Flexible metal conduit is used in many cases for "ease of installation," and an equipment grounding conductor may not be required by Article 250. For example, a lay-in fixture in a drop ceiling could be supplied with a direct connection to EMT. In most cases, the fixture would be supplied from a junction box with a flexible metal conduit or a cable assembly for ease of installation.

352.10(I)

Articles 352 (PVC), 353 (HDPE), 355 (RTRC) and 362 (ENT)
Part II Installation

Insulation Temperature Limitations

Code Language

352.10 Uses Permitted. The use of PVC conduit shall be permitted in accordance with 352.10(A) through (I).

(I) Insulation Temperature Limitations. Conductors or cables rated at a temperature higher than the listed temperature rating of PVC conduit shall be permitted to be installed in PVC conduit, provided the conductors or cables are not operated at a temperature higher than the listed temperature rating of the PVC conduit.

Significance of the Change

This revision is user-friendly, providing clear requirements for the application of conductor operating temperatures with respect to the temperature rating of the conduit. The revision was made to the 3XX.10 sections of Articles 352 (PVC), 353 (HDPE), 355 (RTRC), and 362 (ENT). The 2008 *NEC* located the conductor operating temperatures in the 3XX.12 sections of these articles under "Uses Not Permitted," followed by an exception permitting the use of conductors rated at temperatures higher than the conduit rating, provided they were not operated at higher temperatures than the rating of the conduit.

The revised text is clear and provides the requirements for conductor operating temperatures for PVC, HDPE, RTRC, and ENT in positive text, eliminating the exceptions. A conductor with a temperature rating higher than that listed for the conduit is permitted to be used, provided the conductors or cables are not operated at a temperature higher than the listed temperature rating of the conduit.

Change Summary

The conductor operating temperature requirements for PVC, HDPE, RTRC, and ENT have been clarified and placed in the 3XX.10 Uses Permitted sections of each article. Conductors rated at temperatures higher than the conduit rating are permitted, provided they are not operated higher than the temperature rating of the conduit.

Comment: None

Proposals: 8-67a, 85a, 96a, 139a

392.12

Article 392 Cable Trays
Part II Installation

Uses Not Permitted

Code Language

392.12 Uses Not Permitted. Cable tray systems shall not be used in hoistways or where subject to severe physical damage.

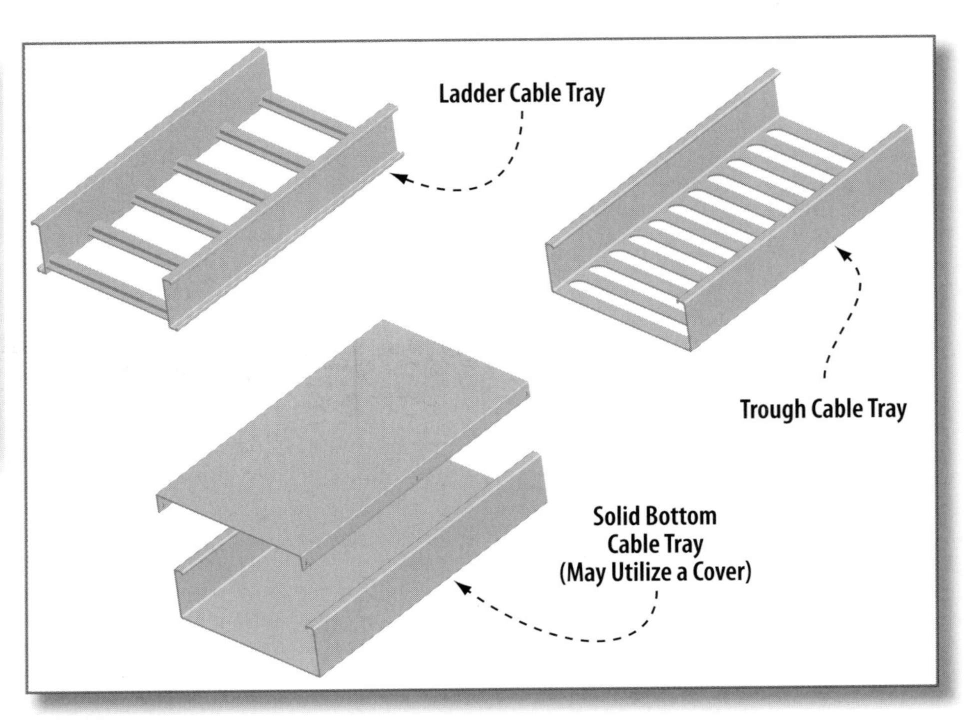

Change Summary

The second sentence of this section, containing a general prohibition of cable trays in air-handling spaces and a reference to 300.22, has been deleted.

Significance of the Change

The "Uses Not Permitted" section in Article 392 has been relocated from 392.4 to 392.12, to correlate with the common numbering system used throughout Chapter 3. In the 2008 *NEC*, the second sentence addressed the use of cable tray in air-handling spaces as follows:

Cable tray systems shall not be used in ducts, plenums, and other air-handling spaces, except as permitted in 300.22, to support wiring methods recognized for use in such spaces.

This sentence has been deleted. Instead, cable tray is covered as follows:

— Cable tray is prohibited in 300.22(A) Ducts for Dust, Loose Stock, or Vapor Removal.

— Cable tray is prohibited in 300.22(B) Ducts Specifically Fabricated for Environmental Air.

— Cable tray is permitted in (C) Other Spaces Used for Environmental Air (Plenums).

See coverage of the changes in 300.22.

Comment: None

Proposal: 8-253

392.18(H)

Article 392 Cable Trays
Part II Installation

Marking

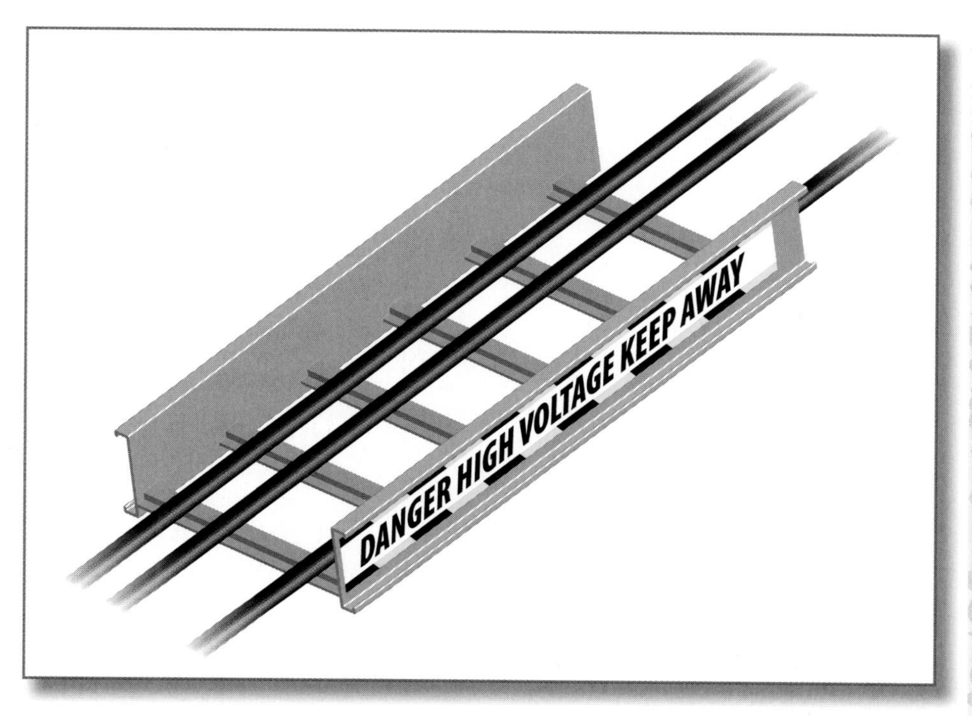

Code Language

392.18 Cable Tray Installation

(H) Marking. Cable trays containing conductors rated over 600 volts shall have a permanent, legible warning notice carrying the wording "DANGER-HIGH VOLTAGE-KEEP AWAY" placed in a readily visible position on all cable trays, with the spacing of warning notices not to exceed 3 m (10 ft).

Significance of the Change

This new section requires that all cable tray installations containing conductors rated over 600 volts be marked in a readily visible manner to notify all persons, whether qualified or unqualified, of the presence of high-voltage conductors. This requirement mandates that the marking be "readily visible" and have a maximum spacing of 10 feet. This revision is safety driven; it is designed to warn all persons of the presence of high-voltage conductors, along with all other existing high-voltage warning requirements throughout the *NEC*.

Change Summary

All cable tray installations containing conductors rated over 600 volts are now required to have permanent, legible warning notices with the words "DANGER-HIGH VOLTAGE-KEEP AWAY" at a minimum of every 10 feet and in a readily visible position.

Comment: 8-100a

Proposal: 8-260

392.60(A)

Article 392 Cable Trays
Part II Installation

Metallic Cable Trays

Code Language

392.60 Grounding and Bonding.

(A) Metallic Cable Trays. Metallic cable trays shall be permitted to be used as equipment grounding conductors where continuous maintenance and supervision ensure that qualified persons service the installed cable tray system and the cable tray complies with provisions of this section. Metallic cable trays that support electrical conductors shall be grounded as required for conductor enclosures in accordance with 250.96 and Part IV of Article 250. Metal cable trays containing only non-power conductors shall be electrically continuous through approved connections or the use of a bonding jumper not smaller than a 10 AWG.

Informational Note: Examples of non-power conductors include nonconductive optical fiber cables and Class 2 and Class 3 Remote Control Signaling and Power Limiting Circuits.

Change Summary

The grounding and bonding requirements of Article 392 have been combined in a single section. A new requirement has been added to this section to mandate that all metal cable trays containing non-power conductors, such as optical fiber, Class 2, and Class 3 conductors be electrically continuous.

Significance of the Change

This revision combines the 2008 *NEC* requirements found in 392.3(C) and 392.7, in a new section created to correlate with the common numbering system used throughout Chapter 3. This new section, 392.60 Grounding and Bonding, contains the text found previously in 392.7, and the first-level subdivisions remain the same. Text found previously in 392.3(C) has been relocated into 392.60(A).

A new requirement for metal cable tray containing non-power conductors (communication, data, signal, etc.) has been added to 392.60(A). All metal cable tray containing non-power conductors is now required to be electrically continuous, which may be achieved through the use of approved connections or the use of a bonding jumper not smaller than 10 AWG. Many of the non-power systems installed in cable tray are required to be bonded to the existing electrical system, for which 250.94 provides requirements for accessible termination points. For example, 800.100 requires bonding for communications systems and 820.100 requires bonding for CATV systems. The *NEC* now requires that all sections of metal cable tray containing these non-power conductors be electrically continuous.

Comments: 8-114, 8-100a

Proposal: 8-263

Cable Trays

Code Language

Article 392 Cable Trays
I. General

II. Installation
392.10 Uses Permitted
392.12 Uses Not Permitted
392.22 Number of Conductors or Cables
392.30 Securing and Supporting
392.60 Grounding and Bonding

III. Construction Specifications
392.100 Construction

Significance of the Change

Article 392 has been editorially revised to comply with the *NEC Style Manual* and to correlate with the common numbering system used in the raceway and cable articles. This revision is a good example of how the *NEC* has been evolving over the last few cycles to improve usability and to logically group information in a common numbering format where applicable. The article has been divided into three parts to logically separate its requirements — I General, II Installation, and III Construction Specifications — in a similar format to those of the raceway and cable articles.

The common numbering system, originally implemented in the 2002 *NEC*, provides the *Code* user with a consistent numbering system and logical separation of information. Examples of that system include 392.10 Uses Permitted, 392.12 Uses Not Permitted, 392.22 Number of Conductors or Cables, 392.30 Securing and Supporting, 392.60 Grounding and Bonding, 392.100 Construction.

It should be noted that cable tray is a support system, not a raceway.

Change Summary

Article 392 has been completely reorganized and separated into parts for clarity and usability. Section numbers are revised to correlate with the common numbering system used in the raceway and cable articles.

Comment: None
Proposal: 8-235a

Article 399

Article 399 Outdoor, Overhead Conductors, Over 600 Volts

Outdoor Overhead Conductors Over 600 Volts

Code Language

399.1 Scope. This article covers the use and installation for outdoor overhead conductors over 600 volts, nominal.

399.2 Definition.

Outdoor Overhead Conductors. Single conductors, insulated, covered, or bare, installed outdoors on support structures.

399.10 Uses Permitted. Outdoor overhead conductors over 600 volts, nominal, shall be permitted only for systems rated over 600 volts, nominal, as follows:

(1) Outdoors

(2) For service conductors, feeders, or branch circuits

Informational Note: For additional information on outdoor overhead conductors over 600 volts, see ANSI/IEEE C2-2007, *National Electrical Safety Code.*

399.12 Uses Not Permitted. Overhead conductors, over 600 volts, nominal shall not be permitted to be installed indoors.

399.30 Support

(A) Conductors

(B) Structures

(C) Insulators

Change Summary

This new article addresses outdoor overhead conductors rated over 600 volts. Its requirements apply to overhead conductors installed as services, feeders, or branch circuits.

Comments: 7-63, 7-65, 7-67

Proposal: 7-162

Significance of the Change

This new article is part of an effort to improve requirements in the *NEC* where services, feeders, and branch circuits over 600 volts are installed outdoors. All conductors on the load side of the "service point" are "premises wiring," installations over which the *NEC* has jurisdiction.

The installation of conductors on the load side of the service point must be installed in accordance with the *NEC.* These installations include (1) service conductors, which may be switched at the service point without an overcurrent device installed; (2) feeders, which may be used for distribution to multiple buildings or structures; and (3) branch circuits, which may be used to supply large industrial equipment. The 2008 *NEC* did not provide requirements for these overhead conductors. Section 399.30 now lists performance-based requirements for support, spacing, and installation of overhead conductors, their supporting structures, and insulators used in the installation. Each of these requirements mandates documentation of design by *licensed professional engineers*. The designing engineer may use his or her own requirements or refer to another industry standard such as the *NESC*. See the informational note in 399.10.

IBEW Code-Making Panel Members

TECHNICAL CORRELATING COMMITTEE
Palmer L. Hickman, [Principal]
James T. Dollard, [Principal]

CODE–MAKING PANEL NO. 1
Palmer L. Hickman, [Principal]
Mark Christian, [Alternate]

CODE–MAKING PANEL NO. 2
Donald M. King, [Principal]
Jacob G. Benninger, [Alternate]

CODE–MAKING PANEL NO. 3
Paul J. Casparro, [Chair]
Marty L. Riesberg, [Alternate]

CODE–MAKING PANEL NO. 4
Todd W. Stafford, [Principal]
Brian L. Crise, [Alternate]

CODE–MAKING PANEL NO. 5
Dan Hammel, [Principal]
Paul J. LeVasseur, [Alternate]

CODE–MAKING PANEL NO. 6
William F. Laidler, [Principal]
James R. Weimer, [Alternate]

CODE–MAKING PANEL NO. 7
Samuel R. La Dart, [Principal]
Keith Owensby, [Alternate]

CODE–MAKING PANEL NO. 8
Joseph Dabe, [Principal]
Gary W. Pemble, [Alternate]

CODE–MAKING PANEL NO. 9
Rodney D. Belisle, [Principal]
Rhett A. Roe, [Alternate]

CODE–MAKING PANEL NO. 10
James T. Dollard, [Principal]
Richard E. Lofton, II, [Alternate]

CODE–MAKING PANEL NO. 11
James M. Fahey, [Principal]
Jebediah J. Novak, [Alternate]

CODE–MAKING PANEL NO. 12
David R. Quave, [Principal]
Jeffrey L. Holmes, [Alternate]

CODE–MAKING PANEL NO. 13
Linda J. Little, [Principal]
James T. Dollard, Jr., [Alternate]

CODE–MAKING PANEL NO. 14
John L. Simmons, [Principal]
Thomas E. Dunne, [Alternate]

CODE–MAKING PANEL NO. 15
Stephen M. Lipster, [Principal]
Gary A. Beckstrand, [Alternate]

CODE–MAKING PANEL NO. 16
Harold C. Ohde, [Principal]
Terry C. Coleman, [Alternate]

CODE–MAKING PANEL NO. 17
Randy J. Yasenchak, [Principal]
Brian Myers, [Alternate]

CODE–MAKING PANEL NO. 18
Paul Costello, [Principal]
Jesse Sprinkle, [Alternate]

CODE–MAKING PANEL NO. 19
Ronald Michaelis, [Principal]
Ronald D. Weaver, Jr., [Alternate]

Chapter 4

Articles 400-490

Equipment for General Use

Significant Changes
TO THE NEC® 2011

404.2(C)

Article 404 Switches
Part I Installation

Switches Controlling Lighting Loads

Code Language

404.2 Switch Connections

(C) Switches Controlling Lighting Loads. Where switches control lighting loads supplied by a grounded general-purpose branch circuit, the grounded circuit conductor for the controlled lighting circuit shall be provided at the switch location.

Exception: The grounded circuit conductor shall be permitted to be omitted from the switch enclosure where either of the conditions in (1) or (2) apply:

(1) Conductors for switches controlling lighting loads enter the box through a raceway. The raceway shall have sufficient cross-sectional area to accommodate the extension of the grounded circuit conductor of the lighting circuit to the switch location whether or not the conductors in the raceway are required to be increased in size to comply with 310.15(B)(2)(a).

(2) Cable assemblies for switches controlling lighting loads enter the box through a framing cavity that is open at the top or bottom on the same floor level, or through a wall, floor, or ceiling that is unfinished on one side.

Informational note: The provision for a (future) grounded conductor is to complete a circuit path for electronic lighting control devices.

Change Summary

In general, all switches installed to control lighting loads must now be supplied with the grounded circuit conductor for the controlled lighting circuit in addition to the ungrounded conductor(s) being switched.

Comment: 9-45

Proposal: 9-95

Significance of the Change

Many electronic lighting control devices on the market and in use today require a standby current to maintain the ready state and detection capability of the device. Standby current flows at all times, even when the lighting load being controlled is off, to allow "the brain" of the device to stay active at all times to facilitate immediate switching of the load to the "on" position.

In almost every dwelling unit and most commercial installations, a grounded conductor is not provided in the switch box for switches controlling lighting loads. This forces the design and installation of these control devices to utilize the equipment grounding conductor to conduct the standby current. These occupancy sensors are permitted by UL 773A to have a current of up to 0.5 mA on the equipment grounding conductor.

Escalating energy costs and the increased recognition and adoption of energy-saving codes will promote, and in many cases legislate, the use of these products. This revision clearly portrays the *NEC*'s recognition of an unsafe practice by mandating the inclusion of a grounded conductor. An occupancy sensor may be installed initially or as an improvement at any time and at any switch location.

404.9(B) (Exc.)

Article 404 Switches
Part I Installation

Grounding

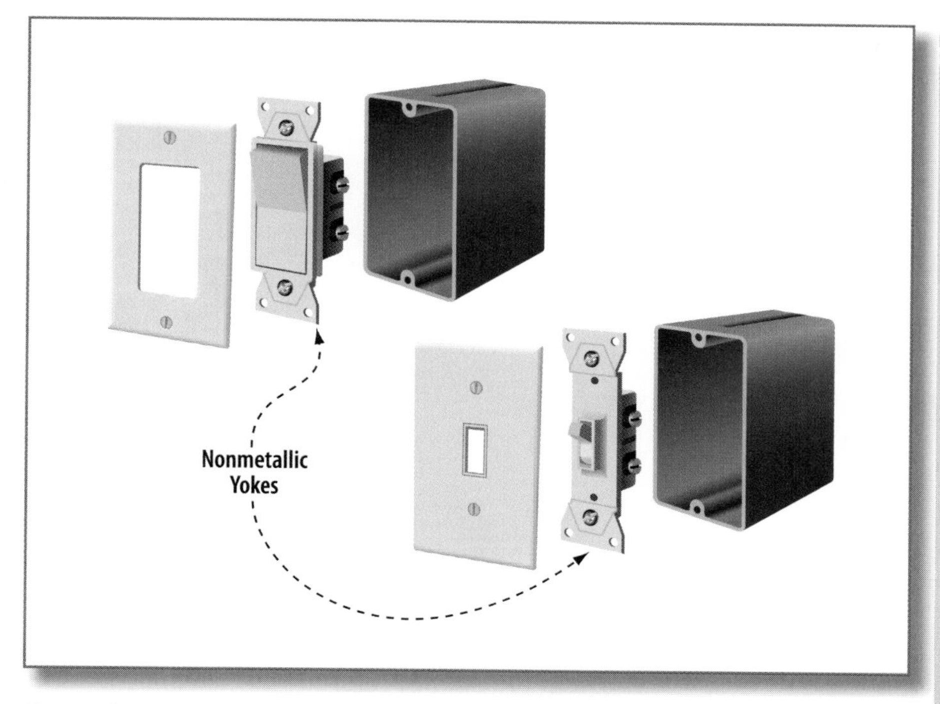

Code Language

Exception No. 1 to (B): ***(Revise second sentence.)*** *A snap switch wired under the provisions of this exception and located within 2.5 m (8 ft) vertically or 1.5 m (5 ft) horizontally, of ground or exposed grounded metal objects shall be provided with a faceplate of nonconducting noncombustible material with nonmetallic attachment screws, unless the switch mounting strap or yoke is nonmetallic or the circuit is protected by a ground-fault circuit interrupter.*

Exception No. 2 to (B): Listed kits or listed assemblies shall not be required to be connected to an equipment grounding conductor if all of the following conditions are met:

(1) The device is provided with a nonmetallic faceplate that cannot be installed on any other type of device,

(2) The device does not have mounting means to accept other configurations of faceplates,

(3) The device is equipped with a nonmetallic yoke, and

(4) All parts of the device that are accessible after installation of the faceplate are manufactured of nonmetallic materials.

Exception No. 3 to (B): A snap switch with integral nonmetallic enclosure complying with 300.15(E) shall be permitted without a connection to an equipment grounding conductor.

Change Summary

The existing exception has been revised and two new exceptions have been added.

Comment: 9-50

Proposals: 9-108, 9-110, 9-111

Significance of the Change

The two new exceptions relax the requirement for an equipment grounding conductor connection to the mounting strap of switches that have nonmetallic yokes and for switches that are an integral part of a listed nonmetallic assembly. Based on the requirements of UL 20, *Standard for General Use Snap Switches*, snap switches that have a plastic yoke and do not provide any means of fastening a standard screw-based wall plate (thus not allowing a metal wall plate to be fastened) are considered safe for installation without the use of an equipment grounding conductor. The new exception to 404.9(B) does not recommend and/or suggest that the grounding means required by 250.148 be eliminated, but that the termination of such grounding means not be required for this unique plastic yoke snap switch.

Generally, the grounding means should always be available and accessible in the outlet box and be properly grounded for replacement purposes or installation of metal yoke devices. Snap switches with metallic mounting straps are required to meet the equipment grounding requirements in 404.9(B) regardless of whether a nonmetallic plate is installed. These exceptions are an appropriate revision and address switch installations that do not expose persons to any conductive parts that could become energized.

Child Care Facility

Code Language

Child Care Facility. A building or structure, or portion thereof, for educational, supervisory, or personal care services for more than four children 7 years old or less.

Change Summary

A new definition of *Child Care Facility* has been added to new Section 406.2. The numbering sequence of Article 406 has been adjusted accordingly.

Significance of the Change

The term *child care facility* is included with new requirements in 406.14 that specifically apply to facilities intended for child supervision as a principle function. Action by CMP-18 on Proposal 18-90 resulted in the creation of 406.2 and a new definition. (The x.2 section of each article is reserved for defined terms used within that article, in accordance with Section 2.2.2.2 of the *NEC Style Manual.*) This new definition incorporates the various aspects of the term's current definitions that appear in the *International Building Code* (Sections 305 and 308) and in the U.S. General Services Administration's *Child Care Design Guide.* The new definition provides users with a clear description of the term *child care facility* to clarify when requirements within Article 406 must be applied. New rules in Article 406 expand the requirements for tamper-resistant receptacles beyond dwelling occupancies. These new rules, in conjunction with the newly defined term, provide clarity for users and for those involved with electrical designs or installations of facilities that provide child care.

Comment: None

Proposal: 18-90

406.4(D)(4)

Article 406 Receptacles, Cord Connectors, and Attachment Plugs (Caps)

Arc-Fault Circuit Interrupter Protection

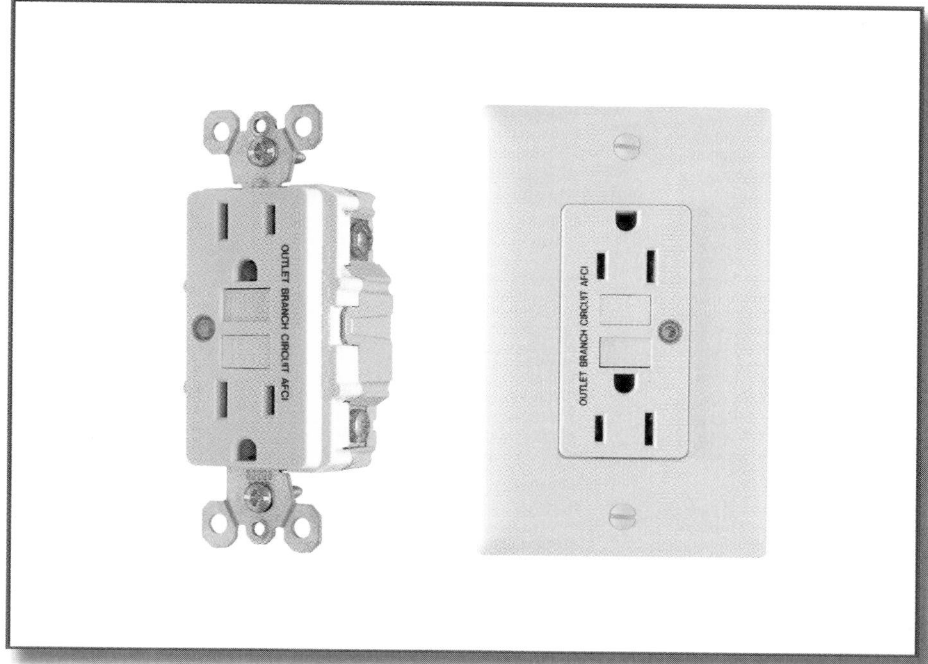

Prototype Courtesy of Pass and Seymour Legrand

Code Language

(4) Arc-Fault Circuit-Interrupter Protection.

Where a receptacle outlet is supplied by a branch circuit that requires arc-fault circuit interrupter protection as specified elsewhere in this *Code*, a replacement receptacle at this outlet shall be one of the following:

(1) A listed outlet branch-circuit-type arc-fault circuit interrupter receptacle.

(2) A receptacle protected by a listed outlet branch circuit-type arc-fault circuit interrupter type receptacle.

(3) A receptacle protected by a listed combination type arc-fault circuit interrupter type circuit breaker.

This requirement becomes effective January 1, 2014.

Significance of the Change

Code Panel 18 acted favorably to expanding arc-fault circuit-interrupter protection requirements beyond new installations, thus extending the requirements for replacement receptacles to include AFCI protection where such protection is now required as provided by 210.12. As a result, if an existing receptacle is replaced in a location where AFCI protection is normally required for the branch circuit, the replacement device must include AFCI protection. This new requirement results in a practical application of AFCI protection in older dwellings.

The existing requirement of 406.4(D)(3) mandates GFCI-protected receptacles where replacements are made at receptacle outlets that are required to be so protected elsewhere in the *NEC*. Substantiation indicated that there is no real practical reason to limit the level of safety and protection provided by AFCI to new homes. Additionally, the benefits of the AFCI requirements in 210.12 have been well substantiated over the last few *NEC* development cycles. This change results in extra protection for older homes by requiring the gradual replacement, over time, of non-AFCI-protected receptacles with new AFCI-protected units. An effective date has been included in this rule to allow manufacturers time to incorporate AFCI technology into an outlet-style device.

Change Summary

Section 406.4(D) is the new location for receptacle replacement requirements formerly found in 406.3(D). This numbering change is related to the acceptance of a new definition of the term *child care facility*, which now is provided in 406.2. A new list item (4) has been added to 406.4(D) to address arc-fault circuit interrupters.

Comments: 18-19, 18-20

Proposal: 18-30

406.4(D)(5) & (6)

Article 406 Receptacles, Cord Connectors, and Attachment Plugs (Caps)

Tamper-Resistant Receptacles and Weather-Resistant Receptacles

Code Language

(5) Tamper-Resistant Receptacles. Listed tamper-resistant receptacles shall be provided where replacements are made at receptacle outlets that are required to be tamper-resistant elsewhere in this *Code*.

(6) Weather-Resistant Receptacles. Weather-resistant receptacles shall be provided where replacements are made at receptacle outlets that are required to be so protected elsewhere in the *Code*.

Change Summary

Section 406.4(D) is the new location for the receptacle replacement requirements formerly found in 406.3(D). The section numbering change is related to the acceptance of a new definition of the term *child care facility*, which now is provided in 406.2. Two new list items (5) and (6) covering receptacle replacements have been added to 406.4(D).

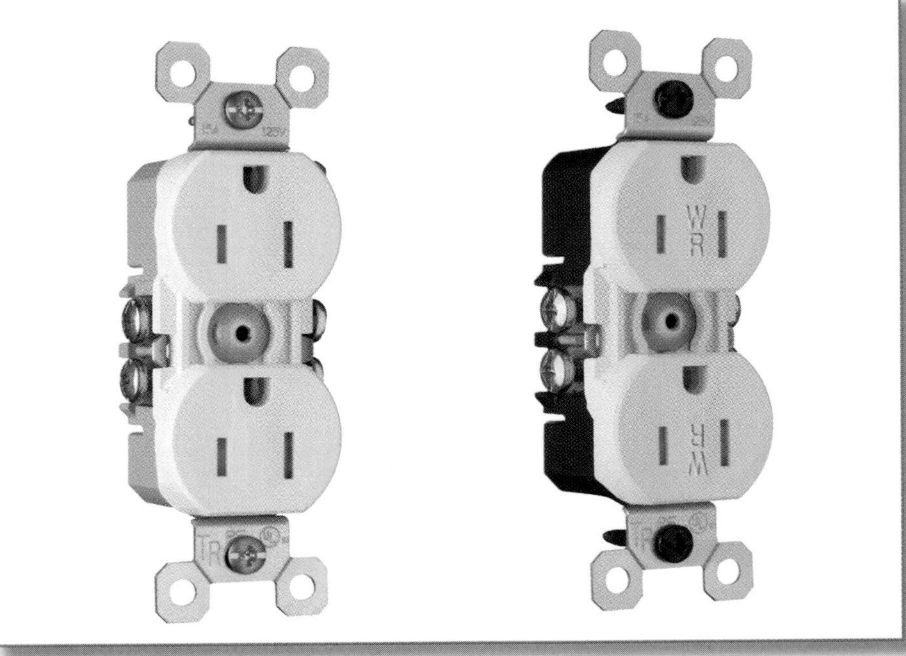

Courtesy of Pass and Seymour Legrand

Significance of the Change

This revision adds new requirements that apply to receptacle replacements covered in 406.4(D). The changes to this section address receptacle replacements in two new list items. New list item (5) requires that where receptacle replacements are made at outlets required to provide tamper-resistant receptacles, the replacement device must be a tamper-resistant type. New list item (6) requires that where receptacle replacements are made at outlets that require weather-resistant receptacles, weather-resistant receptacles must be installed. The net result is that if a receptacle is replaced in a location that now requires a tamper-resistant type or weather-resistant type, the type of replacement device required must be as specified in this section and must provide the level of protection required by the current *Code* requirements for receptacles installed in these locations. This is one of only a few instances in the *NEC* where an existing device is required to be upgraded upon replacement in order to afford equal and effective safety and protection provided by the new technology.

Comment: 18-10

Proposals: 18-24, 18-33

406.12 (Exc.)

Article 406 Receptacles, Cord Connectors, and Attachment Plugs (Caps)

Tamper-Resistant Receptacles in Dwelling Units

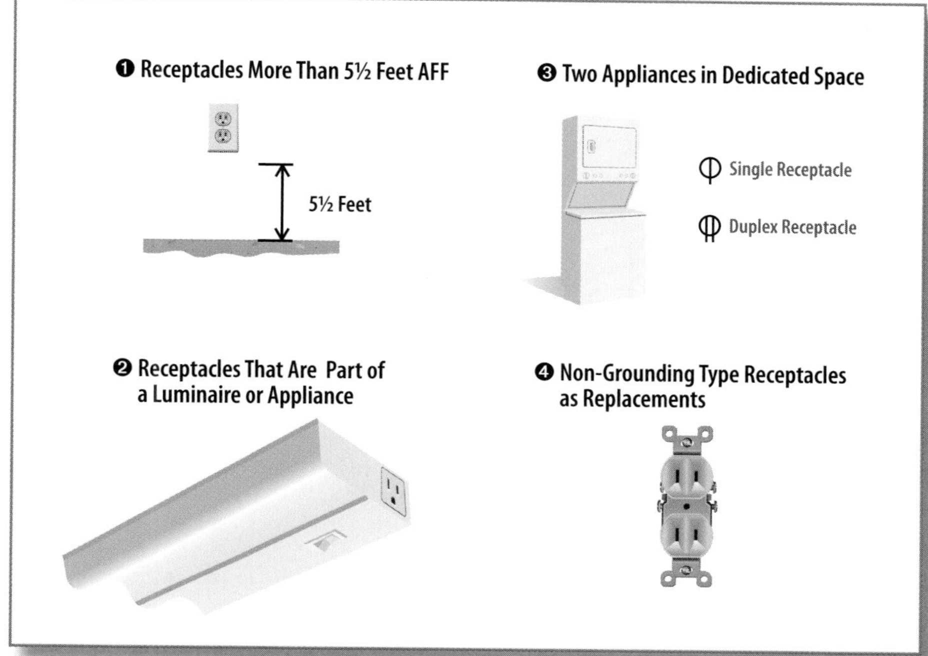

Code Language

406.12 Tamper-Resistant Receptacles in Dwelling Units. In all areas specified in 210.52, all nonlocking type 125-volt, 15- and 20-ampere receptacles shall be listed tamper-resistant receptacles.

Exception: Receptacles in the following locations shall not be required to be tamper-resistant:

(1) Receptacles located more than 1.7 m (5 ½ ft) above the floor.

(2) Receptacles that are part of a luminaire or appliance.

(3) A single receptacle or a duplex receptacle for two appliances located within dedicated space for each appliance that, in normal use, is not easily moved from one place to another and that is cord-and-plug connected in accordance with 400.7(A)(6), (A)(7), or (A)(8).

(4) Nongrounding receptacles used for replacements as permitted in 406.4(D)(2)(a).

Change Summary

The title of this section has been renumbered from 406.11 to 406.12. Additionally, the word *nonlocking* has been added to the rule. A new exception now provides locations or conditions where the receptacles do not have to be tamper-resistant types.

Significance of the Change

This revision results in more specific applicability to nonlocking type receptacles in the voltage and ampere configurations covered by the previous requirement in 406.11. The new exception results from affirmative action by CMP-18 to relax the tamper-resistant requirement for some receptacle outlet locations covered in 210.52. Substantiation with the proposal indicated that receptacles installed above 1.7 m (5 ½ ft.), as provided in list item (1) of the exception, are not accessible and are typically well out of reach of small children. Allowing the exception in list item (3) for a single receptacle or duplex receptacle located within dedicated space eliminates the need for installing tamper-resistant receptacles behind dishwashers, refrigerators, washing machines, and the like, where they are not likely to be accessed by children.

Section 406.4(D)(2)(a) continues to allow the replacement of a nongrounding receptacle with another nongrounding type. This relief in exception list item (4) provides the needed correlation. Exception list item (2) provides practical relief from this requirement, which would otherwise impose tamper-resistant requirements to such electrical products as luminaires that include a nonlocking, 15- or 20-ampere receptacle as part of a listed assembly that is manufactured to current product safety standards.

Comment: 18-33

Proposals: 18-71, 18-82

406.13

Article 406 Receptacles, Cord Connectors, and Attachment Plugs (Caps)

Tamper-Resistant Receptacles in Guest Rooms and Guest Suites

Code Language

406.13 Tamper-Resistant Receptacles in Guest Rooms and Guest Suites. All nonlocking-type, 125-volt, 15- and 20-ampere receptacles located in guest rooms and guest suites shall be listed tamper-resistant receptacles.

Change Summary

A new section, 406.13 Tamper-Resistant Receptacles in Guest Rooms and Guest Suites, has been added to Article 406.

Significance of the Change

This new section extends the tamper-resistant receptacle requirements of Article 406 to the guest rooms and guest suites of hotels and motels. Under the previous edition of the *NEC*, if an occupancy qualified as a dwelling unit by definition, tamper-resistant receptacles were required in areas specified in 210.52. However, those guest rooms and guest suites that did not qualify as dwelling units, specifically transient occupancies frequented by children who were exposed to hazards without the additional level of protection, were exempt from this requirement. This significant change means that the requirements for tamper-resistant receptacles are more consistently applied across occupancies likely to accommodate children and that protection of children is greatly increased.

Comment: 18-35

Proposal: 18-87

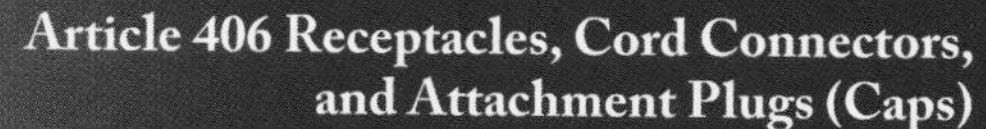

Tamper-Resistant Receptacles in Child Care Facilities

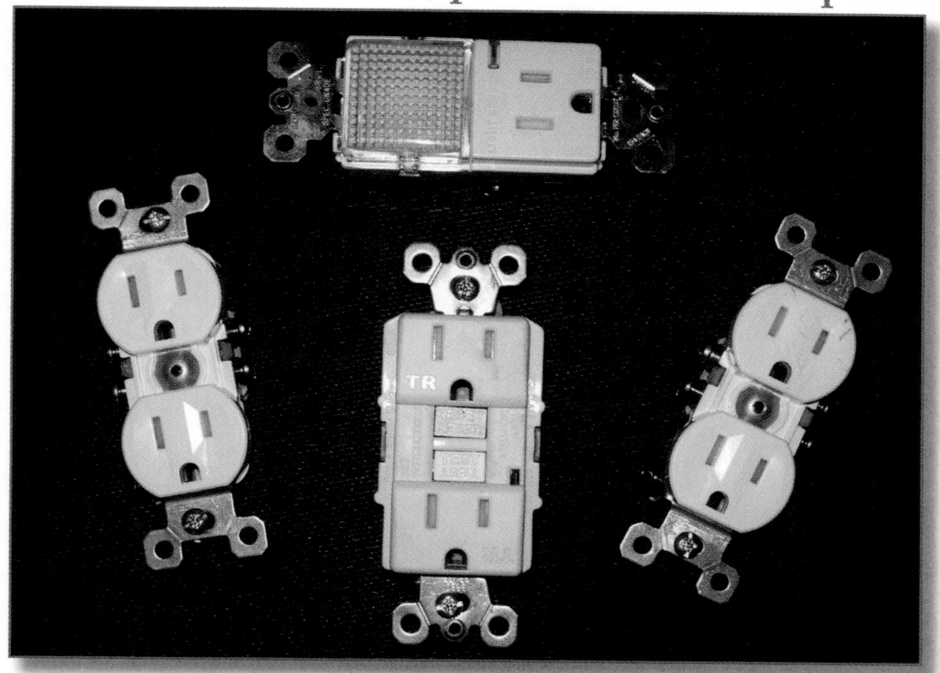

Code Language

406.14 Tamper-Resistant Receptacles in Child Care Facilities. In all child care facilities, all nonlocking-type, 125-volt, 15- and 20-ampere receptacles shall be listed tamper-resistant receptacles.

Significance of the Change

This new rule expands the requirement for tamper-resistant receptacles to facilities beyond dwelling occupancies. The new requirement applies to all nonlocking, 125-volt, 15- and 20-ampere receptacles in child care facilities. Additionally, a new definition of the term *child care facility* has been added to 406.2. Substantiation for the proposal indicated that tamper-resistant receptacles are necessary to prevent incidents of electrical burns and shock that result when children insert conductive objects into receptacles, and so protection should be extended beyond residential and health care facilities. Sections 406.11 and 517.18(C) previously recognized that listed tamper-resistant receptacles provide the most effective means of preventing children from inserting foreign objects into receptacles.

The use of tamper-resistant receptacles is also recognized in the U.S. *General Services Administration Child Care Center Design Guide* as a critical design feature for child care areas. This guide contains the criteria for planning and designing child care centers. Action by CMP-18 applies the requirement for tamper-resistant receptacles to facilities that are dedicated to providing child care and increases consistency in the *NEC* regarding the level of protection provided for unsuspecting children who would otherwise be exposed to similar hazards and risks if this requirement were not applied to child care facilities.

Change Summary

Article 406 now includes a new section, 406.14, titled "Tamper-Resistant Receptacles in Child Care Facilities."

Comment: None

Proposal: 18-90

408.3(F)

Article 408 Switchboards and Panelboards
Part I General

Switchboard or Panelboard Identification

Code Language

408.3 Support and Arrangement of Busbars and Conductors

(F) Switchboard or Panelboard Identification

(2) Ungrounded Systems. A switchboard or panelboard containing an ungrounded electrical system as permitted in 250.21 shall be legibly and permanently field marked as follows:

"Caution Ungrounded System Operating- _____ Volts Between Conductors"

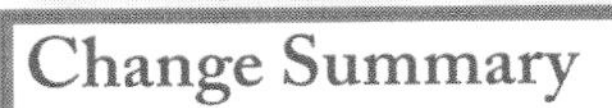

Change Summary

A switchboard or panelboard containing an ungrounded system is now required to be marked in a manner similar to that for systems with a high leg.

Comment: None

Proposal: 9-140

Significance of the Change

This new marking requirement is a safety-driven alert to installers, maintainers, and enforcers to the presence of an ungrounded system. The title of this first-level subdivision has been changed to Switchboard or Panelboard Identification, and the existing requirement for high-leg marking is relocated within.

In the 2008 *NEC*, 408.3(F) was titled High-Leg Identification and contained a marking requirement for 3-phase, 4-wire delta systems where the mid-point of one phase winding is grounded. This requirement is intended to warn installers, maintainers, and enforcers that the system contained a high leg. This requirement still exists as second-level subdivision 408.3(F)(1).

New second-level subdivision 408.3(F)(2) Ungrounded Systems requires that all switchboards or panelboards containing an ungrounded system be marked with the required text and the applicable voltage as follows:

Caution Ungrounded System Operating 480 Volts
Between Conductors

This new marking will ensure that prior to maintaining, expanding, or servicing a switchboard or panelboard with an ungrounded system, all persons, both qualified and unqualified, are alerted to the presence of the ungrounded system.

Source of Supply

Code Language

408.4 Field Identification Required.

(B) Source of Supply. All switchboards and panelboards supplied by a feeder in other than one- or two-family dwellings shall be marked to indicate the device or equipment where the power supply originates.

Significance of the Change

This new first-level subdivision is another marking requirement intended to enhance safety by identifying the source supplying the switchboard or panelboard. This marking serves to aid the installer, maintainer, and designer in all occupancies when a need arises to troubleshoot a problem, perform routine maintenance, or design and install additional loads. It is not uncommon to see this type of marking specified by the design engineer on many installations.

While no requirement presently exists for such marking, many design professionals would include this information on the panelboard schedule or on the switchboard. In the 2008 *NEC*, this section was titled Circuit Directory or Circuit Identification. The new section title is Field Identification Required, and the existing requirement has been merged into first-level subdivision 408.4(A) for circuit directories and marking of all switches and circuit breakers on switchboards.

This marking will also aid the installer/maintainer in performing justified energized work in accordance with NFPA 70E, *Standard for Electrical Safety in the Workplace*. It is imperative that the source of supply and the type of upstream overcurrent device(s) are known when performing an incident energy survey or a hazard risk analysis.

Change Summary

All feeder supplied switchboards and panelboards in other than one- or two-family dwellings are now required to have a marking identifying the source of supply.

Comment: 9-57a

Proposal: 9-142

409.22

Article 409 Industrial Control Panel
Part II Installation

Short-Circuit Current Rating

Code Language

409.22 Short-Circuit Current Rating. An industrial control panel shall not be installed where the available fault current exceeds its short-circuit current rating as marked in accordance with 409.110(4).

Change Summary

Industrial control panels are now prohibited where the available fault current exceeds the short circuit current rating of the industrial control panel.

Significance of the Change

Industrial control panels are required by 409.110 to be marked with a short-circuit current rating. This requirement mandates that the making be based on the short-circuit current rating of a listed and labeled assembly or on the short-circuit current rating established utilizing an approved method. This requirement now prohibits the application of an industrial control panel where the available fault current exceeds the short-circuit current rating of the industrial control panel. Prior to this requirement, the provisions of 110.3(B) would have prevented such misapplication. This new prescriptive requirement in Article 409 provides clarity and usability for the *Code* user.

Comment: 11-2

Proposal: 11-9

409.110

Article 409 Industrial Control Panel
Part III Construction Specifications

Marking

Code Language

409.110 Marking. An industrial control panel shall be marked with the following information that is plainly visible after installation:

(3) Industrial control panels supplied by more than one power source such that more than one disconnecting means is required to disconnect all power within the control panel shall be marked to indicate that more than one disconnecting means is required to de-energize the equipment.

(Remainder of list items unchanged)

Change Summary

Where industrial control panels are fed from more than one source and more than one disconnecting means is required to deenergize the equipment, the industrial control panel must be marked to convey that information.

Significance of the Change

This new marking requirement is designed to provide the installer/maintainer with critical information alerting them that multiple disconnects must be opened to remove all voltage in the industrial control panel. Together with the other marking requirements in this section, this new requirement provides the following information:

(1) Manufacturer's name and trademark
(2) Supply voltage, number of phases, frequency, and full-load currents
(3) More than one disconnecting means is required to de-energize
(4) Short-circuit current rating of the industrial control panel
(5) If intended as service equipment, it must be marked as suitable
(6) Electrical wiring diagram
(7) An enclosure type number

The new requirement does not provide an example of the exact marking text, but the following examples would comply:

Warning: Two sources of power supply this equipment
Warning: Multiple power sources
Warning: Power sources supplying this equipment are located in:
Panelboard LP1A: circuits 3, 5 and 7
MCC-B4: Sewage Ejection Pump 14

Comment: 11-4
Proposal: 11-23

Luminaires as Raceways

Code Language

410.64 Luminaires as Raceways. Luminaires shall not be used as a raceway for circuit conductors unless they comply with 410.64(A), (B), or (C).

(A) Listed. Luminaires listed and marked for use as a raceway shall be permitted to be used as a raceway.

(B) Through-Wiring. Luminaires identified for through wiring, as permitted by 410.21, shall be permitted to be used as a raceway.

(C) Luminaires Connected Together. Luminaires designed for end-to-end connection to form a continuous assembly, or luminaires connected together by recognized wiring methods, shall be permitted to contain the conductors of a 2-wire branch circuit, or one multiwire branch circuit, supplying the connected luminaires and shall not be required to be listed as a raceway. One additional 2-wire branch circuit separately supplying one or more of the connected luminaires shall also be permitted.

Informational Note: See Article 100 for the definition of Multiwire Branch Circuit.

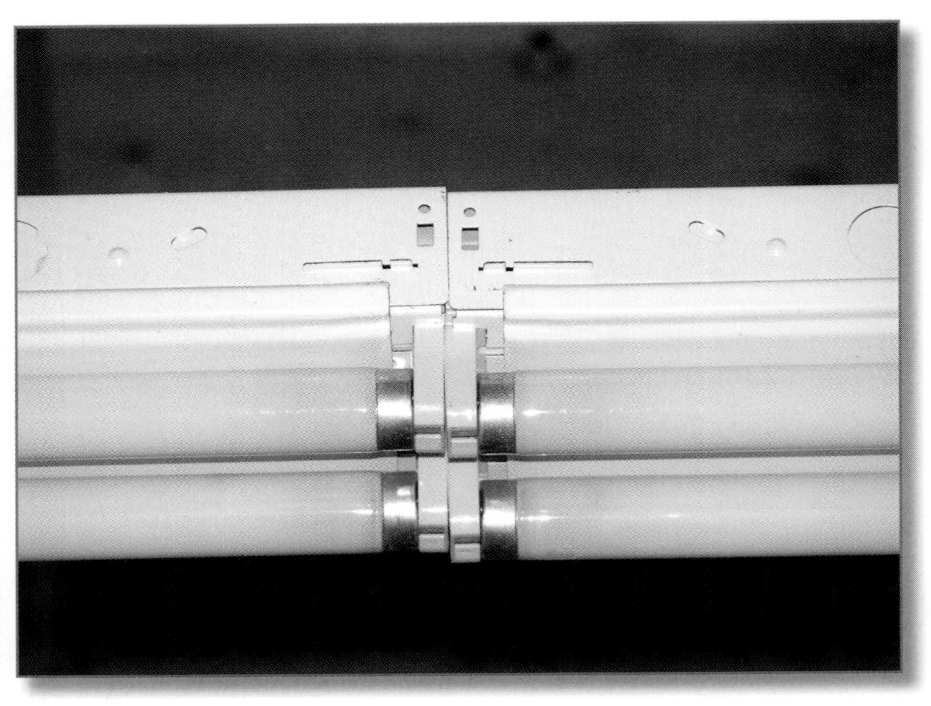

Significance of the Change

In the 2008 *NEC* the requirements of 410.64 and 410.65 conflicted with each other. Section 410.64 required that all luminaires being used as a raceway for circuit conductors be listed and marked for the purpose, while 410.65 permitted otherwise. This revision deletes 410.65 and rolls those requirements into 410.64, then separates 410.64 into three user-friendly first-level subdivisions. Where *Code* requirements can be grouped into a single section with each rule in an individual and titled subdivision, the *NEC* becomes much easier to use.

This revision is editorial in nature, yet it provides the *Code* user with a logical separation of requirements in an easy-to-use format.

Change Summary

NEC requirements for luminaires used as a raceway have been logically reorganized into a single section to improve usability.

Comment: None
Proposal: 18-136a

410.110

Article 410 Luminaires, Lampholders, and Lamps
Part X Special Provisions for Flush and Recessed Luminaires

General

Code Language

410.110 General. Luminaires installed in recessed cavities in walls or ceilings, including suspended ceilings, shall comply with 410.115 through 410.122.

Change Summary

Luminaires installed in suspended ceilings are now subject to the requirements of 410.115 through 410.122.

Significance of the Change

A formal interpretation has been in place since 1982 to clarify that suspended ceilings are subject to the requirements of Part XI of Article 410. The Formal Interpretation is posted on the NFPA website for the 2008 *NEC*. The Formal Interpretation states that fixtures installed in suspended ceilings are subject to the requirements of Part XI of Article 410.

While this formal interpretation has been in place for almost 30 years, the reader of the *NEC* was unaware of its existence unless a problem arose with an AHJ who knew of the interpretation. This revision now requires that all luminaires installed in suspended ceilings comply with Article 410, Part XI:

410.115 Temperature
410.116 Clearance and Installation
410.117 Wiring
410.118 Temperature
410.120 Lamp Wattage Marking
410.121 Solder Prohibited
410.122 Lampholders

Comment: None
Proposal: 18-166

410.116(B)

Article 410 Luminaires, Lampholders, and Lamps
Part X Special Provisions for Flush and Recessed Luminaires

Installation

Code Language

410.116 Clearance and Installation

(B) Installation. Thermal insulation shall not be installed above a recessed luminaire or within 75 mm (3 in.) of the recessed luminaire's enclosure, wiring compartment, ballast, transformer, LED driver, or power supply unless the luminaire is identified as Type IC for insulation contact.

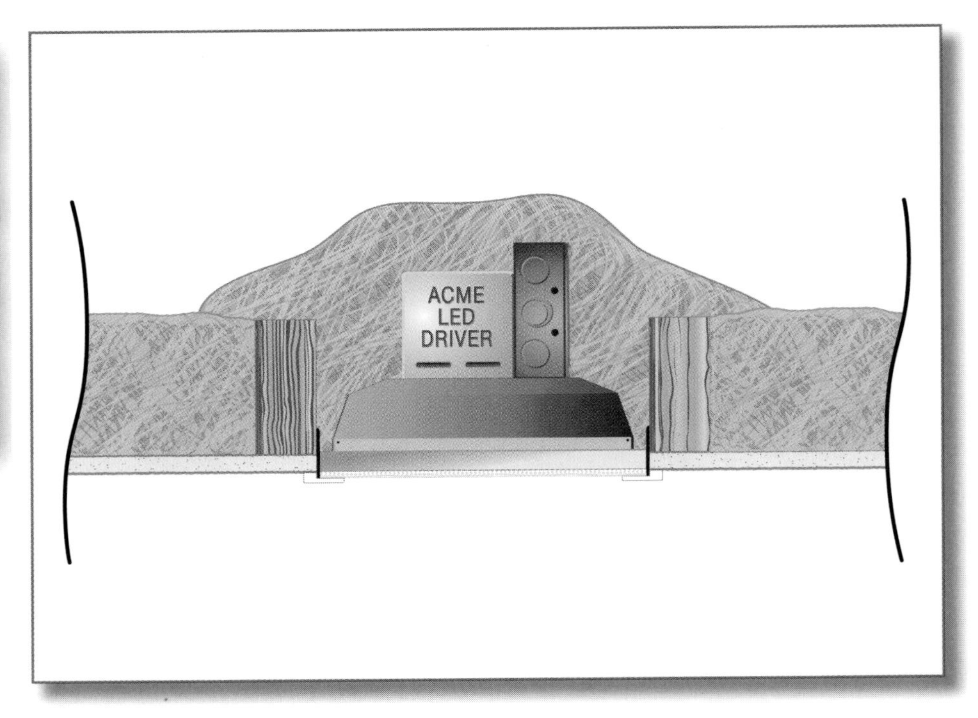

Change Summary

LED drivers and power supplies have been included in 410.116(B) to prohibit the installation of thermal insulation above or within 3 inches of the luminaire unless it is marked as Type IC.

Significance of the Change

Throughout the 2011 *NEC*, revisions have been made to address the increasing use of LED technology. Article 410 contains requirements for luminaires, lampholders, and lamps, and many sections have incorporated LED technology into existing requirements, including but not limited to, 410.16, 410.24, 410.62, 410.68, 410.74, 410.116, 410.136, and 410.137. This revision also incorporates LED drivers and power supplies in the installation requirement in order to prohibit insulation on top of or within 3 inches of the luminaire. The term *Type IC Luminaire* is defined as follows in the *UL White Book*:

TYPE IC LUMINAIRE — Luminaires marked "**TYPE IC**" may be installed such that insulation and other combustible materials are in contact with, and over the top of, the luminaire.

Section 410.116, located in Part X of Article 410 to address specific requirements of that article, is titled Special Provisions for Flush and Recessed Luminaires. This requirement applies to any luminaire that is recessed or flush mounted. For example, the installation of a high hat type fixture in any type of insulated ceiling must be marked Type IC.

Comment: None
Proposal: 18-168

410.130(G)

Article 410 Luminaires, Lampholders, and Lamps
Part XII Special Provisions for Electric-Discharge Lighting Systems of 1000 Volts or Less

Disconnecting Means

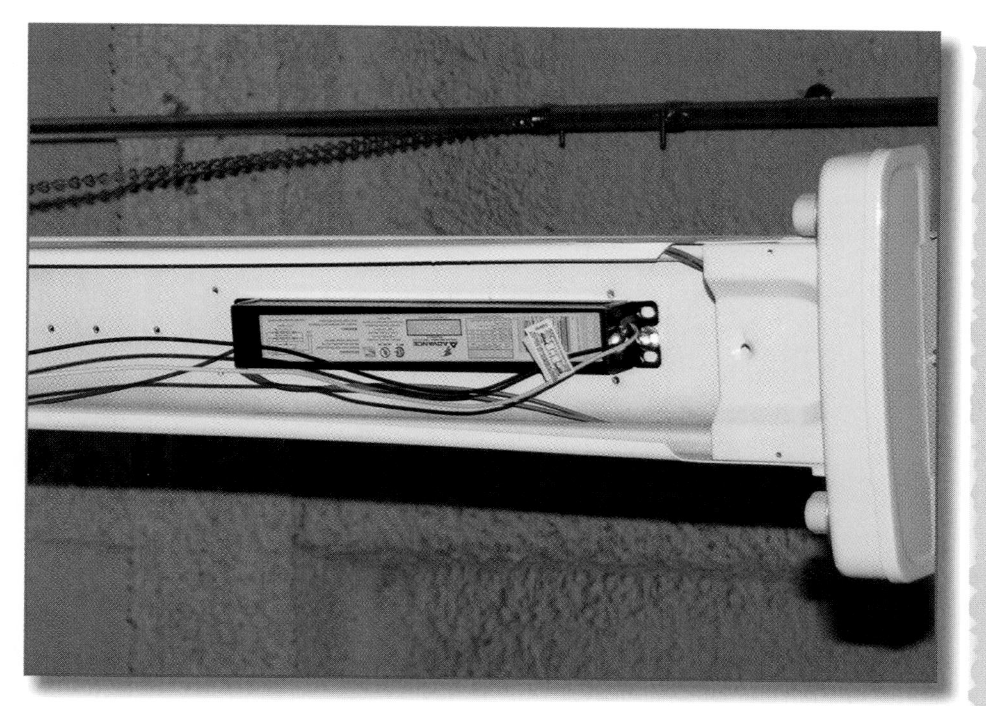

Code Language

410.130 General

(G) Disconnecting Means

(1) General. In indoor locations other than dwellings and associated accessory structures, fluorescent luminaires that utilize double-ended lamps and contain ballast(s) that can be serviced in place shall have a disconnecting means either internal or external to each luminaire. For existing installed luminaires without disconnecting means, at the time a ballast is replaced, a disconnecting means shall be installed. The line side terminals of the disconnecting means shall be guarded.

Exceptions 1 through 5 ***(No change.)***

Change Summary

A disconnecting means is now required to be installed during the replacement of a ballast if a disconnecting means in accordance with 410.130(G) or any of the exceptions does not exist.

Significance of the Change

Section 410.130(G) requires all fluorescent luminaires that utilize double-ended lamps and contain ballast(s) that can be serviced in place to be provided with a means to disconnect the ballast from the source of supply. This requirement is limited to indoor locations other than dwellings and associated accessory structures. The intent of this requirement is to allow the installer/maintainer to disconnect the ballast from the source of supply before the task begins and to therefore perform the task in a deenergized state and eliminate a potential shock situation.

This revision requires that all ballast replacements include a disconnecting means if one is not already in place. Due to the lack of a disconnecting means, the installer/maintainer in many cases would choose to perform the task in an energized state, creating the potential for serious injury or death.

Section 410.130(G)(2) requires that all the supply conductors to the ballast, including the grounded conductor, be simultaneously disconnected. A ballast replacement in most cases that uses a disconnect that breaks both the ungrounded and grounded conductors. The general requirement in 410.130(G) has driven luminaire manufacturers to include a disconnect to open both the ungrounded and grounded conductors in fixtures for compliance.

Comment: None
Proposal: 18-175

422.2

Article 422 Appliances
Part I General

Vending Machine

Code Language

422.2 Definitions.

Vending Machine. Any self-service device that dispenses products or merchandise without the necessity of replenishing the device between each vending operation and is designed to require insertion of coin, paper currency, token, key, or receipt of payment by other means.

Change Summary

Section 422.2 has been added to include definitions that apply only in Article 422. A definition of *vending machine* has been added for application throughout Article 422.

Comment: None

Proposals: 17-3, 17-32

Significance of the Change

A definition is included in the *NEC* only where the term or phrase must be defined for clarity and usability. The *NEC Style Manual* requires that, in general, terms and phrases that are defined and are used in more than one article must be located in Article 100. General industry terms are considered to be understood and so are not defined. A term or phrase that is defined and used in only one article is located in the second section of that article.

The definition of *vending machine* comes directly from the description that existed in 422.51 of the 2008 *NEC*. Relocating this text into a new definition in 422.2 provides the *Code* user with the necessary information to determine whether an appliance is indeed a vending machine. For example, a washing machine, while it is an appliance, is not a vending machine because it does not dispense products or merchandise. Typical vending machines include those that dispense soda, candy, cigarettes, lottery tickets, coffee, and the like. A photo booth would be a vending machine because it does dispense a product.

422.31(C)

Article 422 Appliances
Part III Disconnecting Means

Motor-Operated Appliances Rated over ⅛ Horsepower

Code Language

422.31 Disconnection of Permanently Connected Appliances.

(C) Motor-Operated Appliances Rated over ⅛ Horsepower. For permanently connected motor-operated appliances with motors rated over ⅛ horsepower, the branch-circuit switch or circuit breaker shall be permitted to serve as the disconnecting means where the switch or circuit breaker is within sight from the appliance. The disconnecting means shall comply with 430.109 and 430.110.

Exception: If an appliance of more than ⅛ hp is provided with a unit switch that complies with 422.34(A), (B), (C), or (D), the switch or circuit breaker serving as the other disconnecting means shall be permitted to be out of sight from the appliance.

Change Summary

Disconnect requirements for motor-operated appliances with motors rated over ⅛ hp, previously located in 422.31(B) and 422.32, have been revised for clarity in a new 422.31(C).

Significance of the Change

This revision provides clarity and eliminates conflicting requirements. The title Motor-Operated Appliances Rated over ⅛ Horsepower clearly conveys that this requirement is intended to provide an increased level of safety for persons installing or maintaining appliances that are capable of sudden movement. Sections 422.31(B) and 422.32 were conflicting in the 2008 *NEC*, in that 422.31(B) permitted the branch-circuit switch or circuit breaker disconnecting means for an appliance rated over ⅛ hp to be locked in the open position, yet 422.32 required that for a motor-driven appliance rated over ⅛ hp, the disconnecting means must be within sight from the motor controller.

The revised text in 422.31(C) now clarifies that a disconnecting means is required within sight from the appliance. The exception remains and permits a disconnecting means to be out of sight from an appliance of more than ⅛ hp where a unit switch that complies with 422.34(A), (B), (C), or (D) is provided for the appliance. Section 422.31(B) no longer applies to an appliance rated over ⅛ hp, and 422.32 has been deleted.

Comment: 17-14
Proposal: 17-21

424.19(A)

Article 424 Fixed Electric Space-Heating Equipment
Part III Control and Protection of Fixed Electric Space-Heating Equipment

Heater with a Motor Rated Over ⅛ Horsepower

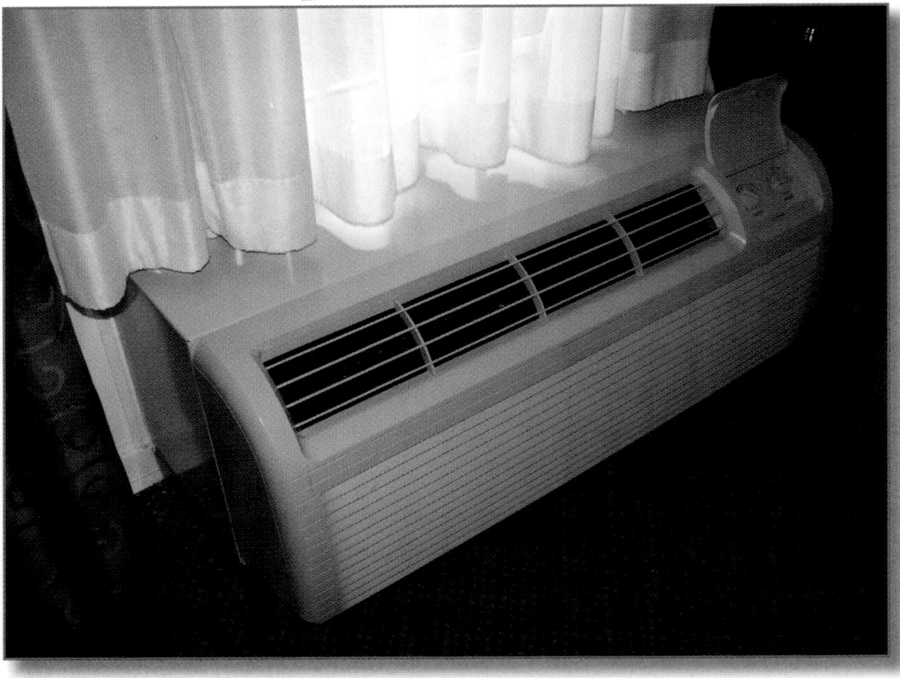

Code Language

424.19 Disconnecting Means

(A) Heating Equipment with Supplementary Overcurrent Protection

(2) Heater Containing a Motor(s) Rated over ⅛ Horsepower. The above disconnecting means shall be permitted to serve as the required disconnecting means for both the motor controller(s) and heater under either of the following conditions:

(1) Where the disconnecting means is in sight from the motor controller(s) and the heater and complies with Part IX of Article 430.

(2) Where a motor(s) of more than ⅛ hp and the heater are provided with a single unit switch that complies with 422.34(A), (B), (C), or (D), the disconnecting means shall be permitted to be out of sight from the motor controller.

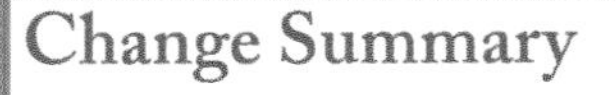

Change Summary

The disconnect requirement for a heater containing a motor over ⅛ horsepower is clarified for usability purposes. This section now mirrors 422.31(C) because the disconnect requirement should be the same whether it is a motor-driven appliance or part of fixed space-heating equipment.

Significance of the Change

This revision of 424.19(A)(2) provides uniformity in the disconnect requirements of a heater containing a motor rated over ⅛ horsepower, and in 422.31(C) (422.32 in 2008 *NEC*) for motor-driven (over ⅛ horsepower) appliances. This revision now requires that the disconnecting means comply with 424.19(A)(1) or (A)(2).

Section 424.19(A)(1) requires that the disconnecting means to be in sight from the motor controller(s) and the heater and to comply with Part IX of Article 430.

In accordance with 424.19(A)(2), the disconnecting means is permitted to be out of sight from the motor as long as the motor and heater are provided with a single unit switch that complies with 422.34. This section permits a unit switch with a marked "off" position that is part of an appliance that disconnects all ungrounded conductors as long as another disconnect is provided in accordance with 422.34(A), (B), (C), or (D). Section 424.19(A)(2) permits a disconnecting means to be out of sight where a unit switch in compliance with 422.34 exists. This requires a marked "off" position that is a part of the appliance and disconnects all ungrounded conductors.

Comment: None

Proposal: 17-44

424.44(G)

Article 424 Fixed Electric Space-Heating Equipment
Part V Electric Space-Heating Cables

Ground-Fault Circuit-Interrupter Protection

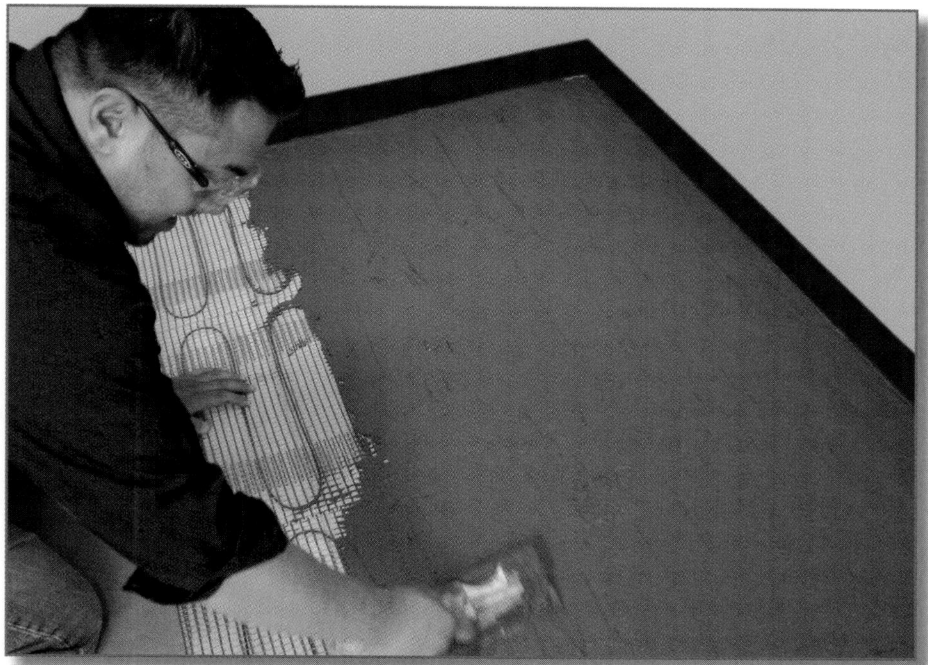

Courtesy of Raychem Quicknet, Tyco Thermal Controls

Code Language

424.44 Installation of Cables in Concrete or Poured Masonry Floors.

(G) Ground-Fault Circuit-Interrupter Protection. Ground-fault circuit-interrupter protection for personnel shall be provided for cables installed in electrically heated floors of bathrooms, kitchens, and in hydromassage bathtub locations.

Change Summary

Where heating cable in installed in a kitchen floor, the circuit is now required to be GFCI-protected. The previous requirement for GFCI protection was limited to bathrooms and hydromassage bathtub locations.

Significance of the Change

The previous requirement for GFCI protection of electrical heating cable installed in concrete or poured masonry floors was limited to bathrooms and hydromassage bathtub locations, for the obvious reason of standing water. The GFCI requirement now governs kitchens. Concrete and masonry are considered conductive, and the presence of water only enhances the conductivity of these surfaces. In areas where water is likely or will possibly be present on the floor, it is prudent to protect persons from shock by providing GFCI protection. Note that this is not an all-inclusive requirement, in that heating cables installed in a concrete or poured masonry floor of a family room or den would not be required to be GFCI-protected.

Comment: None

Proposal: 17-58

Ground-Fault Protection of Equipment

Courtesy of Tyco Thermal Controls

Code Language

426.28 Ground-Fault Protection of Equipment. Ground-fault protection of equipment shall be provided for fixed outdoor electric deicing and snow-melting equipment.

Change Summary

All fixed outdoor electric deicing and snow-melting equipment installed outdoors for deicing or snow-melting must now be provided with ground-fault protection of equipment (GFPE) regardless of where or how it is installed.

Significance of the Change

The title of this section has been revised from Equipment Protection to Ground-Fault Protection of Equipment, to more clearly convey the content of this section. The second half of this section, which excluded Type MI cable installed in concrete or other noncombustible medium, has been deleted. All Type MI (mineral-insulated) cable installed outdoors for deicing or snow-melting must now be provided with ground-fault protection of equipment (GFPE).

GFPE protection is not equivalent to GFCI protection. The latter is intended for protection of persons. A Class A GFCI device will trip when the current level to ground exceeds 6 mA and will not trip below 4 mA. A GFPE device trips at levels above 6 mA. A device used in this application will trip in the 30 mA range. It is important to note that this is well above the let-go threshold of 10 mA and is intended only for protection of equipment, not for protection of personnel.

Comment: None

Proposals: 17-75, 17-76

430.6(D)

Article 430 Motors, Motor Circuits, and Controllers
Part I General

Valve Actuator Motor Assemblies

Code Language

430.6 Ampacity and Motor Rating Determination

(D) Valve Actuator Motor Assemblies. For valve actuator motor assemblies (VAMs), the rated current shall be the nameplate full-load current, and this current shall be used to determine the maximum rating or setting of the motor branch-circuit short-circuit and ground-fault protective device and the ampacity of the conductors.

Significance of the Change

This requirement is necessary to provide the *Code* user with prescriptive text clarifying that the nameplate current is used on a VAM for selection of the maximum rating or setting of the motor branch-circuit short-circuit and ground-fault protective device and the ampacity of the conductors. Confusion abounds in the industry regarding the installation of valve actuator motors, based on the installation of a VAM and whether to use table, locked rotor, or nameplate values of current. In the 2008 *NEC*, 430.6 did not address a VAM, so many *Code* users have applied 430.6(B), which mandated the use of locked rotor current values, thinking that a VAM is a torque motor. A VAM is not a torque motor.

Section 430.6 is the starting point for all motor installations, where the *Code* user determines which value of current to use when sizing the maximum rating or setting of the motor branch-circuit short-circuit and ground-fault protective device and the ampacity of the conductors. Current values for overload protection are addressed in Part III Motor and Branch-Circuit Overload Protection of Article 430.

Change Summary

A first-level subdivision has been added to 430.6 to clarify which value of current must be used for a valve actuator motor assembly (VAM). When sizing the maximum rating or setting of the motor branch-circuit short-circuit and ground-fault protective device and the ampacity of the conductors for a VAM, the nameplate current is now required to be used.

Comment: None

Proposal: 11-26

430.22(G)

Article 430 Motors, Motor Circuits, and Controllers
Part II Motor Circuit Conductors

Conductors for Small Motors

Code Language

430.22 Single Motor

(G) Conductors for Small Motors. Conductors for small motors shall not be smaller than 14 AWG unless otherwise permitted in 430.22(G)(1) or (G)(2).

(1) 18 AWG copper…

(2) 16 AWG Copper…

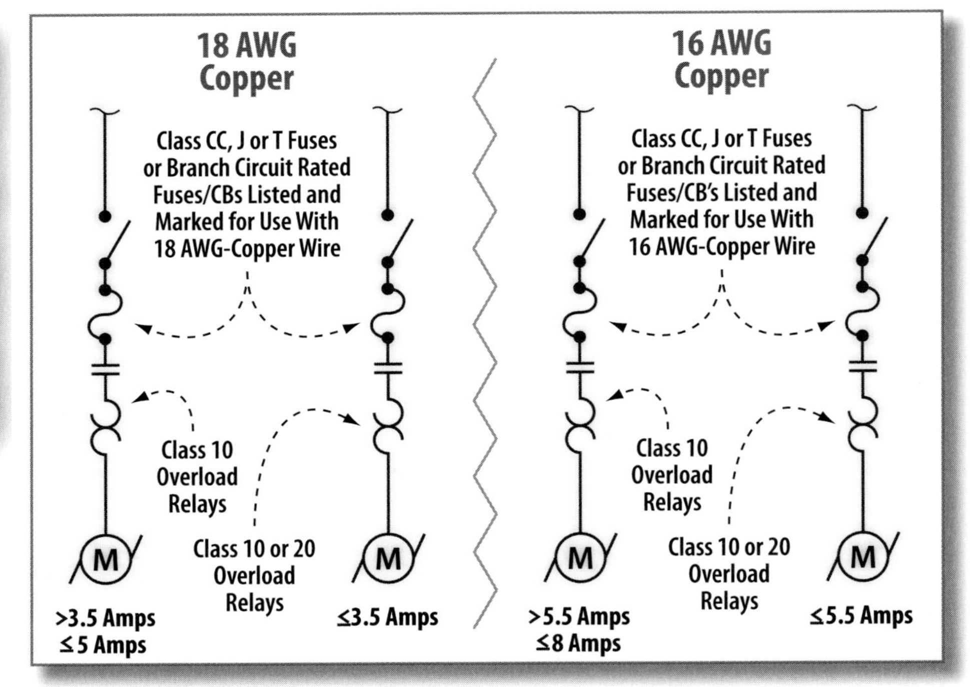

Change Summary

A new first-level subdivision has been added to 430.22 to permit 16 AWG and 18 AWG copper conductors in specific applications. This revision will allow U.S. machinery manufacturers to be more competitive in the international market, where the allowable minimum IEC motor circuit conductors are smaller than the minimum 14 AWG conductors allowed by the *NEC*.

Comments: 11-17a, 18, 19, 20

Proposal: 11-50

Significance of the Change

This new text correlates the detailed requirements for applying small motor circuit conductors in Article 430 with those covered by NFPA 79. The size and specific installation are limited to protected areas such as cabinets or enclosures. These conductors may be applied only with specific classes of overload relays and specific sizes of overcurrent protective devices, based on the motor full-load ampacity. Fuses and circuit breakers must be specifically listed for use with these smaller conductors, or they may be class CC, J, or T type fuses. This revision mirrors the present requirements of NFPA 79 for industrial machinery. In the 2008 cycle, 240.4(D) was revised in a similar fashion to correlate with NFPA 79 and to help U.S. machinery manufacturers compete in the international market.

This revision applies only where conductors are installed in a cabinet or enclosure. The conductors may be individual copper conductors, copper conductors that are part of a jacketed multiconductor cable assembly, or copper conductors in a flexible cord. In all cases, protection must comply with 430.52 and 240.4(D)(1)(2), and overload protection is specified according to ampacity to be a Class 10 or Class 20. Overload protection "Class" is categorized by NEMA Standard ICS 2.

430.52(C)(7)

Article 430 Motors, Motor Circuits, and Controllers
Part IV Motor Branch-Circuit Short-Circuit and Ground-Fault Protection

Motor Short-Circuit Protector

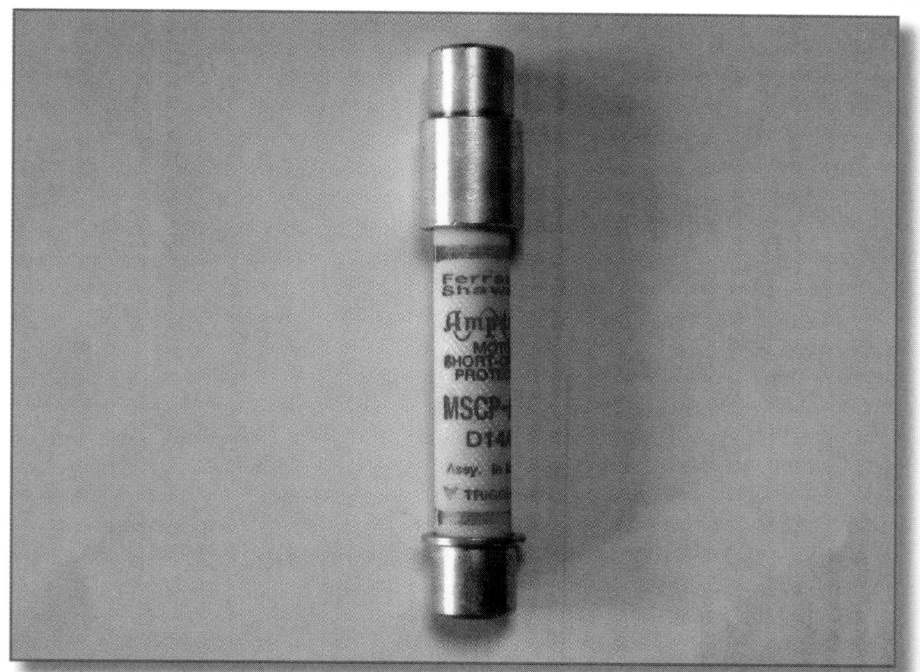

Courtesy of National Electric Fuse Association

Code Language

430.52 Rating or Setting for Individual Motor Circuit

(C) Rating or Setting

(7) Motor Short-Circuit Protector. A motor short-circuit protector shall be permitted in lieu of devices listed in Table 430.52 if the motor short-circuit protector is part of a listed combination motor controller having coordinated motor overload protection and short-circuit and ground-fault protection in each conductor and it will open the circuit at currents exceeding 1300 percent of motor full-load current for other than Design B energy-efficient motors and 1700 percent of motor full-load motor current for Design B energy-efficient motors.

Informational Note: A motor short-circuit protector, as used in this section, is a fused device and is not an instantaneous trip circuit breaker.

Significance of the Change

This informational note will help to eliminate confusion in the industry regarding which type of device is addressed by 430.52(C)(7). The type addressed in this second-level subdivision is a fused device, not an instantaneous-type circuit breaker commonly known in these applications as an MCP, or motor circuit protector. Application of 430.52(C)(3), (C)(6), or (C)(7) requires a listed combination motor controller. Section 430.52(C)(3), Instantaneous Trip Circuit Breaker, addresses the use of an MCP, or motor circuit protector, as a device that provides only short-circuit and ground-fault protection but not any level of overload protection. An MCP is not a branch circuit overcurrent protective device as defined in Article 100, because it does not provide a full range of overcurrent protection.

According to the *UL White Book*, a combination motor controller provides the motor branch-circuit functions of motor controller, disconnect means, short-circuit and ground-fault protection, and motor overload protection. These functions may be provided by individual discrete components or be combined in a single controller unit.

Change Summary

An informational note has been added to 430.52(C)(7) to clarify that a "motor short-circuit protector" is a fused device, not an MCP-type, or "motor circuit protector-type," circuit breaker.

Comment: None

Proposal: 11-67

430.53

Article 430 Motors, Motor Circuits, and Controllers
Part IV Motor Branch-Circuit Short-Circuit and Ground-Fault Protection

Several Motors or Loads on One Branch Circuit

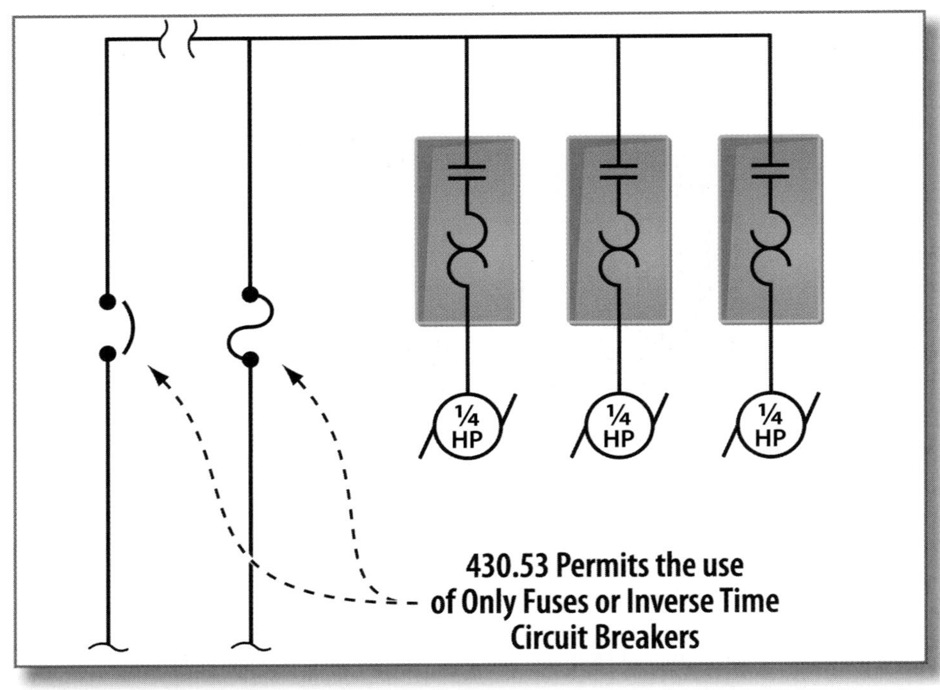

Code Language

430.53 Several Motors or Loads on One Branch Circuit. Two or more motors or one or more motors and other loads shall be permitted to be connected to the same branch circuit under conditions specified in 430.53(D) and in 430.53(A), (B), or (C). The branch-circuit protective device shall be fuses or inverse time circuit breakers.

Change Summary

A new last sentence has been added to 430.53 to clarify that inverse-time circuit breakers and fuses are the only permitted means for providing group motor branch short-circuit and ground-fault protection.

Significance of the Change

Where several motors or loads are placed on one branch circuit, the only permitted means for providing group motor branch short-circuit and ground-fault protection is fuses or inverse-time circuit breakers. An instantaneous-type device such as a motor circuit protector (MCP) type circuit breaker is not permitted.

Overcurrent protection of motor installations can be via a single overcurrent device that meets the requirement of Part IV with requirements for motor branch-circuit short-circuit and ground-fault protection and Part III with requirements for motor and branch-circuit overload protection. In many cases two devices are used.

Overload protection is provided in accordance with Part III to protect the motor and the branch circuit conductors from damage. The overloads will not protect against short circuits or ground faults. Motor branch short-circuit and ground-fault protection is provided in accordance with Part IV. This device is then sized larger than would be permitted to protect the motor branch-circuit conductors to allow the motor to start.

Comment: None

Proposal: 11-68

430.53(C)

Article 430 Motors, Motor Circuits, and Controllers
Part IV Motor Branch-Circuit Short-Circuit and Ground-Fault Protection

Several Motors or Loads on One Branch Circuit

Code Language

430.53 Several Motors or Loads on One Branch Circuit.

(C) Other Group Installations ...

(1) Each motor overload device is either (a) listed for group installation with a specified maximum rating of fuse, inverse time circuit breaker, or both, or (b) selected such that the ampere rating of the motor branch short-circuit and ground-fault protective device does not exceed that permitted by 430.52 for that individual motor overload device and corresponding motor load.

(2) Each motor controller is either (a) listed for group installation with a specified maximum rating of fuse, circuit breaker, or both, or (b) selected such that the ampere rating of the motor branch short-circuit and ground-fault protective device does not exceed that permitted by 430.52 for that individual controller and corresponding motor load.

Significance of the Change

The previous text of 430.53(C)(1) and (C)(2) in the 2008 *NEC* mandated that each motor overload device be listed for group installation with a specified maximum rating of fuse, inverse time circuit breaker, or both, and each motor controller be listed for group installation with a specified maximum rating of fuse, circuit breaker, or both.

This new text will now clearly permit an individual overload device or individual motor controller to be applied to a single motor that is part of a group motor installation provided that the requirements of 430.52 are met.

There are four additional list items with requirements for 430.53(C):

(3) Each circuit breaker must be listed and of the inverse time type

(4) The calculation of the branch circuit rating of the required fuses or inverse time circuit breakers is explained in detail. Where that calculation results in a rating less than the ampacity of the supply conductors, the provisions of 240.4(B) are permitted.

(5) The branch-circuit fuses or inverse time circuit breakers can not be larger than allowed by 430.40 for the overload relay protecting the smallest rated motor of the group

(6) Overcurrent protection for loads other than motor loads shall be in accordance with Parts I through VII of Article 240.

Change Summary

Text added to 430.53(C)(1) and (C)(2) now allows overload devices and motor controllers for single motors that are not listed for "group installation" for single motors in the group, provided they individually meet the requirements of 430.52.

Comment: None
Proposal: 11-69

Disconnecting Means

Code Language

450.14 Disconnecting Means. Transformers, other than Class 2 or Class 3 transformers, shall have a disconnecting means located either in sight of the transformer or in a remote location. Where located in a remote location, the disconnecting means shall be lockable, and the location shall be field marked on the transformer.

Change Summary

A new section has been added to clearly require a disconnecting means for transformers. The required disconnect may be either within sight of the transformer or in a remote location. A means to lock the disconnect is required where the disconnect it is not within sight of the transformer. Where the disconnect is not within sight, the location of the remote disconnect must be field marked on the transformer.

Significance of the Change

Transformers, other than Class 2 and Class 3 transformers, are now required to have a disconnecting means. The new text permits the location of the disconnecting means to be within sight of the transformer in an electrical closet, for example. The new text also permits the required disconnect to be located remote from the transformer. For example, a feeder could leave an electrical closet and supply a transformer in another equipment room. Where the disconnecting means is located remote from the transformer this new section requires that it must be capable of being locked in the open position and the transformer must be field marked to denote the location of the remote disconnect.

This safety driven requirement will now mandate: (1) a disconnect means within sight or (2) a disconnecting means capable of being locked in the open position for disconnects located remote from the transformer and field marking of the transformer with the location of the remote disconnect.

Comment: None
Proposal: 9-176

NECA Code-Making Panel Members

TECHNICAL CORRELATING COMMITTEE
Stanley J. Folz, [Principal]
Larry D. Cogburn, [Alternate]

CODE–MAKING PANEL NO. 1
Harry J. Sassaman, [Principal]

CODE–MAKING PANEL NO. 2
Thomas H. Wood, [Principal]

CODE–MAKING PANEL NO. 3
Stanley D. Kahn, [Principal]

CODE–MAKING PANEL NO. 4
Ronald J. Toomer, [Chair]
Larry D. Cogburn, [Alternate]

CODE–MAKING PANEL NO. 5
Michael J. Johnston, [Chair]
Larry D. Cogburn, [Alternate]

CODE–MAKING PANEL NO. 6
Scott Cline, [Chair]
Phillip J. Huff, [Alternate]

CODE–MAKING PANEL NO. 7
Michael W. Smith, [Chair]
Wesley L. Wheeler, [Alternate]

CODE–MAKING PANEL NO. 8
Stephen P. Poholski, [Principal]

CODE–MAKING PANEL NO. 9
Monte Szendre, [Principal]

CODE–MAKING PANEL NO. 10
Richard Sobel, [Principal]

CODE–MAKING PANEL NO. 11
Wayne Brinkmeyer, [Chair]
Stanley J. Folz, [Alternate]

CODE–MAKING PANEL NO. 12
Thomas L. Hedges, [Principal]
Charles M. Trout, [Alternate]

CODE–MAKING PANEL NO. 13
Martin D. Adams, [Principal]

CODE–MAKING PANEL NO. 14
Marc J. Bernsen, [Principal]

CODE–MAKING PANEL NO. 15
Bruce D. Shelly, [Principal]

CODE–MAKING PANEL NO. 16
W. Douglas Pirkle, [Principal]

CODE–MAKING PANEL NO. 17
Don W. Jhonson, [Chair]
Bobby J. Gray, [Alternate]

CODE–MAKING PANEL NO. 18
Charles M. Trout, [Principal]

CODE–MAKING PANEL NO. 19
Howard D. Hughes, [Principal]

Chapter 5

Articles 500-590

Special Occupancies

Significant Changes
TO THE NEC® 2011

Definition of Combustible Dust

Code Language

500.2 Definitions.

Combustible Dust. Any finely divided solid material that is 420 microns (0.017 in.) or smaller in diameter (material passing a U.S. No. 40 Standard Sieve) and presents a fire or explosion hazard when dispersed and ignited in air. [499, 2008]

Change Summary

A definition of the term *combustible dust* has been added to 500.2. It was extracted from NFPA 499, *Recommended Practice for the Classification of Combustible Dusts and of Hazardous (Classified) Locations for Electrical Installations in Chemical Process Areas.*

Significance of the Change

The definition for *combustible dust* has been revised in a number of NFPA documents in an effort to achieve harmony across various codes and standards. This revision eliminates previous inconsistencies about dust material size within the *NEC* and within NFPA 499 regarding considerations of a combustible dust. This new definition in Article 500 establishes consistent criteria of what constitutes a combustible dust within the context of how the term is expressed in the requirements in which the term appears. In NFPA 654, a combustible dust is now defined as "a combustible particulate solid that presents a fire or deflagration hazard when suspended in air or some other oxidizing medium over a range of concentrations, regardless of particle size or shape." Inclusion of the proper historic definition for combustible dust upon which these *NEC* articles and NFPA 499 are based will retain this important term without adding confusion between the same term used and defined in other standards beyond NFPA 499 and NFPA 70.

Comment: None

Proposals: 14-8, 14-9

501.10(A)(1)

Article 501Class I Locations
Part II Wiring

Wiring Methods

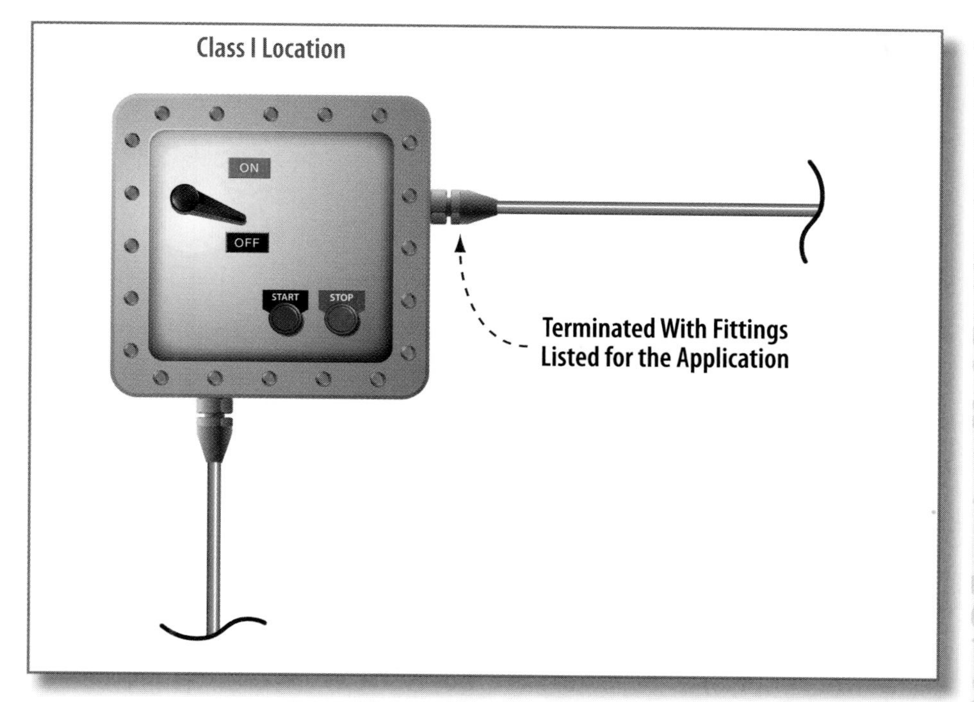

Code Language

501.10(A)(1)

(A) Class I, Division 1.

(1) General...

(a) *Threaded rigid metal conduit...*

Exception... ***(No changes)***

(b) *Type MI cable terminated with fittings listed for the location.*

(c) *In industrial establishments with restricted public access*, where the conditions of maintenance and supervision ensure that only qualified persons service the installation, Type MC-HL cable...

(d) *In industrial establishments with restricted public access*, where the conditions of maintenance and supervision ensure that only qualified persons service the installation, Type ITC-HL cable...

(See NEC for full text of this section.)

Change Summary

The concept of cables being "terminated with" listed fittings has been incorporated consistently throughout this section. The informational note about MC-HL cable following paragraph (c) has been included in the requirements of this section. The informational note about Type ITC cable following (d) has been included in the requirements of this section. The 2008 FPNs have been incorporated into the rule.

Significance of the Change

Section 501.10(A)(1) covers wiring methods for Class I locations. One of the features necessary for qualifying as an acceptable wiring method for Class I, Division 1 locations is that the fittings for cable assemblies be listed for terminating the type of cable involved. The editorial revisions provide consistent wording of that requirement throughout the section. The two informational notes (these were FPNs in the 2008 *NEC*) to paragraphs (c) and (d) dealt more with restrictions of use, which is more related to installation requirements or restrictions. Both have been incorporated into the text of this section to reference the appropriate articles respectively.

Comment: 14-26

Proposals: 14-37, 14-168

Process Sealing

Code Language

501.17 Process Sealing. This section shall apply to process-connected equipment, which includes, but is not limited to, canned pumps, submersible pumps, flow, pressure, temperature, or analysis measurement instruments. A process seal is a device to prevent the migration of process fluids from the designed containment into the external electrical system. Process connected electrical equipment that incorporates a single process seal, such as a single compression seal, diaphragm, or tube to prevent flammable or combustible fluids from entering a conduit or cable system capable of transmitting fluids, shall be provided with an additional means to mitigate a single process seal failure. The additional means may include, but is not limited to the following:

(1) A suitable barrier... ***(See NEC for full text.)***

(2) A listed Type MI cable assembly... ***(See NEC for full text.)***

(3) A drain or vent... ***(See NEC for full text.)***

Process-connected electrical equipment that does not rely on a single process seal or is listed and marked "single seal" or "dual seal" shall not be required to be provided with an additional means of sealing.

Informational Note: For construction and testing requirements for process sealing for listed and marked "single seal" or "dual seal" requirements, refer to ANSI/ISA-12.27.01-2003, *Requirements for Process Sealing Between Electrical Systems and Potentially Flammable or Combustible Process Fluids.*

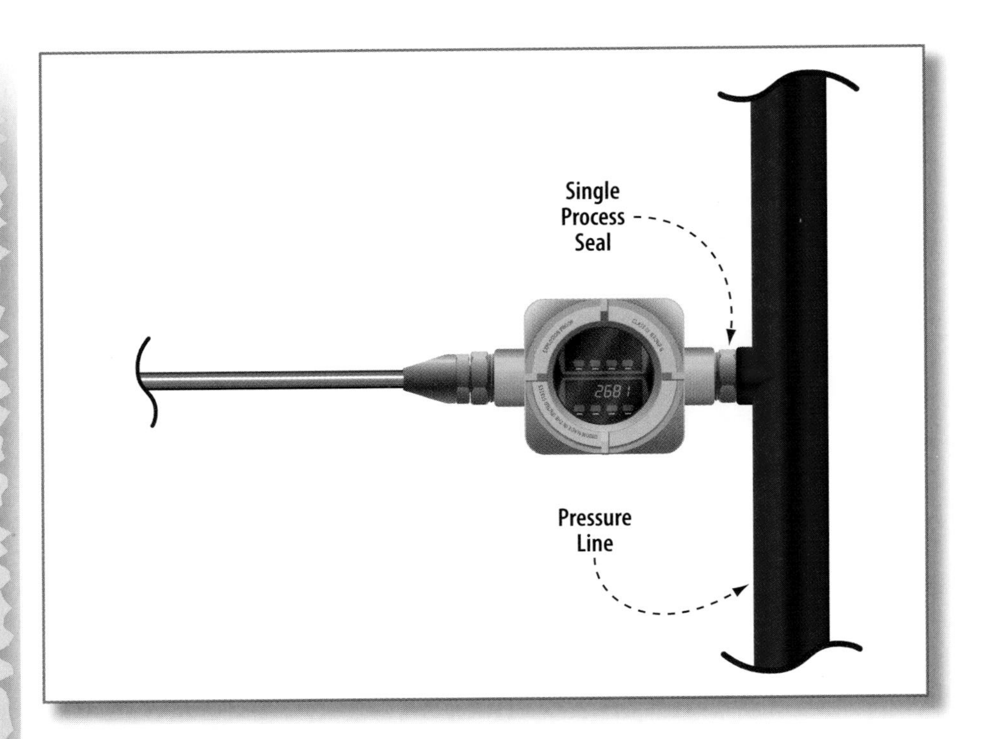

Change Summary

Section 501.15(F)(3) has been relocated and renumbered as 501.17. It has been expanded to provide more detail about process sealing methods. This section now provides three examples of additional means that can be used to mitigate a single process seal failure. Section 505.26 has also been revised to include consistent requirements for process sealing in applications under the "Zone" system of hazardous (classified) locations.

Process Sealing (continued)

Courtesy of Cogburn Brothers, Inc.

Significance of the Change

Process seals are devices that are intended to prevent the migration of process fluids from a designed containment into the external electrical system. This section has been rewritten into a more user-friendly format and expanded to describe what process sealing is intended to accomplish. As in the previous requirements, both single-seal and dual-seal processes are described. The revision also lists three alternatives to achieve the additional means to mitigate a single process seal failure.

These additional requirements for single-seal equipment address both pressure and temperature cycling, followed by a leakage and burst test. The three additional methods permitted are a suitable barrier, use of listed MI cable, and a drain or vent as fully described in list items (a), (b), and (c). As before, if a process seal is listed and marked "single seal" or "dual seal," the additional mitigation methods for sealing or drainage are not required. The provisions of ANSI/ISA-12.27.01 include both construction and performance requirements for single- and dual-sealed electrical equipment.

Comment: 14-38
Proposal: 14-60

501.30(B), 502.30(B), 503.30(B), 505.25(B), & 506.25(B)

Articles 501, 502, 503, 505, 506
Part II Wiring (in 501, 502, and 503); N/A for Articles 505 and 506

Grounding and Bonding Requirements

Code Language

501.30

(B) Types of Equipment Grounding Conductors. Flexible metal conduit and liquidtight flexible metal conduit shall include an equipment bonding jumper of the wire type in compliance with 250.102.

502.30

(B) Types of Equipment Grounding Conductors. Liquidtight flexible metal conduit shall include an equipment bonding jumper of the wire type in compliance with 250.102.

503.30

(B) Types of Equipment Grounding Conductors. Liquidtight flexible metal conduit shall include an equipment bonding jumper of the wire type in compliance with 250.102.

505.25(B)

(B) Types of Equipment Grounding Conductors. Flexible metal conduit and liquidtight flexible metal conduit shall include an equipment bonding jumper of the wire type in compliance with 250.102.

506.25(B)

(B) Types of Equipment Grounding Conductors. Liquidtight flexible metal conduit shall include an equipment bonding jumper of the wire type in compliance with 250.102.

Flexible metal conduit and liquidtight flexible metal conduit installed in hazardous (classified) locations is not permitted as an equipment bonding jumper.

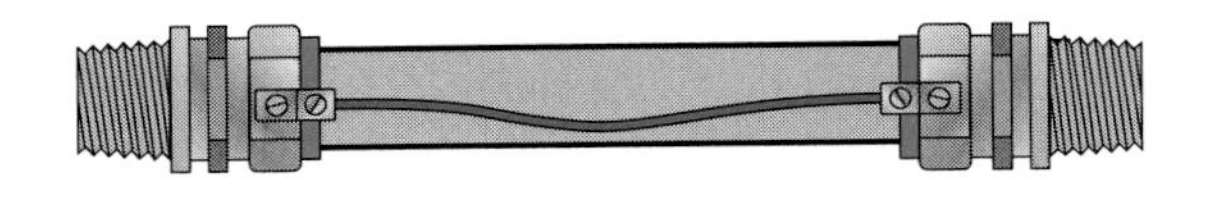

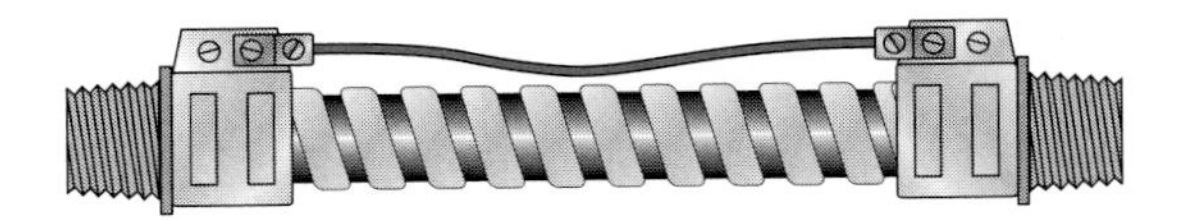

A wire-type equipment bonding jumper or equipment grounding conductor is required to ensure an effective ground-fault current path is provided.

Change Summary

Sections 501.30(B), 502.30(B), 503.30(B), 505.25(B), and 506.25(B) have been revised to include the phrase "include an equipment bonding jumper of the wire type." This revision clarifies that flexible wiring methods are not permitted as an effective ground-fault current path in hazardous (classified) locations.

Grounding and Bonding Requirements (continued)

Significance of the Change

The previous text in each of these sections could have been interpreted as permissive by addressing equipment bonding jumpers "where installed" and by not requiring a wire-type equipment bonding jumper. This revision provides clarity about the bonding requirements for flexible metal conduit and liquidtight flexible metal conduit that is installed in Class I and liquidtight flexible metal conduit in Class II locations. Sections 503.30(B) and 506.25(B) only address bonding around liquidtight flexible metal conduit. The interlocking metal tape-type construction of these wiring methods must be supplemented by an equipment bonding jumper of the wire type to minimize arcing effects during ground-fault conditions. The equipment bonding jumper provides an effective path for fault current in addition to the interlocking metal tape construction of the conduit.

The revision to this section provides clear direction for users that the wire-type equipment bonding jumper must be installed and that the sizing requirements for it are provided in 250.102. Similar revisions have been made to 502.30(B), 503.30(B), and 505.25(B) for the same bonding requirements provided for Class II and Class III locations and wiring installed under the zone system as covered in 505.25(B) and 506.25(B).

Comment: None

Proposals: 14-66, 14-104, 14-14-140, 14-215a, 14-245a

501.140(B)(4) & (5)

Article 501 Class I Locations
Part III Equipment

Flexible Cords, Class I, Divisions 1 and 2

Code Language

501.140(B) Installation... *(No changes)*

(1) *(No changes)*

(2) *(No changes other than editorial)*

(3) *(Previous (3) has been deleted.)*

(4) In Division 1 locations or in Division 2 locations where the boxes, fittings, or enclosures are required to be explosionproof, the cord shall be terminated with a cord connector or attachment plug listed for the location or a cord connector installed with a seal listed for the location. In Division 2 locations where explosionproof equipment is not required, the cord shall be terminated with a listed cord connector or listed attachment plug.

(5) Be of continuous length. Where 501.140(A)(5) is applied, cords shall be of continuous length from the power source to the temporary portable assembly and from the temporary portable assembly to the utilization equipment.

(No change to the informational note)

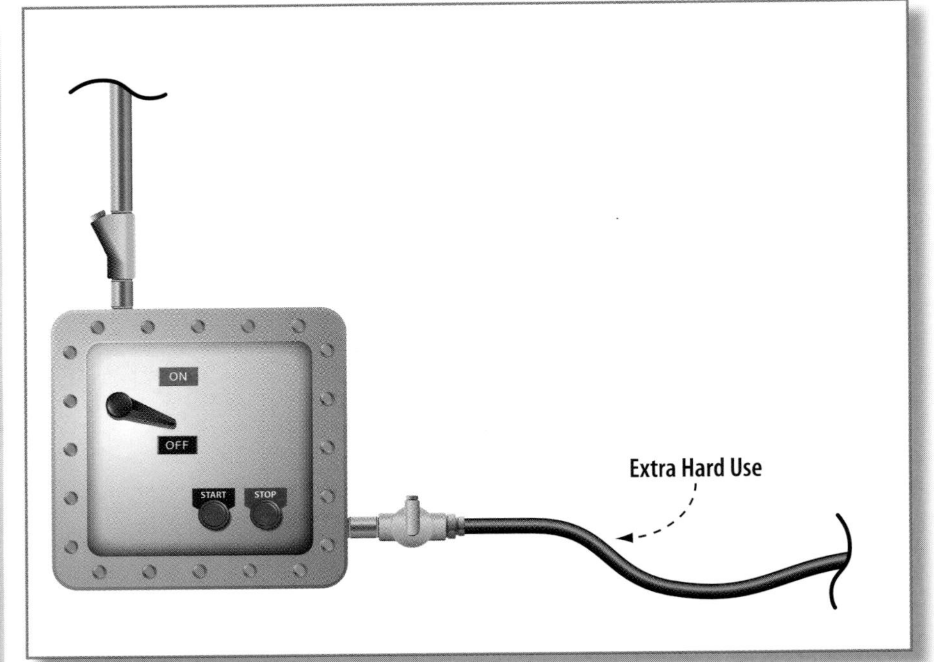

Change Summary

List item (4) in this section has been revised to clearly distinguish between Division 1 and Division 2 requirements. The exception to former list item (5) has been deleted. List item (5) addresses continuous lengths of cords installed in accordance with the provisions of this section.

Significance of the Change

Cord is typically terminated to enclosures with listed fittings, but their fittings, as with sealing fittings, are usually not provided with the cord. Instead, they are separate products. The revision to this section clarifies that where cord is used in Class I, Division 1 or Division 2 locations, it must be terminated to enclosures using fittings listed for the location. Where cords are used with explosionproof enclosures, the integrity (explosionproof characteristics) of the enclosure must be maintained by a listed seal installed at the point of cord entry to such enclosures. In Division 2 locations, where enclosures are not required to be explosionproof, cords must be terminated with a listed cord connector or listed attachment plug. The exception to this section has been deleted because neither 501.10(B) nor 501.105(B)(6) gives specific cord-sealing requirements and no exception is given to completing an enclosure that is required to be explosionproof with a seal on the entry.

Comments: 14-54, 14-53a

Proposal: 14-88

Zone Equipment

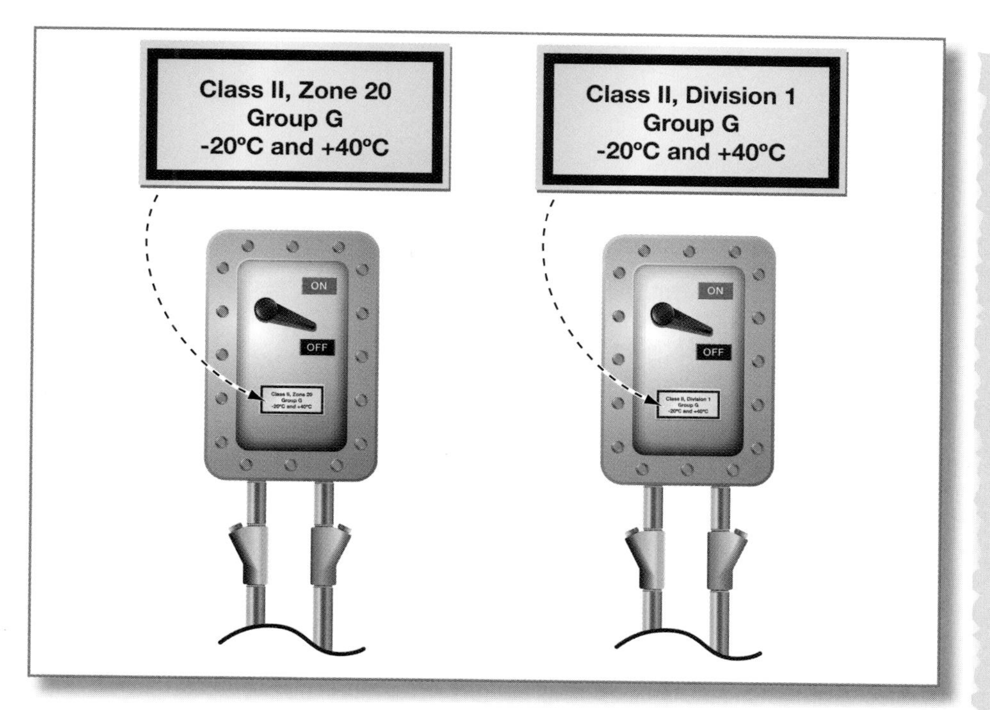

Code Language

502.6 Zone Equipment. Equipment listed and marked in accordance with 506.9(C)(2) for Zone 20 locations shall be permitted in Class II, Division 1 locations for the same dust atmosphere; and with a suitable temperature class.

Equipment listed and marked in accordance with 506.9(C)(2) for Zone 20, 21, or 22 locations shall be permitted in Class II, Division 2 locations for the same dust atmosphere and with a suitable temperature class.

Significance of the Change

Section 502.6 addresses equipment with zone suitability markings for use in Class II, Division 1 or 2 locations. This new section provides users with *Code* text that allows equipment identified for a Zone 20 location to be installed and used in Class II, Division 1 locations as long as the type of dust atmosphere is the same and the equipment is provided with a suitable temperature classification. Where equipment is marked in accordance with 506.9(C)(2) for Zone 20, 21, or 22 locations, it can be installed and used within a Class II, Division 2 location for the same dust atmosphere and a suitable temperature classification marking. Essentially what this revision does is provide some interchangeability for equipment that carries specific markings for zone classification applications that are in similar environments classified under the division system of classification. The key to proper application of this equipment is to verify the temperature suitability.

Change Summary

A new section 502.6 titled Zone Equipment has been added to Part I of Article 502. The new text addresses equipment that is listed and marked for use in Zone classified areas as being suitable for either Class II, Division 1 (Zone 20) or Class II Division 2 (Zones 20, 21, and 22) if the equipment has a suitable temperature classification.

Comment: None

Proposal: 14-93

502.100, 115, 120, 125, 130, 135, & 150

Article 502 Class III Locations
Part III Equipment

Equipment

Code Language

Part III Equipment

502.100 Transformers and Capacitors
502.115 Switches, Circuit Breakers, Motor Controllers, and Fuses
502.120 Control Transformers and Resistors
502.125 Motors and Generators
502.130 Luminaires
502.135 Utilization Equipment
502.150 Signaling, Alarm, Remote-Control, and Communications Systems; and Meters, Instruments, and Relays

(See Part III of Article 502 for full text including the revisions.)

Change Summary

Part III of Article 502 has been revised to include the phrase *identified for the location* in multiple sections, and to replace the term *metal dusts* with the term *Group E* in 502.100(A)(3). The effective date of January 1, 2011, has been deleted from 502.120(B)(2).

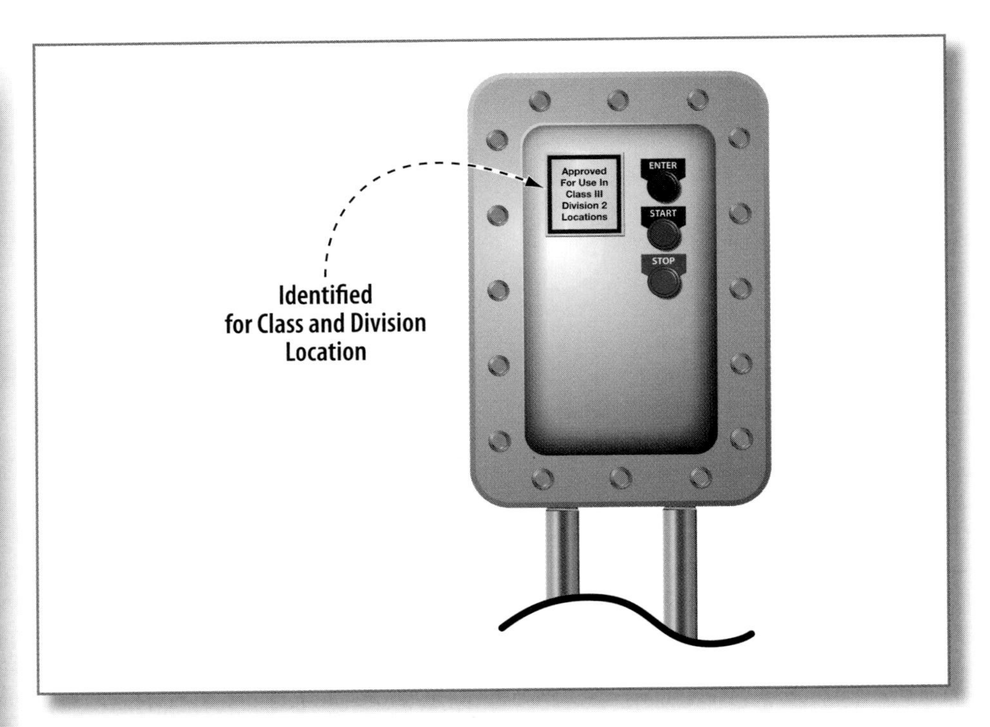

Significance of the Change

Part III of Article 502 provides specific requirements for equipment installed in Class II, Division 1 and 2 locations. Sections in Part III of Article 502 have been revised to improve clarity and usability and to remove unnecessary, redundant text. The significance of these revisions is more specific references to equipment that is required to be suitable for use in dust environments. The specific *Group E* has replaced the term *metal dusts* to clarify the grouping requirements and provide consistent *Code* text that aligns with how equipment enclosures are identified for these environments. The phrase *identified for the location* is important within each of these requirements because it is directly related to safe operation of these equipment types installed in Class II (dust) environments.

Comment: 14-55
Proposal: 14-90

Zone Equipment

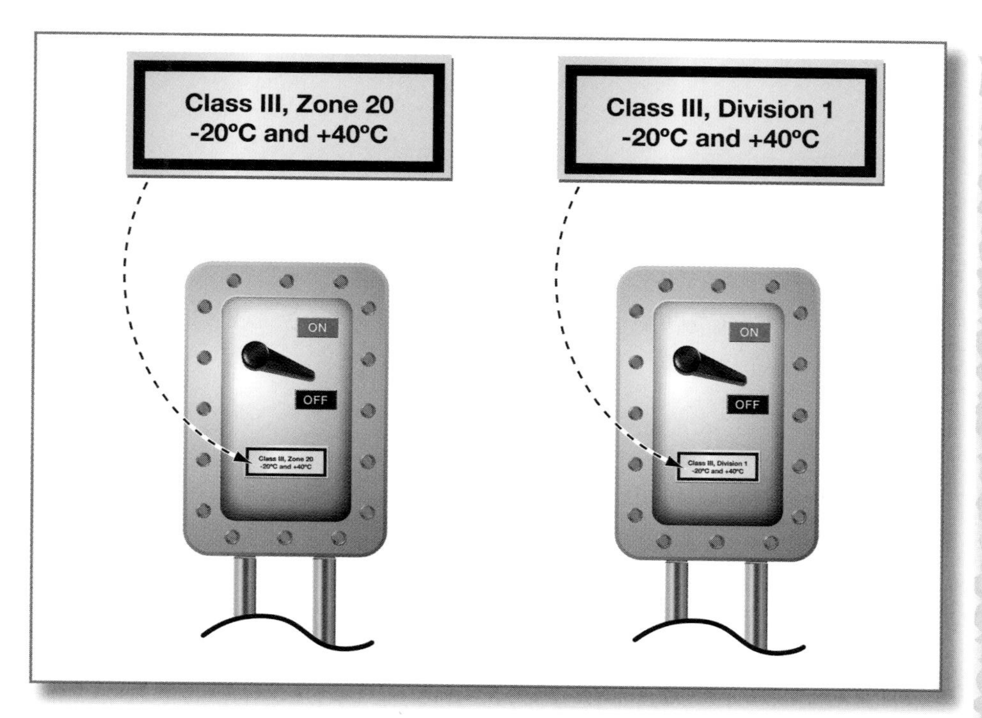

Code Language

503.6 Zone Equipment. Equipment listed and marked in accordance with 506.9(C)(2) for Zone 20 locations and with a temperature class of not greater than T120°C (for equipment that may be overloaded) or not greater than T165°C (for equipment not subject to overloading) shall be permitted in Class III, Division 1 locations.

Equipment listed and marked in accordance with 506.9(C)(2) for Zone 20, 21, or 22 locations and with a temperature class of not greater than T120°C (for equipment that may be overloaded) or not greater than T165°C (for equipment not subject to overloading) shall be permitted in Class III, Division 2 locations.

Significance of the Change

Section 503.6 is new and addresses equipment with zone suitability markings that can be used in Class III, Division 1 or 2 locations. This new section provides users with *Code* text that allows equipment identified for a Zone 20 location to be installed and used in a Class III, Division 1 location as long its temperature class is not greater than T120°C (for equipment that may be overloaded) or not greater than T165°C (for equipment not subject to overloading). Where equipment is marked in accordance with 506.9(C)(2) for Zone 20, 21, or 22 locations, it can be installed and used within a Class III, Division 2 location as long as it is identified with a temperature class of not greater than T120°C (for equipment that may be overloaded) or not greater than T165°C (for equipment not subject to overloading).

Essentially what this revision does is provide some interchangeability for equipment that carries specific marks for zone classification applications that are of similar environments classified under the division system of classification. The key to proper application of this equipment is verification of temperature suitability.

Change Summary

A new section, 503.6 Zone Equipment, has been added to Part I of Article 503. The new provisions in this section address suitability of Zone equipment in Class III locations as long as the equipment carries suitable temperature classifications for the environment involved.

Comment: None

Proposal: 14-130

503.10(A)

Article 503 Class III Locations
Part II Wiring

Class III, Division 1

Code Language

(A) Class III, Division 1.

(1) General. In Class III, Division 1 locations, the wiring method shall be in accordance with (1) through (4):

(1) Rigid metal conduit, Type PVC conduit, Type RTRC conduit, intermediate metal conduit, electrical metallic tubing, dust-tight wireways, or Type MC or MI cable with listed termination fittings.

(2) Type PLTC and Type PLTC-ER cable in accordance with the provisions of Article 725 including installation in cable tray systems. The cable shall be terminated with listed fittings.

(3) Type ITC and Type ITC-ER cable as permitted in 727.4 and terminated with listed fittings.

(4) Type MC, MI, or TC cable installed in ladder, ventilated trough, or ventilated channel cable trays in a single layer, with a space not less than the larger cable diameter between the two adjacent cables, shall be the wiring method employed.

Exception to (4): Type MC cable listed for use in Class II, Division 1 locations shall be permitted to be installed without the spacings required by 503.10(A)(1)(4).

(2) Boxes and Fittings. All boxes and fittings shall be dusttight.

(3) Flexible Connections. Where necessary to employ flexible connections, one or more of the following shall be permitted:

(1) Dusttight flexible connectors

(2) Liquidtight flexible metal conduit with listed fittings,

(3) Liquidtight flexible nonmetallic conduit with listed fittings,

(4) Interlocked armor Type MC cable having an overall jacket of suitable polymeric material and installed with listed dusttight termination fittings

(5) Flexible cord in compliance with 503.140

Class III, Division 1 (continued)

Additional wiring methods have been added to Section 503.10(A).

503.10(A) Class III, Division 1

(1) General

(1) RMC, PVC, RTRC, IMC, EMT, MC or MI cable, and dusttight wireways

(2) Type PLTC and Type PLTC-ER cables with listed fittings

(3) Type ITC and Type ITC-ER cables with listed fittings

(4) Type MC, MI, or TC cables in single layer installed in ladder, ventilated trough or ventilated channel cable tray with spacings not less than the larger cable diameter of adjacent cables.

The exception to (A)(1)(4) *relaxes the spacing requirement for MC cable listed for use in Class II, Division 1 locations*

(2) Boxes and Fittings. (No changes)

(3) Flexible Connections

(1) (No changes)

(2) (No changes)

(3) (No changes)

(4) Interlocking armor Type MC cable having an overall jacket of suitable polymeric material and installed with listed dusttight termination fittings.

Significance of the Change

Article 503 provides requirements for Class III locations that are hazardous due to the presence of ignitible fibers or accumulations of combustible flyings or materials in sufficient quantities to create a fire hazard. Part II of Article 503 covers wiring methods that are suitable in Class III locations. Section 503.10(A) has been revised and expanded to clarify the types of conduit wiring methods — for example, PVC conduit and RTRC conduit — that are permitted in Class III, Division 1 locations. This section has also been revised to include two new list items [(2) and (3)] covering cable-type wiring methods such as Types PLTC, PLTC-ER, ITC, ITC-ER, MC, MI, or TC cable. Flexible connections for wiring methods installed in Class III, Division 1 locations are addressed in (A)(3), which has been expanded to include a new list item (4) that addresses Type MC cable with an overall polymeric jacket and terminated in listed dusttight fittings. List item (5) of 503.10(A)(3) has been revised to clarify that cords installed for flexibility in Class III, Division 1 locations must be listed for extra-hard use, terminated with listed cord connectors, and meet the applicable requirements in 503.140. The revisions to this section also align the formatting with the structure and sequence of Articles 501 and 502.

Change Summary

In subdivision (A), the wiring method "rigid nonmetallic conduit" has been revised to "PVC conduit," and "RTRC conduit" has been added to list item (1). Additionally, list items (3) and (4) are new to (A)(1), and under list item (A)(2)(3) Flexible Connections, list item (4) is new and list item (5) has been revised.

Comment: 14-69

Proposal: 14-131

505.7(E)

Article 505 Class I, Zone 0, 1, and 2 Locations

Simultaneous Presence of Flammable Gases and Combustible Dusts...

Code Language

505.7 Special Precaution.

(E) Simultaneous Presence of Flammable Gases and Combustible Dusts or Fibers/Flyings. Where flammable gases, combustible dusts, or fibers/flyings are or may be present at the same time, the simultaneous presence shall be considered during the selection and installation of the electrical equipment and the wiring methods, including the determination of the safe operating temperature of the electrical equipment.

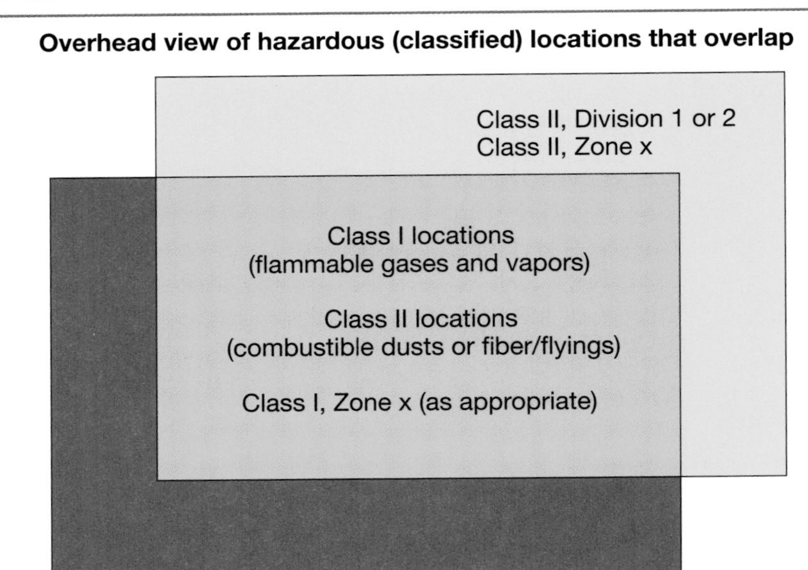

The equipment and wiring methods installed must be suitable for both types of hazardous atmospheres that are classified and overlay one another.

Change Summary

A new subdivision (E) titled Simultaneous Presence of Flammable Gases and Combustible Dusts or Fibers/Flyings has been added to 505.7. This new subdivision addresses the selection of equipment for use in areas with different classifications that overlap and are simultaneously present in the same location.

Significance of the Change

This new subdivision incorporates a provision that addresses simultaneous area classifications for hazardous (classified) locations where flammable gases and combustible dusts or fibers and flyings are present at the same time. This important new requirement is directly related to the selection and installation of equipment and wiring methods suitable for both types of hazardous atmospheres. Information in the proposal's substantiation pointed out that 500.8(C)(4) coordinates between Class I (divisions only, since Class I zones stand alone in Article 505) and Class II (divisions only since Article 506 is not referred to as Class II). Section 506.6(D) ties Zones 20, 21, and 22 to Class I zones, but nothing ties the Class I Zones to Class II Divisions, which is a situation not precluded by reclassification as permitted in 505.7(C) or 506.6(C). This new provision would allow areas previously classified as Class I, Division x and Class II, Division y simultaneously to reclassify the Class I, Division x location as Class I, Zone x as would be appropriate.

Comment: None

Proposal: 14-175

Encapsulation "m"

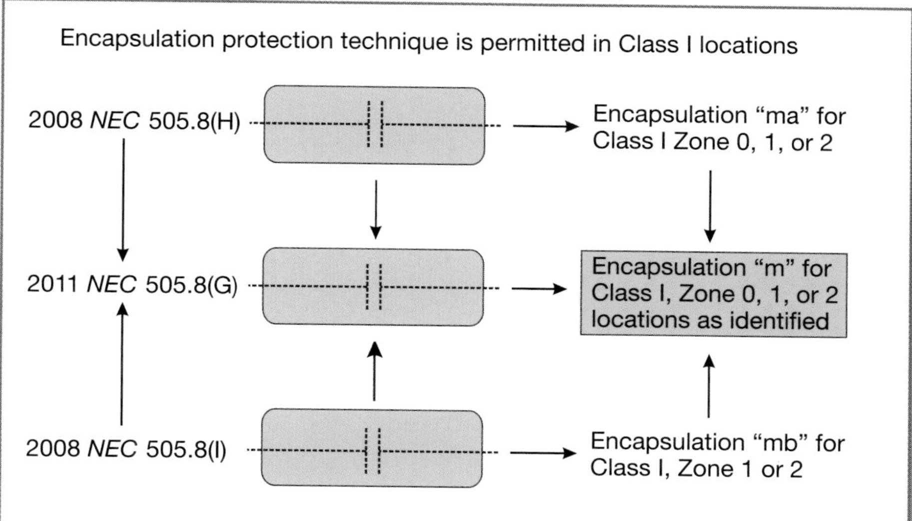

Code Language

(G) Encapsulation "m". This protection technique shall be permitted for equipment in Class I, Zone 0, Zone 1, or Zone 2 locations for which it is identified.

Informational Note: See Table 505.9(C)(2)(4) for the descriptions of subdivisions for encapsulation.

Significance of the Change

This revision results in coverage of all encapsulation protection techniques "ma" and "mb" for zone applications under subdivision (G) only. Previously, the encapsulation protection techniques for Zone 0 applications were provided in subdivisions (H) and (I), in two encapsulation categories. By including the requirement for identifying the zone for which the encapsulation method is suitable, one subdivision in this section is all that is necessary in the *Code*. The new informational note provides an important reference to Table 505.9(C)(2)(4), which includes the descriptions of encapsulation subdivisions. This new format is also consistent with the format used for Intrinsic Safety and Protection Type "n."

Change Summary

Former subdivisions (H) (Encapsulation "ma") and (I) (Encapsulation "mb") have been deleted, and the term *Zone 0* has been added to subdivision (G) (Encapsulation "m"). The phrase *for which it has been identified* has been added to the end of subdivision (G). The remaining subdivisions have been re-identified in alphanumeric sequence accordingly. A new informational note has been added to this section.

Comment: None

Proposal: 14-177

Combustible Gas Detection Systems

Code Language

505.8(I) Combustible Gas Detection System...

Informational Note No. 1 ***(No changes)***

Informational Note No. 2: For further information, see ANSI/ISA-60079-29-2, *Explosive Atmospheres - Part 29-2: Gas detectors - Selection, installation, use and maintenance of detectors for flammable gases and oxygen.*

Informational Note No. 3: *For further information, see ISA-TR12.13.03, Guide for Combustible Gas Detection as a Method of Protection.*

Change Summary

Former 505.8(K) has been renumbered as 505.8(I) as a result of action by CMP-14 on Proposal 14-177. Informational Note No. 2 has been revised, and a new informational note referencing other standards related to combustible gas detection protection techniques has been included.

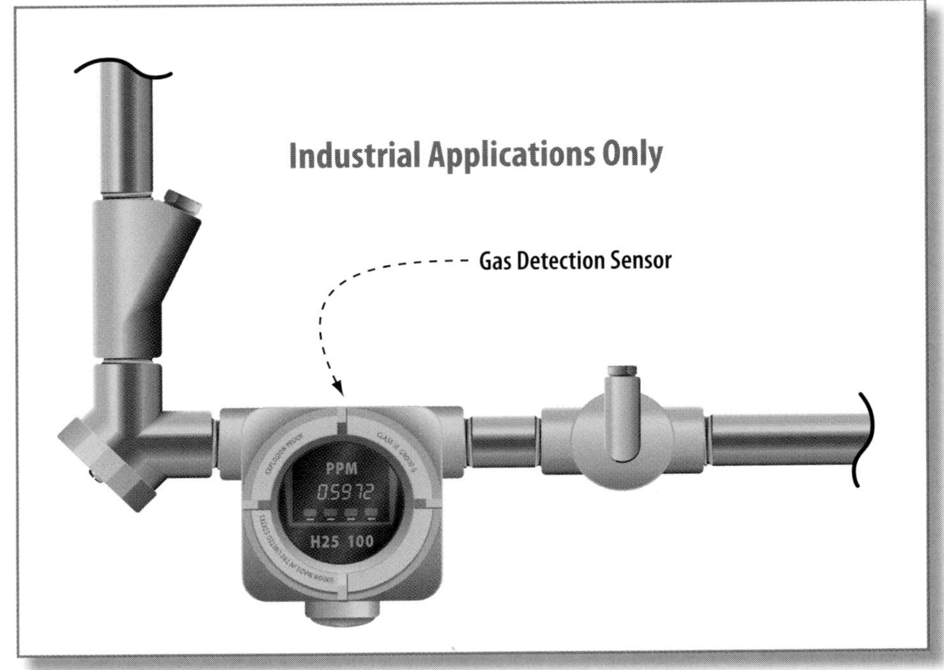

Significance of the Change

Combustible gas detection systems are protection techniques whose use is limited to industrial establishments with restricted public access. Other limiting use factors relate to conditions of maintenance and supervision and ensure that only qualified persons service the system. The type of detection equipment, its listing information, locations, alarms and shutdown criteria, and calibration frequency are all required to be documented. The title of Informational Note No. 2 has been revised to match the current title of the ISA standard. Informational Note No. 3 is new and references ISA-TR12.13.03, *Guide for Combustible Gas Detection as a Method of Protection*. Note No. 3 provides a useful reference to guidelines for installing and operating gas detection systems but only in compliance with the controlled conditions described in this section.

Comment: None

Proposals: 14-176, 14-178

Marking Zone Equipment

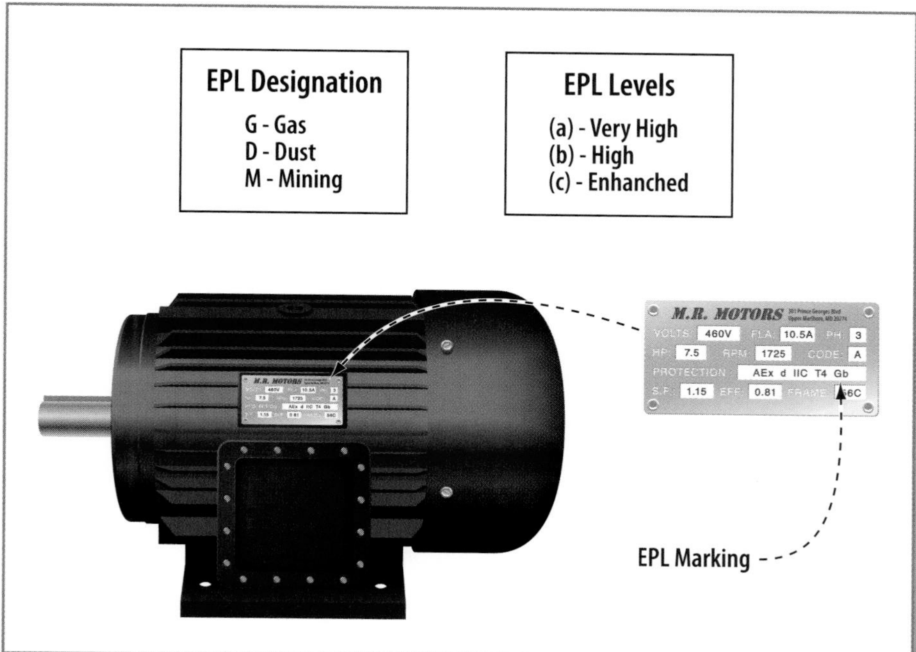

Code Language

505.9 Equipment

(C) Marking...

(2) Zone Equipment...

Informational Note No. 4: The EPL (or equipment protection level) may appear in the product marking. EPLs are designated as G for gas, D for dust, or M for mining and are then followed by a letter (a, b, or c) to give the user a better understanding as to whether the equipment provides either (a) a "very high," (b) a "high," or (c) a "enhanced" level of protection against ignition of an explosive atmosphere. For example, an AEx d IIC T4 motor (which is suitable by protection concept for application in Zone 1) may additionally be marked with an EPL of "Gb" to indicate that it was provided with a high level of protection, such as AEx d IIC T4 Gb.

Change Summary

New Informational Note No. 4 has been added to 505.9(C)(2). This new informational note provides information about equipment protection levels that may appear on products that are suitable for use in hazardous (classified) locations.

Significance of the Change

This new informational note indicates that equipment protection levels (EPLs) may be marked on electrical equipment in addition to the currently required markings under the product safety standards. Equipment protection levels were introduced into IEC standards in 2006 and are now also included in ANSI/ISA-60079-0 and ANSI/UL 60079-0. This new informational note serves to alert users about the additional equipment protection level markings that may appear on some products that meet applicable IEC and North American electrical safety standards. Examples of such EPL markings would include an (a) "very high," (b) "high," or (c) "enhanced" level of protection against ignition of an explosive atmosphere. These additional EPL markings have no impact on any currently required product markings and would be in addition to such markings.

Comment: None

Proposal: 14-179

505.16(C)(1)(b)

Article 505 Class I, Zone 0, 1, and 2 Locations

Conduit Seals

Code Language

505.16 Sealing and Drainage...

(C) Zone 2 ...

(1) Conduit Seals...

(b) In each conduit run passing from a Class I, Zone 2 location... ***(remainder is the same except for two new last sentences as follows:)***

...Conduits shall be sealed to minimize the amount of gas or vapor within the Class I, Zone 2 portion of the conduit from being communicated to the conduit beyond the seal. Such seals shall not be required to be flameproof or explosionproof but shall be identified for the purpose of minimizing passage of gases under normal operating conditions and shall be accessible.

Change Summary

Two new sentences have been added to 505.16(C)(1)(b) as follows: "Conduits shall be sealed to minimize the amount of gas or vapor within the Class I, Zone 2 portion of the conduit from being communicated to the conduit beyond the seal. Such seals shall not be required to be flameproof or explosionproof but shall be identified for the purpose of minimizing passage of gases under normal operating conditions and shall be accessible."

Seals required in each conduit passing from a Class I, Zone 2 location to an unclassified location.

Boundary

Class I, Zone 2 location

Unclassified location

Seal permitted on either side of boundary but within 10 feet of that boundary

The seal is not required to be explosion-proof or flameproof.

The seal shall be identified for minimizing the passage of gases or vapors under normal operating conditions, and is required to be accessible.

Significance of the Change

This revision provides performance language that describes how the required seals are intended to perform at the boundary of a Class I, Zone 2 location. This descriptive text is very similar to the same performance language included in 501.16 dealing with Class I locations. The new text indicates that seals are necessary to minimize the passage of gas or vapor across the seal at the boundary and that the seal does not have to be a flameproof or explosionproof type. Where not of an explosionproof or flameproof type, the sealing fitting does have to be identified for the purpose of minimizing gas or vapor passage across the seal during normal operation. The sealing fitting also must be accessible, just as other seal fittings are required to be accessible.

Comment: None
Proposal: 14-202

514.11(A)

Article 514 Motor Fuel Dispensing Facilities

Circuit Disconnects

Code Language

514.11 Circuit Disconnects.

(A) General. Each circuit leading to or through dispensing equipment, including all associated power, communications, data, and video circuits, and equipment for remote pumping systems, shall be provided with a clearly identified and readily accessible switch or other approved means, located remote from the dispensing devices, to disconnect simultaneously from the source of supply, all conductors of the circuits, including the grounded conductor, if any.

Single-pole breakers utilizing handle ties shall not be permitted.

Change Summary

The words "all associated power, communications, data, and video circuits, and" have been incorporated in the first sentence of this section. This section now contains an additional requirement for disconnecting associated communications, video, and data circuits.

Significance of the Change

Section 514.11 provides requirements for emergency means of disconnect for fuel dispensing equipment. The revision to this section results in a requirement for disconnecting all systems routed to fuel dispensing equipment, including power, data, communications, and video, in addition to equipment for remote pumping. The substantiation provided with the proposal for this change clearly indicated that the number of circuits and systems supplying dispensing equipment is increasing and that the number of electrical circuits and systems related to communication and marketing at the pump is on the rise. As such, fuel dispensing equipment is getting more and more sophisticated and so requires more of manufacturers of fuel dispensing products.

The requirement for disconnecting all sources supplying dispensing equipment is now extended to limited energy systems, for example, to data and communications circuitry routed to this equipment. This revision improves worker safety by requiring a disconnecting means or multiple means of removing all voltage sources and other sources of energy leading to or through dispensing equipment. Typically, the disconnecting means required by this section is field-installed. However, this revision provides an opportunity for dispensing equipment manufacturers to incorporate such safety provisions for disconnects to meet the requirements of this section.

Comment: None

Proposal: 14-279

514.13

Maintenance and Service of Dispensing Equipment

Code Language

514.13 Provisions for Maintenance and Service of Dispensing Equipment. Each dispensing device shall be provided with a means to remove all external voltage sources, including power, communications, data, and video circuits and including feedback, during periods of maintenance and service of the dispensing equipment. The location of this means shall be permitted to be other than inside or adjacent to the dispensing device. The means shall be capable of being locked in the open position.

Significance of the Change

Section 514.13 provides disconnecting means requirements for maintenance and service operations associated with fuel dispensing equipment. The revision to this section results in a requirement for disconnecting all systems routed to fuel dispensing equipment, including power, data, communications, and video, in addition to equipment for remote pumping, which may introduce feedback. The substantiation for the proposal for change clearly indicated that the number of circuits and systems supplying dispensing equipment is increasing and that the number of electrical circuits and systems related to communication and marketing at the pump are on the rise. Fuel dispensing equipment is getting more and more sophisticated and requires more from fuel dispensing product manufacturers.

Comment: None
Proposal: 14-281

Maintenance and Service of Dispensing Equipment (continued)

Change Summary

The words "including power, communications, data, and video circuits and" have been incorporated into the first sentence of this section. This section now contains an additional requirement for disconnecting associated communications, video, and data circuits.

The requirement for disconnecting all sources supplying dispensing equipment is now extended to systems such as the data and communications circuits. The revision improves worker safety by requiring a disconnecting means or multiple means of removing all voltage sources and other sources of energy leading to or through dispensing equipment. Typically, the disconnecting means required by this section is field-installed. However, this revision provides an opportunity for dispensing equipment manufacturers to incorporate such safety provisions for disconnects to meet the requirements of this section.

Definition of Battery-Powered Lighting Units

Code Language

517.2 Definitions.

Battery-Powered Lighting Units. Individual unit equipment for backup illumination consisting of the following:

(1) Rechargeable battery
(2) Battery-charging means
(3) Provisions for one or more lamps mounted on the equipment, or with terminals for remote lamps, or both
(4) A relaying device arranged to energize the lamps automatically upon failure of the supply to the unit equipment.

Change Summary

A new definition of the term *battery-powered lighting units* has been added to 517.2. The new definition correlates with new requirements to provide backup illumination for anesthetizing locations and to connect the unit equipment to the critical branch lighting circuit in the area.

Comment: 15-1

Proposal: 15-3a

Significance of the Change

Battery-powered lighting units (unit equipment) are included for use in grounded electrical systems supplying power to anesthetizing locations in health care facilities. Section 517.63(A) provides a requirement to install battery-powered lighting units in anesthetizing locations, to provide minimum lighting levels during procedures in the event of a power failure. The essential backup system in a hospital is typically generators that are allowed up to 10 seconds to restore backup power if normal power is interrupted. These battery-powered emergency lighting units provide the lighting necessary during the transition. They also will provide illumination for a minimum of 90 minutes if the power source for the essential system does not restore power to the essential system loads for any reason.

This new definition describes the characteristics required for a battery-powered lighting unit used for supplying illumination in anesthetizing locations of health care facilities. It is noteworthy that the definition parallels the same description and operational characteristics of unit equipment provided in 700.12(F). These units must have a rechargeable battery, a battery charging means, provisions for one or more lamps mounted on the equipment or with terminals for remote lamps, or both, and a relaying device arranged to energize the lamps automatically upon failure of the supply to the unit equipment.

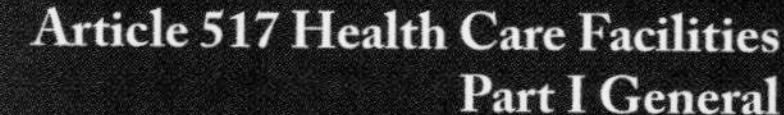

Definition of Transportable X-Ray Installations

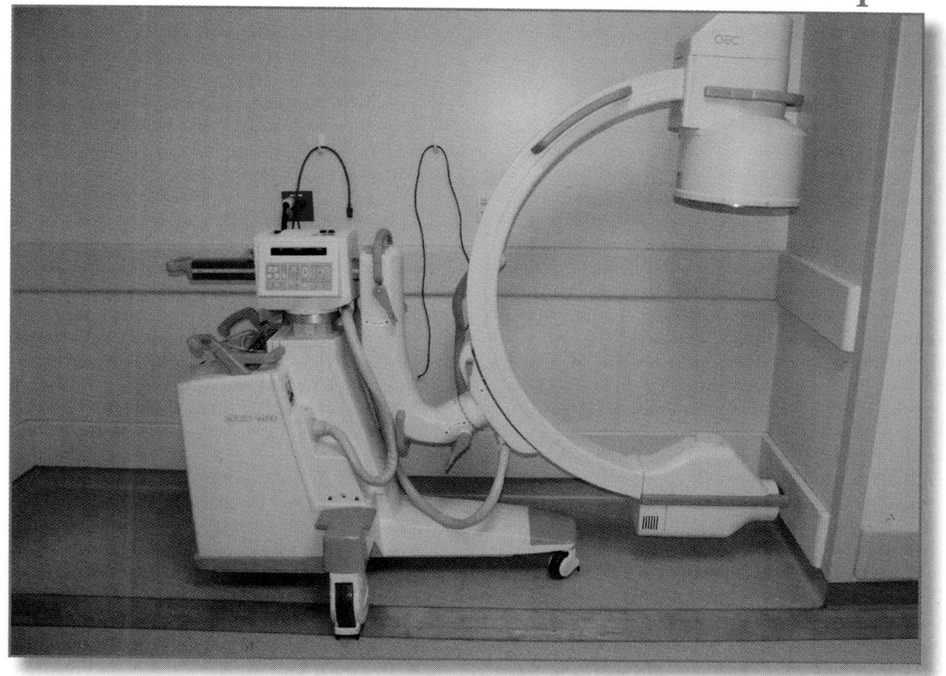

Code Language

517.2 Definitions.

X-Ray Installations, Transportable. X-ray equipment to be conveyed by a vehicle or that is readily disassembled for transport by a vehicle.

Significance of the Change

Transportable X-ray installations are used in health care applications. The definition is almost identical to the definition in Article 660 and applies to a similar type of X-ray equipment that is transportable by a vehicle. This type of equipment is intended to be conveyed by a vehicle or can be readily disassembled for transporting to the point of use. Transportable X-Ray equipment is referred to in Section 517.71(B). This revised definition clarifies the type of equipment covered and offers a slight differentiation between what constitutes mobile or portable X-ray equipment. In some cases, mobile X-ray equipment and transportable X-ray equipment seem to be interchangeable.

Change Summary

The definition of the term *X-Ray Installations, Transportable* has been revised by adding the words "to be conveyed by a."

Comment: 15-31

Proposal: 15-21

517.13(B)

Article 517 Health Care Facilities
Part II Wiring and Protection

Insulated Equipment Grounding Conductor

Code Language

(B) Insulated Equipment Grounding Conductor.

(1) General. The following shall be directly connected to an insulated copper equipment grounding conductor that is installed with the branch circuit conductors in the wiring methods as provided in 517.13(A):

(1) The grounding terminals of all receptacles.

(2) Metal boxes and enclosures containing receptacles.

(3) All non-current-carrying conductive surfaces of fixed electrical equipment likely to become energized that are subject to personal contact, operating at over 100 volts.

Exception: An insulated equipment bonding jumper that directly connects to the equipment grounding conductor is permitted to connect the box and receptacle(s) to the equipment grounding conductor.

Exception No. 1 to (3): Metal faceplates shall be permitted to be connected to the equipment grounding conductor by means of a metal mounting screw(s) securing the faceplate to a grounded outlet box or grounded wiring device.

Exception No. 2 to (3): Luminaires more than 2.3 m (7 ½ ft) above the floor and switches located outside of the patient care vicinity shall be permitted to be connected to an equipment grounding conductor return path complying with 517.13(A).

(2) Sizing. Equipment grounding conductors and equipment bonding jumpers shall be sized in accordance with 250.122.

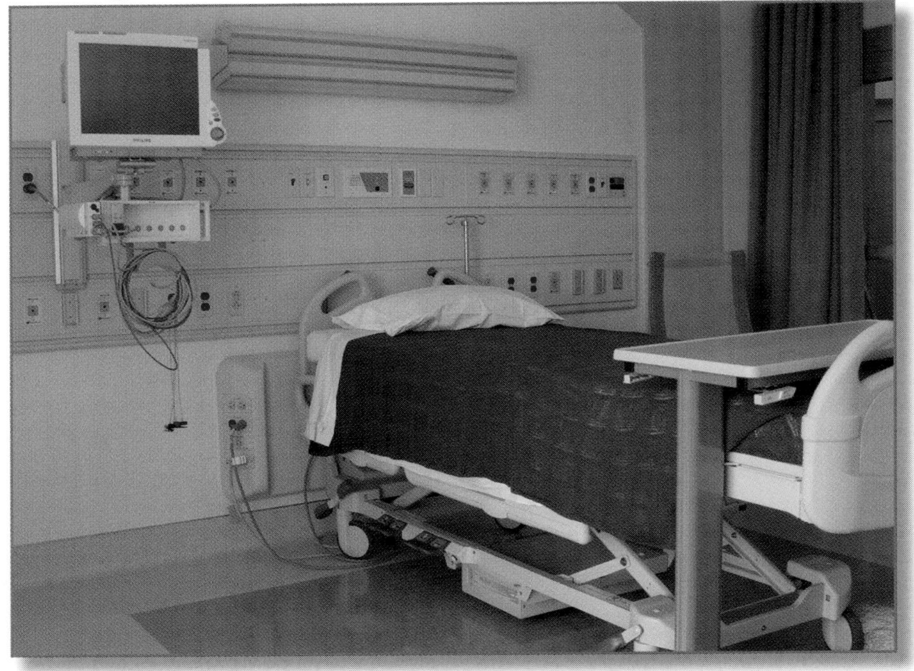

Change Summary

Section 517.13(B) has been restructured into a list format in accordance with the *NEC Style Manual*, and a new exception now follows the main requirement in 517.13(B)(1). Exceptions Nos. 1 and 2 now follow list item (3) and still apply only to that list item. The sizing requirement for equipment grounding conductors and equipment bonding jumpers is included as new list item (2).

Insulated Equipment Grounding Conductor (continued)

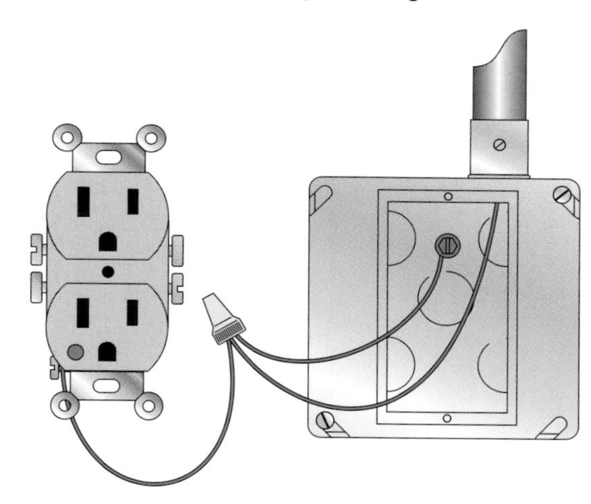

An insulated equipment bonding jumper that directly connects to the equipment grounding conductor is permitted to connect the box and receptacle(s) to the equipment grounding conductor.

Equipment bonding jumpers and equipment grounding conductors must be sized according to 250.122.

Significance of the Change

This revision restructures 517.13(B) into a list format to improve clarity and usability of the requirements. As restructured, this rule is now clear on the three permissible locations at which the required insulated copper equipment grounding conductor must terminate. It can terminate on the grounding terminals of all receptacles, on metal boxes and enclosures containing receptacles, or on all non-current-carrying conductive surfaces of fixed electrical equipment likely to become energized that are subject to personal contact and operating at over 100 volts.

A new exception to 517.13(B)(1) clarifies that a single equipment bonding jumper can connect to the equipment grounding conductors of the circuit and that only one connection between the equipment grounding conductor is necessary to the metal box in this case. This results in an exception recognizing that only one connection between the equipment grounding conductors and the box is required without stacking multiple conductors to the same connection point in the box. The reference to Table 250.122 has been changed to reference all of Section 250.122, which will include some additional sizing requirements for equipment grounding conductors that may be necessary in certain installations, for example, where the ungrounded circuit conductors are increased in size for any reason. The two existing exceptions apply only to list item (3) and now follow that list item as restructured.

Comment: 15-35
Proposal: 15-25

Receptacles with Insulated Grounding Terminals

Code Language

517.16 Receptacles with Insulated Grounding Terminals. Receptacles with insulating grounding terminals, as described in 250.146(D), shall not be permitted.

(The FPN has been deleted.)

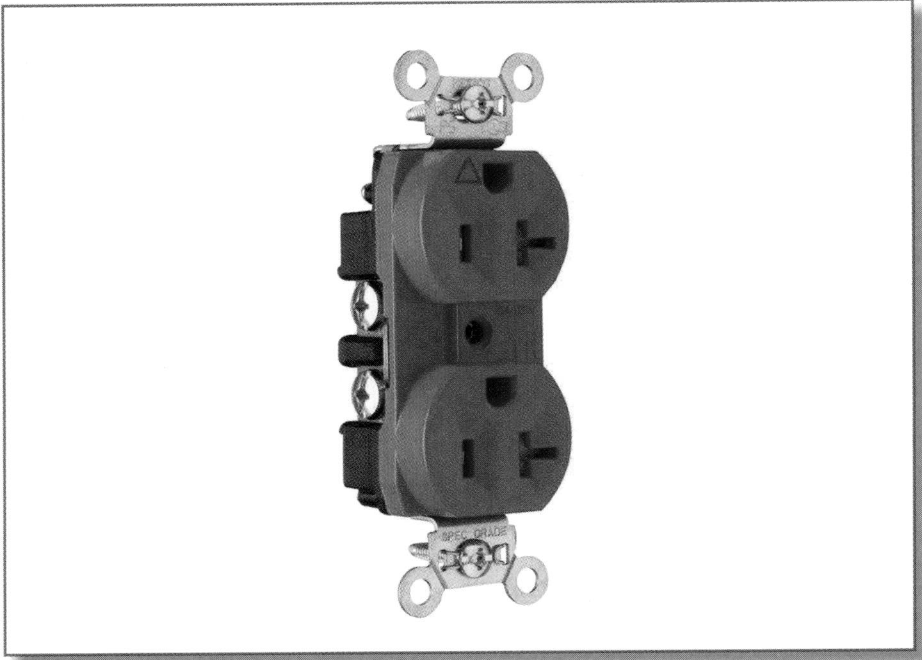

Courtesy of Pass and Seymour Legrand

Change Summary

Section 517.16 has been revised, and the existing informational note has been deleted. Isolated grounding receptacles and circuits are no longer permitted in patient care locations.

Significance of the Change

This revision results in a restriction against the installation of any isolated grounding receptacles and circuits in branch circuits serving patient care locations. Substantiation provided with the proposal pointed out that the existing informational note warned users that the redundant equipment grounding conductor paths could be compromised by the installation of isolated grounding receptacles. Some significant consequences result from this revision, most notably regarding the medical appliances and equipment that recommend installation of an isolated grounding receptacle for the reduction of electrical noise on the grounding circuit.

Another problem created by this revision is the inconsistency introduced between NFPA 99, *Standard for Health Care Facilities*, and this *Code*. NFPA 99 continues to address isolated grounding circuits in health care facilities. This revision in the *NEC* makes clear that no isolated grounding receptacles or circuits are permitted in the branch circuits serving patient care locations. The proposal substantiation also pointed out that the reduction of electrical noise on the grounding circuit should not be taken more seriously than protecting patients, by ensuring redundant grounding paths required by 517.13.

Comments: 15-35, 15-39, 15-41, 15-42

Proposal: 15-29

517.17(C)

Article 517 Health Care Facilities
Part II Wiring and Protection

GFP Selectivity

Code Language

517.17 Ground-Fault Protection

(C) Selectivity. Ground-fault protection for operation of the service and feeder disconnecting means shall be fully selective such that the feeder device, but not the service device, shall open on ground faults on the load side of the feeder device. Separation of ground-fault protection time current characteristics shall conform to manufacturer's recommendations and shall consider all required tolerances and disconnect operating time to achieve 100 percent selectivity.

Significance of the Change

The requirement for a six-cycle separation between the tripping bands of multiple GFP devices has been deleted. The revision now requires conformance to the manufacturer's recommendations on achieving the required separation between the tripping times to attain sufficient selective coordination between the multiple levels of GFP equipment installed to comply with 517.17(B). Substantiation provided with the proposal indicated that the six-cycle time frame is not universally required and is an older requirement that applied to electromechanical relays operating on separate switching mechanisms. The equipment of today for the most part includes ground-fault protection that is integral to the disconnection means that the relay operates. It is no longer necessary to include additional time between the curves of these devices, because the manufacturer's time current information includes all applicable tolerances and operating time features.

The revision removes this long-standing requirement for six cycles of time between the tripping bands of ground-fault protection for equipment installed in health care facility applications. Selectivity is still required, but it can be accomplished without specifying the reaction time between the levels of GFPE installed in the system. Note that performance testing is still a requirement when ground-fault protection for equipment is first installed on-site, as indicated in 517.17(D).

Change Summary

The last two sentences of this section have been replaced by a new sentence addressing the selectivity required. There is no longer a minimum number of cycles required between the tripping bands of the equipment ground-fault protection installed to conform with this rule which, as revised, requires 100% selectivity.

Comment: None

Proposal: 15-40

517.18(A) & 517.19(A)

Article 517 Health Care Facilities
Part II Wiring and Protection

Patient Bed Location Branch Circuits

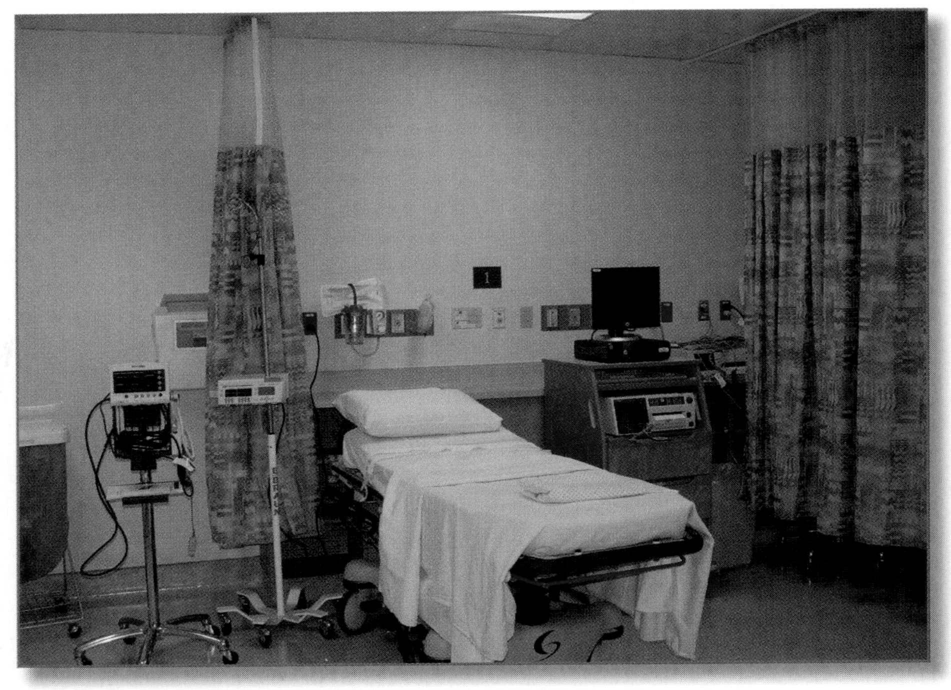

Code Language

517.18 General Care Areas.

(A) Patient Bed Location. Each patient bed location shall be supplied by at least two branch circuits, one from the emergency system and one from the normal system. All branch circuits from the normal system shall originate in the same panelboard.

The branch circuit serving patient bed locations shall not be part of a multi-wire branch circuit.

517.19 Critical Care Areas.

(A) Patient Bed Location Branch Circuits. ***(Unchanged except for new last sentence as follows:)***

The branch circuit serving patient bed locations shall not be part of a multi-wire branch circuit.

Change Summary

A new last sentence has been added to 517.18(A) and 517.19(A), as follows: The branch circuit serving patient bed locations shall not be part of a multi-wire branch circuit.

Significance of the Change

This change restricts the installation of multi-wire branch circuits in patient bed locations in general care areas and critical care areas of health care facilities. Section 210.4(B) requires that multi-wire branch circuits be provided with a means to simultaneously disconnect all ungrounded conductors of the circuit. Disconnection can be accomplished through identified handle ties or multipole breakers. The proposal substantiation pointed out that if one circuit of a multi-wire circuit serving multiple patient care locations tripped, it could simultaneously cause interruption of the other circuits in the multi-wire branch circuit arrangement. The new sentence in each of these sections prohibits multi-wire branch circuits from serving patient bed locations addressed in 517.18(A) and 517.19(A). Branch circuits that do not share a neutral conductor are required for supplying these patient care locations.

Comment: None

Proposal: 15-45

517.18(B)

Article 517 Health Care Facilities
Part II Wiring and Protection

Patient Bed Location Receptacles

Code Language

(B) Patient Bed Location Receptacles. Each patient bed location shall be provided with a minimum of four receptacles. They shall be permitted to be of the single, duplex, or quadruplex type, or any combination of the three. All receptacles, whether four or more, shall be listed "hospital grade" and so identified. The grounding terminal of each receptacle shall be connected to an insulated copper equipment grounding conductor sized in accordance with Table 250.122.

Significance of the Change

Section 517.18(B) addresses the minimum number of receptacles required at a patient bed location in a general care area. The revision includes the word *quadruplex* in recognition of some styles of manufactured receptacle product configurations that would meet this requirement at a patient bed location in a general care area. This section still requires that a minimum of four receptacles be provided but allows multiple alternatives in achieving that minimum quantity. At least one of these four receptacles must be connected to the normal system branch circuit required in 517.18(A), and at least one of these receptacles must be connected to the emergency system. The branch circuits supplied from the normal system must originate from the same panelboard, as required in previous editions of the *Code*.

Note that all of the receptacles installed in this location must be hospital grade and are required to be connected to an insulated copper equipment grounding conductor installed with the branch circuit. The minimum size required for this insulated copper equipment grounding conductor must be in accordance with 250.122.

Change Summary

This section has been revised by adding the word *quadruplex* in the second sentence. This revision includes "quadruplex" receptacle assemblies that could be used to achieve the minimum number of required receptacles in patient bed locations (general care areas).

Comment: 15-54

Proposal: 15-46

Pediatric Locations

Code Language

517.18 General Care Areas...

(C) Pediatric Locations. Receptacles located within the rooms, bathrooms, playrooms, activity rooms, and patient care areas of designated pediatric locations shall be listed tamper resistant or shall employ a listed tamper-resistant cover.

Courtesy of Pass and Seymour Legrand

Change Summary

The word *wards* has been deleted from, and the words *designated* and *locations* have been added, to 517.18(C). The revision results in more specific locations where tamper-resistant receptacles are required to be installed.

Significance of the Change

The substantiation for this revision correctly indicated that the word *wards* was not defined in the *Code* and could lead to inconsistencies in field application. Including the words *designated* and *locations* in this section makes for easier and more accurate user application of this requirement. This revision actually results in identifying more specific locations where the tamper-resistant receptacle is a requirement for pediatric locations designated for patient care. It also makes clear that the governing body of the facility should be involved in the determination of areas that qualify as pediatric patient care locations rather than areas that might be used for treating pediatric patients but might not be so designated. This revision should result in more consistent application of the requirement by inspectors, installers, designers, and other *Code* users.

Comment: 15-58

Proposal: 15-50

517.19(B)(2)

Article 517 Health Care Facilities
Part II Wiring and Protection

Receptacle Requirements

Patient bed location receptacles

Critical branch

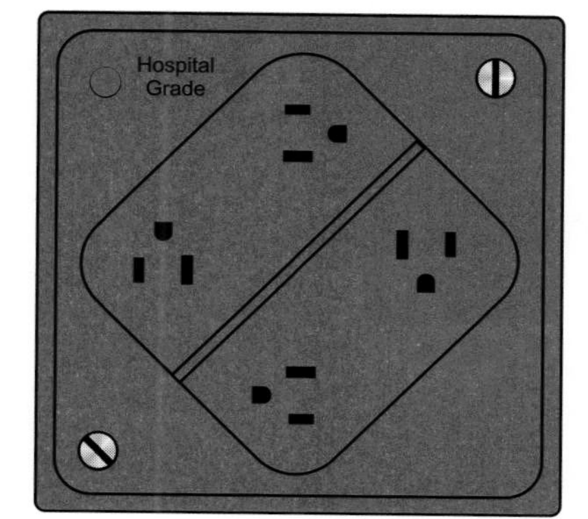

Normal branch

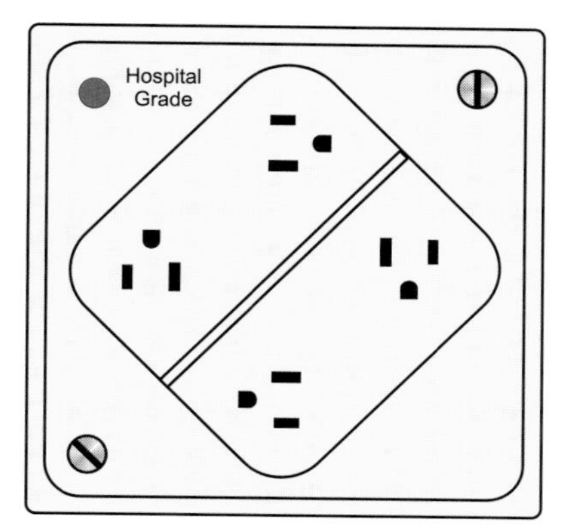

Quadruplex receptacles are permitted to be used to satisfy the receptacle quantity requirements in 517.19(B)(2)

Code Language

(B) Patient Bed Location Receptacles

(2) Receptacle requirements. The receptacles required in 517.19(B)(1) shall be permitted to be single, duplex, or quadruplex type or any combination thereof. All receptacles shall be listed "hospital grade" and shall be so identified. The grounding terminal of each receptacle shall be connected to the reference grounding point by means of an insulated copper equipment grounding conductor.

Change Summary

This section has been restructured to a list format that is numbered to comply with the *NEC Style Manual*. In list item (2) covering receptacle requirements, the word *quadruplex* has been added.

Significance of the Change

Section 517.19(B) addresses the minimum number of receptacles required at a patient bed location in a critical care area. The revision includes the word *quadruplex* in recognition of some styles of manufactured receptacle product configurations that would meet this requirement at a patient bed location in a critical care area. This section still requires that a minimum of six receptacles be provided but allows multiple alternatives in achieving that minimum quantity. At least one of these six receptacles must be connected either to the normal system branch circuit required in 517.19(A) or to an emergency system circuit supplied by a different transfer switch than the other receptacles at the same patient bed location. Note that all of the receptacles installed in this location must be hospital grade and are required to be connected to an insulated copper equipment grounding conductor installed with the branch circuit. The minimum size required for this insulated copper equipment grounding conductor must be in accordance with 250.122.

Comment: 15-60

Proposal: 15-55

517.20(A)

Article 517 Health Care Facilities
Part II Wiring and Protection

Wet Procedure Locations

Code Language

517.20 Wet Procedure Locations.

(A) Receptacles and Fixed Equipment. Wet procedure location patient care areas shall be provided with special protection against electric shock by one of the following means:

(1) A power distribution system that inherently limits the possible ground-fault current due to a first fault to a low value, without interrupting the power supply.

(2) A power distribution system in which the power supply is interrupted if the ground-fault current does, in fact, exceed a value of 6 mA.

Exception: Branch circuits supplying only listed, fixed, therapeutic and diagnostic equipment shall be permitted to be supplied from a grounded service, single- or 3-phase system, provided that:

(a) Wiring for grounded and isolated circuits does not occupy the same raceway, and

(b) All conductive surfaces of the equipment are connected to an insulated copper equipment grounding conductor.

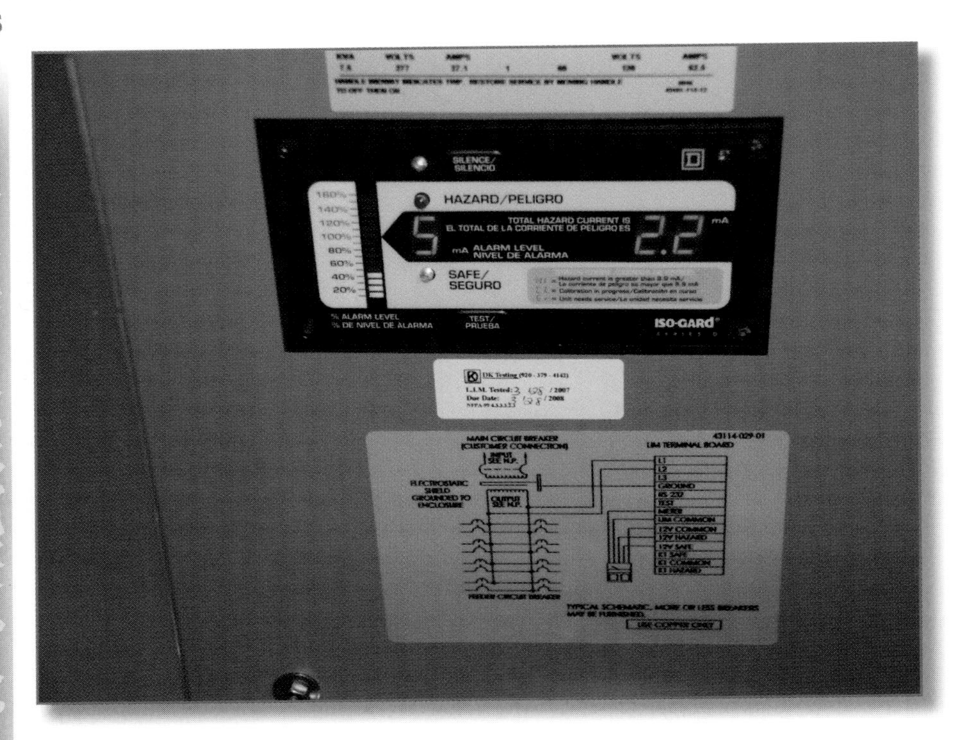

Change Summary

This section has been restructured into a list format to meet the requirements of the *NEC Style Manual*. List items (1) and (2) provide two clear methods of shock protection for wet procedure locations. The words *insulated copper* have been added to Exception No. 2.

Wet Procedure Locations (continued)

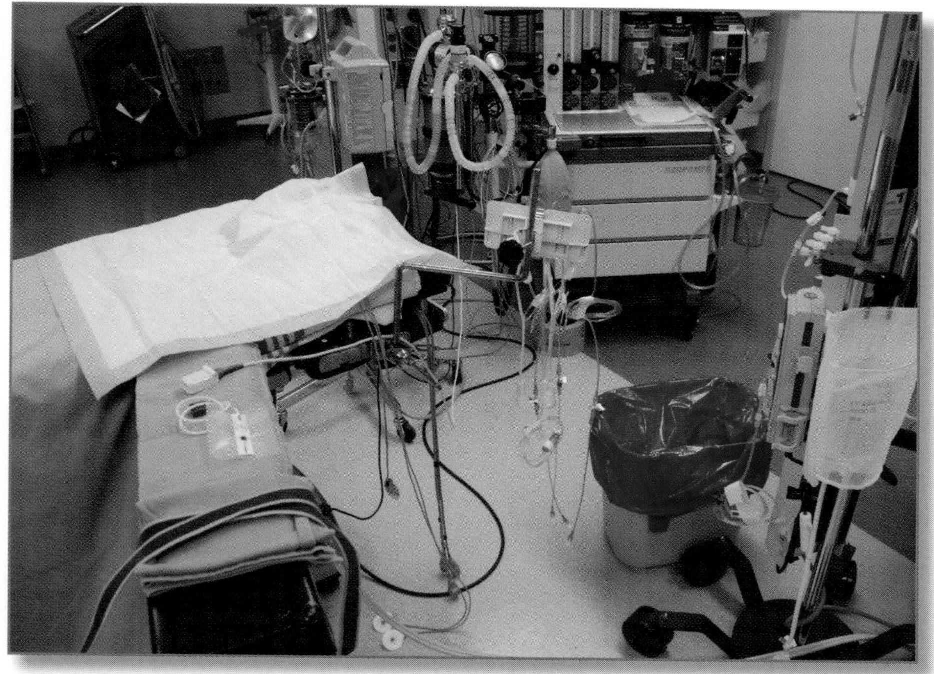

Significance of the Change

The changes to this section negate the concept of installations for which interruption by a GFCI device cannot be tolerated. The revision now recognizes that either type of protection against shock can be provided in accordance with item (1), which addresses isolated power systems, or item (2), which addresses GFCI protective devices. The significance of this change is that according to previous editions of the *Code*, the qualifying factor for installing an isolated power system was that interruption by a GFCI could not be tolerated. Now, however, a choice is provided under the new arrangement of this rule.

Note that it is typically a decision of the governing body of the facility whether to provide isolated power systems in areas such as operating rooms or other critical care areas where interruption of power by the operation of a GFCI device creates a more hazardous situation for patients being treated therein. There remains a need to determine where interruption of a GFCI is not tolerated, but the governing body typically makes this decision when determining the extent of wet procedure locations. Section 210.8(B)(5) is still applicable in areas that include a receptacle within 6 feet of a sink unless they meet the provisions in Exception 2 to 210.8(B)(5).

Comment: 15-69

Proposals: 15-64, 15-65

517.30(C)(5)

Article 517 Health Care Facilities
Part III Essential Electrical System

Mechanical Protection of the Emergency System

Code Language

517.30 Essential Electrical Systems for Hospitals

(C) Wiring Requirements...

(3) Mechanical Protection of the Emergency System. The wiring of the emergency systems in hospitals shall be mechanically protected. Where installed as branch circuits in patient care areas, the installation shall comply with the requirements of 517.13(A) and (B). The following wiring methods shall be permitted:

(1) ***(No change)***
(2) ***(No change)***
(3) ***(No change)***
(4) ***(No change)***
(5) Cables for Class 2 or Class 3 systems permitted by Part VI of this article, with or without raceways.

Change Summary

This list item has been revised by deleting the phrases *secondary circuits* and *signaling systems* and the word *communications* and adding the word *cables*.

Significance of the Change

This section was revised by deleting the phrases *secondary circuits* and *signaling circuits* and the word *communications* and adding the word *cables*. The revision also results in a reference to Part VI of Article 517, which specifically addresses communications, signaling systems, data systems, fire alarm systems, and systems less than 120 volts, nominal. The net result of this revision is a clarification that all Class 2 or Class 3 systems permitted in the essential electrical system of a hospital are allowed to be installed with or without raceways, even if part of an essential electrical system. The 2008 *NEC* revised this section to provide additional clarification about running these systems within or without raceways. This new revision clarifies the specific Class 2 and Class 3 systems permitted without the physical protection of a raceway installation. It also coordinates with a similar revision in 517.80 that also exempts Class 2 and Class 3 circuits and communications systems from being installed in raceway systems.

Comment: 15-101
Proposals: 15-86, 15-87

Article 517 Health Care Facilities
Part VI Communications, Signaling Systems, Data Systems, Fire Alarm Systems, and Systems Less Than 120 Volts, Nominal

Patient Care Areas

Code Language

517.80 Patient Care Areas. Equivalent insulation and isolation to that required for the electrical distribution systems in patient care areas shall be provided for communications, signaling systems, data system circuits, fire alarm systems, and systems less than 120 volts, nominal. Class 2 and Class 3 signaling and communications systems and power-limited fire alarm systems shall not be required to comply with the grounding requirements of 517.13, to comply with the mechanical protection requirements of 517.30(C)(3)(5), or to be enclosed in raceways unless otherwise specified by Chapter 7 or 8.

(Remainder unchanged)

Significance of the Change

This revision provides a new second sentence in 517.80, to clarify that Class 2 and Class 3 signaling and communications systems and power-limited fire alarm systems are not required to comply with the grounding requirements of 517.13 because these systems are not branch circuits by definition. These systems are also not required to comply with the mechanical protection requirements of 517.30(C)(3) or (C)(5), or to be enclosed in raceways unless specifically required by Chapter 7 or Chapter 8. This revision should reduce the misapplication of this section's requirements by enforcement officials and other users, and allow this installation exemption for limited energy systems as intended. Power-limited fire alarm systems are also exempted on the basis of the power limitations of such circuits and the reduction in shock hazards.

The revision to this section also correlates with similar changes in 517.30(C)(5), which recognizes the installation of these systems, with or without raceways, as a means of physical protection whether part of an emergency system or not. Obviously, where these systems are subject to physical damage and the need for physical protection is predicated, protection should be provided as a general requirement. A nurse call system is an example of a limited energy system that would fall under the requirements of 517.80.

Change Summary

A new second sentence has been added to 517.80, as follows: Class 2 and Class 3 signaling and communications systems and power-limited fire alarm systems shall not be required to comply with the grounding requirements of 517.13, to comply with the mechanical protection requirements of 517.30(C)(3)(5), or to be enclosed in raceways unless otherwise specified by Chapter 7 or 8.

Comment: 15-132
Proposal: 15-120

517.160(A)(5)

Article 517 Health Care Facilities
Part VII Isolated Power Systems

Conductor Identification

Code Language

517.160(A) Installations.

(5) Conductor Identification. The isolated circuit conductors shall be identified as follows:

(1) Isolated Conductor No. 1 — Orange with at least one distinctive colored stripe other than white, green, or gray along the entire length of the conductor

(2) Isolated Conductor No. 2 — Brown with at least one distinctive colored stripe other than white, green, or gray along the entire length of the conductor

For 3-phase systems, the third conductor shall be identified as yellow with at least one distinctive colored stripe other than white, green, or gray along the entire length of the conductor. Where isolated circuit conductors supply 125-volt, single-phase, 15- and 20-ampere receptacles, the striped orange conductor(s) shall be connected to the terminal(s) on the receptacles that are identified in accordance with 200.10(B) for connection to the grounded circuit conductor.

Change Summary

The phrase "along the entire length of the conductor" has been added to list items (1) and (2) and in the last paragraph of this rule.

Comment: None
Proposal: 15-125

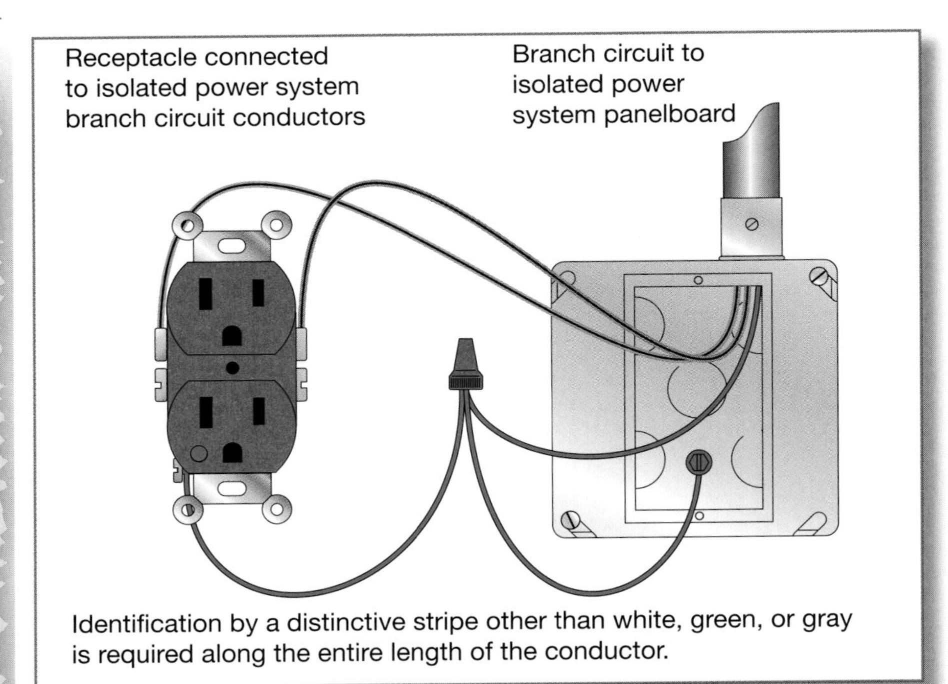

Significance of the Change

This revision clarifies that the single distinctive colored stripe for identification must be along the entire length of the conductor. In other words, the conductor has to be manufactured with this identification means on the insulation. The problem with the previous language was that compliance could be achieved literally by attaching a single stripe of marking tape around the conductor at exposed locations. This was never the intent of this new identification requirement accepted in the 2008 edition of the *NEC*.

The purpose of this new identification requirement is to establish a clear distinction between conductors of a 480Y/277-volt, 3-phase, 4-wire power system that may be installed in the health care facility and identified using brown, orange, and yellow conductor insulation. This common method of identification for system conductors at this voltage conflicts with the required means of conductor identification for isolated power systems as required by 517.160(A)(5). The requirement for this single stripe along the entire length of the conductor should alert installers and maintenance personnel to the more specific purpose and use of such conductors.

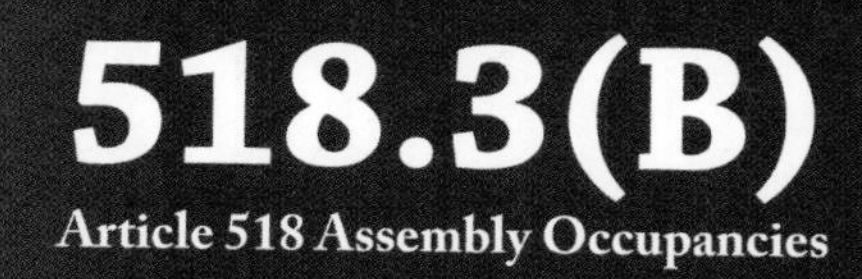

Temporary Wiring

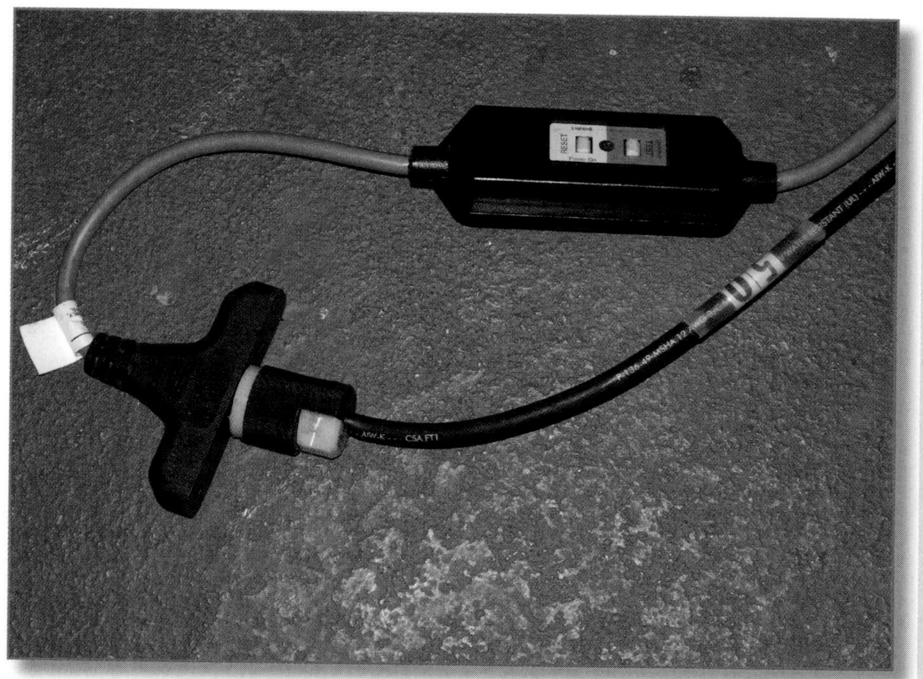

Code Language

518.3 Other Articles

(B) Temporary Wiring. In exhibition halls used for display booths, as in trade shows, the temporary wiring shall be permitted to be installed in accordance with Article 590. Flexible cables and cords approved for hard or extra-hard usage shall be permitted to be laid on floors where protected from contact by the general public. The ground-fault circuit-interrupter requirements of 590.6 shall not apply. All other ground-fault circuit-interrupter requirements of this *Code* shall apply.

Where ground-fault circuit interrupter protection for personnel is supplied by plug-and-cord-connection to the branch circuit or to the feeder, the ground-fault circuit-interrupter protection shall be listed as portable ground-fault circuit-interrupter GFCI protection or provide a level of protection equivalent to a portable ground-fault circuit-interrupter GFCI, whether assembled in the field or at the factory.

Exception: (No change)

Change Summary

The GFCI requirements for exhibition halls used for display booths such as in trade shows have been clarified. Where GFCI protection is derived at an unprotected receptacle outlet by plug-and-cord connection to the branch circuit or to the feeder, the device is required to be listed as portable GFCI protection or to provide a level of protection equivalent to that of a portable GFCI.

Significance of the Change

Section 518.3(B) provides temporary wiring requirements for exhibition halls used for display booths in trade shows. This second-level subdivision amends the GFCI requirements of 590.6, to exempt the temporary power distributed to vendors in a trade show at 15/20/30-amps 125 volts from the requirement to supply GFCI protection. A sentence has been added to mandate that all other ground-fault circuit-interrupter requirements of the *NEC* apply. Therefore, if a vendor in a trade show for swimming pools builds a small version of its pool complete with water, the rules for GFCI protection in Article 680 would apply.

The new paragraph mandates that where the required GFCI protection is supplied by a cord-and-plug connection, the device must be listed as a "portable GFCI" or must provide a level of protection equivalent to that of a portable GFCI. A portable GFCI device provides open neutral protection. To function properly, the brain of a GFCI needs a source of power. If the neutral is opened on a non-portable GFCI, current will flow to ground without the device opening. A portable device will not operate upon loss of power.

Comment: None

Proposals: 15-126, 15-127

547.5(G)

Article 547 Agricultural Buildings

GFCI Protection

Code Language

547.5 Wiring Methods

(G) Receptacles. All 125-volt, single-phase, 15- and 20-ampere general-purpose receptacles installed in the locations listed in (1) through (4) shall have ground-fault circuit-interrupter protection:

(1) Areas having an equipotential plane
(2) Outdoors
(3) Damp or wet locations
(4) Dirt confinement areas for livestock

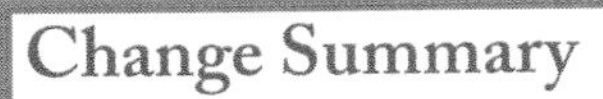

Change Summary

The permission to install a receptacle without GFCI protection to supply a dedicated load within 3 feet of a GFCI protected receptacle has been deleted.

Significance of the Change

In the 2008 *NEC* revision cycle, a new last paragraph was added to this first-level subdivision, permitting a receptacle without GFCI protection to supply a dedicated load, provided it was installed within 3 feet of a GFCI protected receptacle. This decision has been reversed. It is recognized that a dedicated load supplied from a receptacle can and will be easily moved to another location, leaving the unprotected receptacle available for use by persons.

A GFCI will not operate until the circuit has a leakage current of 4 to 6 mA. Current product standards do not permit a leakage current of more than 0.5 mA. There should no tripping of a GFCI device by an appliance or tool due to the very low leakage currents mandated by the current product standards. However, complaints are still heard regarding older farm equipment, including electric fence controllers. This type of older equipment may continue to trip a GFCI-protected outlet due to higher levels of leakage current permitted in an older product standard or the equipment may be damaged.

Comment: None

Proposals: 19-24, 19-25

590.4(D)

Article 590 Temporary Installations

Receptacles

Code Language

590.4 General

(D) Receptacles.

(1) All Receptacles. All receptacles shall be of the grounding type. Unless installed in a continuous metal raceway that qualifies as an equipment grounding conductor in accordance with 250.118 or a continuous metal-covered cable that qualifies as an equipment grounding conductor in accordance with 250.118, all branch circuits shall include a separate equipment grounding conductor, and all receptacles shall be electrically connected to the equipment grounding conductor(s). Receptacles on construction sites shall not be installed on any branch circuit that supplies temporary lighting.

(2) Receptacles in Wet Locations. All 15- and 20-ampere, 125- and 250-volt receptacles installed in a wet location shall comply with 406.9(B)(1).

Significance of the Change

The requirements of 590.4(D) are separated into two subdivisions. 590.4(D)(1) clearly prohibits branch circuits supplying temporary lighting from supplying power to a receptacle for "temporary power." The justification for this requirement is to ensure that an overload or fault on a branch circuit used for temporary power does not leave an area without lighting. For example, if temporary lighting for a small auditorium were wired on the same branch circuit as temporary power, a construction worker on a scaffold could cause the circuit to trip on overload. This would create a very dangerous situation for the worker when descending the scaffold in the dark.

This *NEC* rule mirrors a similar requirement in the OSHA 1926 construction standard, in 1926.405(a)(2)(ii)(C). This *NEC* revision is necessary because 210.4(B) now requires simultaneous disconnection of multiwire branch circuits.

The new requirement in 590.4(D)(2) mandating compliance with 406.9(B)(1) now requires that in wet locations (1) an enclosure be weatherproof whether or not the attachment plug cap is inserted for 15- and 20-ampere, 125- and 250-volt receptacles and (2) all 15- and 20-ampere, 125- and 250-volt nonlocking-type receptacles be listed weather-resistant type.

Change Summary

Section 590.4(D)(1) has been revised to clarify that receptacles on construction sites must not be installed on any branch circuit that supplies temporary lighting. Section 590.4(D)(2) now requires compliance with 406.9(B)(1).

Comment: 3-56

Proposals: 3-122, 3-123

590.6(A)

Article 590 Temporary Installations

Ground-Fault Protection for Personnel

Code Language

590.6 Ground-Fault Protection for Personnel

(A) Receptacle Outlets. Temporary receptacle installations used to supply temporary power to equipment used by personnel during construction, remodeling, maintenance, repair, or demolition of buildings, structures, equipment, or similar activities shall comply with the requirements of 590.6(A)(1) through 590.6(A)(3), as applicable.

Exception: (No Change)

(1) Receptacle Outlets Not Part of Permanent Wiring. All 125-volt, single-phase, 15-, 20-, and 30-ampere receptacle outlets that are not a part of the permanent wiring of the building or structure and that are in use by personnel shall have ground-fault circuit-interrupter protection for personnel.

(2) Receptacle Outlets Existing or Installed as Permanent Wiring. Ground-fault circuit-interrupter protection for personnel shall be provided for all 125-volt, single-phase, 15-, 20-, and 30-ampere receptacle outlets installed or existing as part of the permanent wiring of the building or structure and used for temporary electric power. Listed cord sets or devices incorporating listed ground-fault circuit interrupter protection for personnel identified for portable use shall be permitted.

(... Continued on next page ...)

Significance of the Change

The parent text of 590.6(A) has been revised to clearly and logically separate the requirements for GFCI protection of receptacle outlets used in construction or similar activities where Article 590 is employed. Three new second level subdivisions have been created in 590.6(A) to clarify these GFCI requirements. 590.6(A)(1) requires GFCI protection for all receptacle outlets operating at 125-volts, single-phase, 15-, 20-, and 30-amperes that are not a part of the permanent wiring of the building or structure.

590.6(A)(2) requires GFCI protection for all receptacle outlets operating at 125-volts, single-phase, 15-, 20-, and 30-amperes installed or existing as part of the permanent wiring of the building or structure. This can be achieved by protecting the entire branch circuit with a GFCI type circuit breaker, a GFCI type receptacle at the first outlet, or listed cord sets or devices incorporating listed ground-fault circuit interrupter protection for personnel identified for portable use.

Ground-Fault Protection for Personnel (continued)

Code Language

(... Continued from previous page ...)

(3) Receptacles on 15 kW or less Portable Generators. All 125-volt and 125/250-volt, single-phase, 15-, 20-, and 30-ampere receptacle outlets that are a part of a 15 kW or smaller portable generator shall have listed ground-fault circuit interrupter protection for personnel. All 15- and 20-ampere, 125- and 250-volt receptacles, including those that are part of a portable generator, used in a damp or wet location shall comply with 406.9(A) and (B). Listed cord sets or devices incorporating listed ground-fault circuit interrupter protection for personnel identified for portable use shall be permitted for use with 15 kW or less portable generators manufactured or remanufactured prior to January 1, 2011.

Section 590.6(A)(3) contains two new requirements. The first is that GFCI protection is now required for all receptacle outlets operating at 125-volts and 125/250-volts, single-phase, 15-, 20-, and 30-amperes that are a part of a 15-kW or smaller portable generator. Older generators manufactured or remanufactured prior to January 1, 2011 are permitted to be used where listed cord sets or devices incorporating listed ground-fault circuit interrupter protection as personnel identified for portable are used. The second new requirement in 590.6(A)(3) is that all 15- and 20-ampere, 125- and 250-volt receptacles, that are part of a portable generator used in a damp or wet location must comply with section 406.9(A) and (B). This will require these generator receptacles to be listed for damp and/or wet locations and enclosures that are weather proof in accordance with the requirements of 406.9(A) & (B).

Change Summary

GFCI requirements for 125-volt, single-phase, 15-, 20-, and 30-ampere receptacle outlets used by personnel during construction, remodeling, maintenance, repair, or demolition of buildings, structures, equipment, or similar activities have been clarified.

Comment: 3-81

Proposals: 3-139, 3-140, 3-141

Chapter 6

Articles 600-695
Special Equipment

600.1

Article 600 Electric Signs and Outline Lighting
Part I General

Scope

Code Language

600.1 Scope. This article covers the installation of conductors, equipment, and field wiring for electric signs and outline lighting, regardless of voltage. All installations and equipment using neon tubing, such as signs, decorative elements, skeleton tubing, or art forms, are covered by this article.

Informational Note: Sign and outline lighting illumination systems include, but are not limited to, cold cathode neon tubing, high intensity discharge lamps (HID), fluorescent or incandescent lamps, light emitting diodes (LEDs), and electroluminescent and inductance lighting.

Courtesy of YESCO

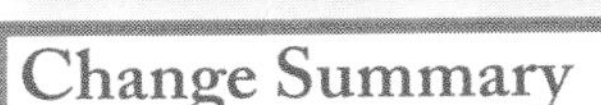

Change Summary

The phrases *and field wiring* and *regardless of voltage* have been added to the scope of Article 600. The revision expands the scope of this article to cover additional types of signs and outline lighting. A new informational note has been added following 600.1.

Significance of the Change

The scope of Article 600 has been revised to clarify that field wiring of electric signs and outline lighting systems is covered within its requirements. Electrical signs and outline lighting systems are supplied by a variety of voltages, ranging from low voltage levels for Class 2 circuits and systems up to 15,000 volts (7,500 to ground) for use with neon tubing and cold cathode installations. The phrase *regardless of voltage* underscores the requirement that Article 600 applies to sign wiring installations at any voltage. Many newer installations are incorporating limited energy circuits that drive light-emitting diode (LED) technology used for signs and outline lighting. Additional requirements have also been added to Article 600 to specifically address LED systems. The new informational note lists the various types of lighting systems commonly used for electric signs and outline lighting systems to help clarify the equipment and systems covered by Article 600.

Comment: None

Proposal: 18-197

Definition of LED Sign Illumination System

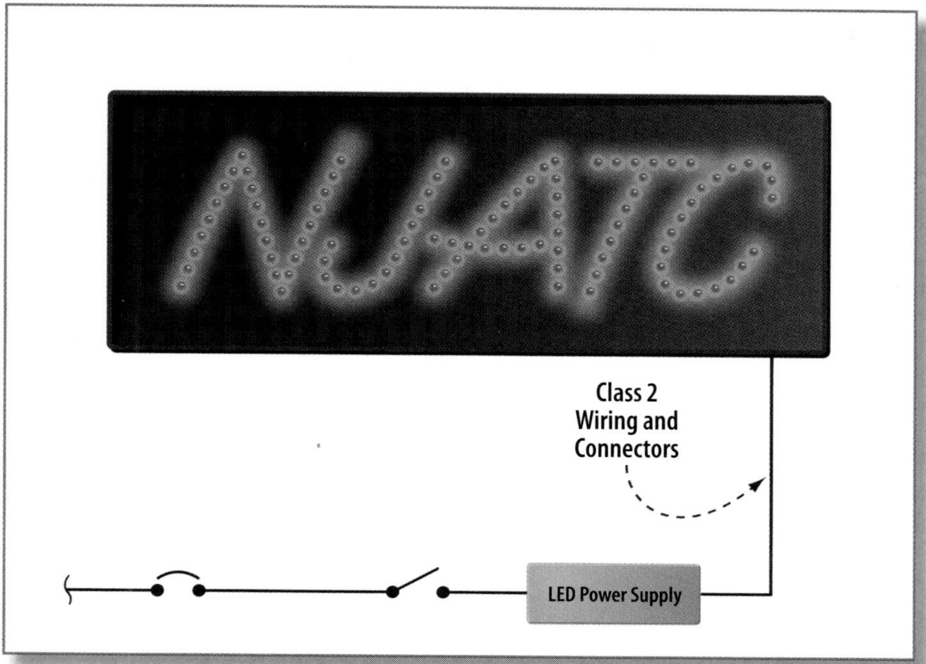

Code Language

600.2 Definitions.

LED Sign Illumination System. A complete lighting system for use in signs and outline lighting consisting of light emitting diode (LED) light sources, power supplies, wire, and connectors to complete the installation.

Significance of the Change

Advancement in LED technology is progressive and deliberate to the extent that new rules in Articles 410 and 600 of the *NEC* are necessary. New requirements for LED sign illumination systems have been incorporated into Article 600, and new 600.33 addresses specific rules for secondary wiring supplying LED illumination equipment. The new definition addresses the entire system, consisting of light emitting diodes as the light source and the power supply, conductors, connectors, and so forth, that are required for the complete LED system installed for a sign or outline lighting system. Note that 600.3 applies to these LED systems. In other words, they are required to be listed systems that include components that have been evaluated for use together in a complete system. The applicable product standards govern the manufacturing of LED signs and outline lighting systems. In this case, UL 48, *Electric Signs*, covers listed electric signs as well as LED sign illumination systems.

Change Summary

A definition of the term *LED Sign Illumination System* has been added to 600.2. Such systems include the LEDs, power supplies, conductors and connectors required to complete the installation.

Comment: None

Proposal: 18-198

600.4(C) & (D)

Article 600 Electric Signs and Outline Lighting
Part I General

Visibility and Durability

Code Language

600.4 Markings.

(Subdivisions (A) and (B) remain unchanged.)

(C) Visibility. The markings required in 600.4(A) and listing labels shall not be required to be visible after installation but shall be permanently applied in a location visible during servicing.

(D) Durability. Marking labels shall be permanent, durable and, when in wet locations, shall be weatherproof.

(E) Section Sign. ***[unchanged, just re-identified as (E)]***

Change Summary

Two new subdivisions, (C) and (D), have been added to 600.4, and former subdivision (C) has been re-identified as (E) to accommodate these new subdivisions. These new subdivisions clarify the location requirements for required markings on signs and outline lighting systems.

NECA Photo - © 2010 Rob Colgan

Significance of the Change

This revision provides additional clarification about the marking requirements for electric signs and outline lighting systems. New subdivision (C) addresses the visibility requirements for the listing markings and the other marking information required under (A), including the manufacturer's name or trademark, input voltage, and current ratings. These markings are now required to be visible only during servicing procedures, not after the installation is complete. New subdivision (D) requires that the markings addressed in this rule be durable for the environment in which they are installed. In wet locations, the markings must now be weatherproof. This change affects manufacturers of electrical signs and outline lighting systems because the markings covered by this section are generally not applied by the installer in the field.

The marking requirements in this section are important for inspectors and sign service personnel, both during the installation and periodically while service is performed after this installation is complete and operational. The markings are required on the equipment, but they are not required to be visible after the installation is completed.

Comment: None

Proposal: 18-206a

600.5(B)

Article 600 Electric Signs and Outline Lighting
Part I General

Rating

I-Stock Photo, Courtesy of NECA

Code Language

(B) Rating. Branch circuits that supply signs shall be rated in accordance with 600.5(B)(1) or (B)(2) and shall be considered to be continuous loads for the purposes of calculations.

(1) Neon Signs. Branch circuits that supply neon tubing installations shall not be rated in excess of 30 amperes.

(2) All Other Signs. Branch circuits that supply all other signs and outline lighting systems shall be rated not to exceed 20 amperes.

Significance of the Change

This revision clarifies that the load associated with electric signs and outline lighting systems must be considered as a continuous load for calculation purposes. Previously, this section indicated only that the branch circuit had to have a minimum rating based on the provisions of 600.5(B)(1) or (B)(2) requiring a rating not to exceed a 20-ampere circuit for an incandescent, fluorescent type of illumination, or any other type of electric sign or outline lighting system, and a rating not to exceed a 30-ampere branch circuit for circuits supplying neon tubing installations. New list item (2) sets that maximum branch circuit rating for circuits supplying signs and outline lighting systems other than neon types.

The revision provides a requirement to include the sign or outline lighting load at 125 percent within the total load calculation of the system. The definition of *Continuous Load* in Article 100 is clear that this is a load at maximum current for 3 hours or more. Electric signs and outline lighting systems generally fall into that use category and thus qualify as continuous loads. This revision clarifies how these loads must be counted in the overall calculation of the system load. No changes were made to the maximum rating of the branch circuit supplying electric sign and outline lighting system loads.

Change Summary

The phrase *and shall be considered to be continuous loads for the purposes of calculations* has been added to 600.5(B). Former list item (2) covering neon signs has been renumbered as (B)(1), and a new (B)(2) addresses all other electric signs and outline lighting system branch circuit ratings.

Comment: None

Proposals: 18-206b, 18-207

600.6

Article 600 Electric Signs and Outline Lighting
Part I General

Disconnects

Code Language

600.6 Disconnects. Each sign and outline lighting system, feeder circuit or branch circuit supplying a sign, outline lighting system, or skeleton tubing shall be controlled by an externally operable switch or circuit breaker that opens all ungrounded conductors and controls no other load. The switch or circuit breaker shall open all ungrounded conductors simultaneously on multi-wire branch circuits in accordance with 210.4(B). Signs and outline lighting systems located within fountains shall have the disconnect located in accordance with 680.12.

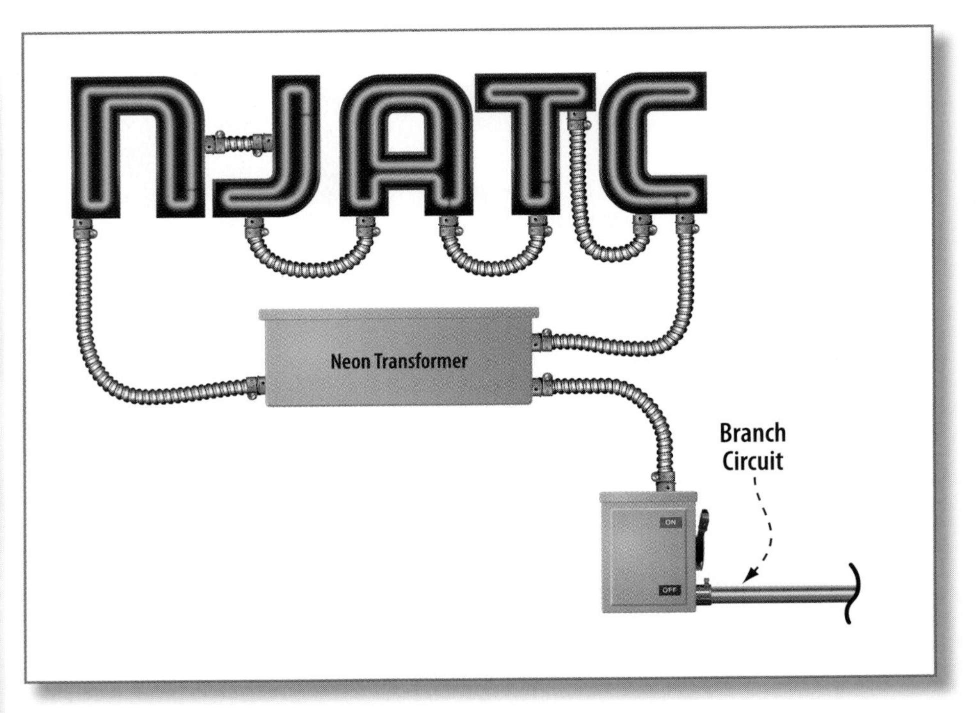

Change Summary

The word *each*, the phrase *or skeleton tubing*, and the phrase *and controls no other load* have been added to 600.6. A new second sentence addresses the requirements for handle-ties or multipole circuit breakers to meet the requirements for disconnecting multi-wire branch circuits.

Comment: 18-72

Proposals: 18-210, 18-212

Significance of the Change

This revision makes clear that the disconnecting means for electric signs and outline lighting systems must supply only those loads, and that they are not permitted to supply any other load. The previous language of this section did not clearly indicate that the disconnecting means was limited to the sign or outline equipment supplied. This revision provides that clarification and is expected to be a safety enhancement for service persons working on this type of installation. Another clarification provided by this revision is the inclusion of skeleton tubing installations. It was generally understood that this section applied to skeleton tubing systems, but it was not previously addressed specifically within this requirement. The revision makes it clear that the required disconnect for each feeder or branch circuit supplying a sign or outline lighting system is not permitted to supply another load.

Some electric signs require feeders because of the number of branch circuits required to supply the entire assembly. The feeder supplying the panelboard for this type of installation can supply only the branch circuits for the sign, no other loads. In this case, the feeder is required to be equipped with a disconnecting means. The second sentence referencing 210.4(B) has been added to address the requirements for handle-ties or multipole circuit breakers installed for multi-wire branch circuits supplying this equipment.

600.7(B)(1)

Article 600 Electric Signs and Outline Lighting
Part I General

Bonding

NECA Photo - © 2010 Rob Colgan

Code Language

600.7(B)(1) ***(Unchanged)***

Exception: Remote metal parts of a section sign or outline lighting system only supplied by a remote Class 2 power supply shall not be required to be bonded to an equipment grounding conductor.

Significance of the Change

The new exception to 600.7(B)(1) relaxes the requirements for bonding metal parts of equipment supplied by Class 2 circuits. This section provides correlation with the provisions of 250.112(G) that generally require all metal parts of signs and outline lighting systems to be bonded to the equipment grounding conductor of the supply circuit. Previously, what the sign or outline lighting system voltage was did not matter — all metal parts had to be grounded and bonded. This new exception recognizes the limits of low-voltage systems supplying signs and outline systems. The new exception applies only to the remote metal parts associated with this type of equipment — in other words, to the parts that are exposed only to the low-voltage circuit supplied by the Class 2 transformer or power supply.

The revision aligns with similar information that is being incorporated in UL Standard 48, *Standard for Electric Signs*. It should be noted that the equipment grounding conductor is still required with the branch circuit supplying this type of equipment, as specified in 600.7(A)(1). The equipment grounding conductor is required to be connected to the metal enclosure where the Class 2 transformer or power supply is installed for such listed systems.

Change Summary

A new exception has been added to 600.7(B)(1). The new exception provides a relaxation of the bonding requirements for remote metal parts as long as the sign is supplied by a Class 2 power supply.

Comment: None

Proposal: 18-223

Field-Installed Skeleton Tubing and Outline Lighting

Code Language

II. Field-Installed Skeleton Tubing, Outline Lighting, and Secondary Wiring

600.30 Applicability. Part II of this article shall apply to all of the following:

(1) Field-installed skeleton tubing
(2) Field-installed secondary circuits
(3) Outline lighting

These requirements are in addition to the requirements of Part I.

Change Summary

The words *outline lighting* and *secondary* have been added to the title of Part II of Article 600 and have been incorporated in 600.30(2) and (3).

Courtesy of YESCO

Significance of the Change

This revision incorporates secondary wiring and outline lighting system wiring under the coverage of Part II of Article 600. Previously, this part of the article applied only to field-installed skeleton tubing systems and associated wiring. The scope of Article 600 has been expanded to include light-emitting diode technology in electric signs and outline lighting systems. Therefore, Part II has been expanded to include the secondary circuit wiring for such systems, thus requiring the revision to the title of Part II of the article.

The revision clarifies that requirements for outline lighting systems and field-installed secondary circuits for such systems, in addition to field-installed skeleton tubing systems, are covered in this part of Article 600. This includes secondary circuit wiring for neon tubing systems and light-emitting diode systems. Additionally, a new section, 600.33, covering light-emitting diode (LED) sign illumination systems has been added to this part of the article, driving the need for revising the title and applicability of Part II.

Comment: None
Proposals: 18-246, 18-247

LED Sign Illumination Systems, Secondary Wiring

Courtesy of YESCO

Change Summary

Section 600.33 LED Sign Illumination Systems, Secondary Wiring has been added to Part II of Article 600. This new section includes four subdivisions. These new requirements are specific to the locations, conductors, and grounding/bonding requirements that must be applied to Class 2 circuits of LED sign illumination systems.

Code Language

600.33 LED Sign Illumination Systems, Secondary Wiring. The wiring methods and materials shall be installed in accordance with the sign manufacturer's installation instructions using any applicable wiring methods from Chapter 3 and the requirements for Class 2 circuits contained in Part III of Article 725.

(A) Insulation and Sizing of Class 2 Conductors. Listed Class 2 cable that complies with Table 725.154(G) shall be installed on the load side of the Class 2 power source. The conductors shall have an ampacity not less than the load to be supplied and shall not be sized smaller than 22 AWG.

(1) Wet Locations. Class 2 cable used in a wet location shall be identified for use in wet locations or have a moisture-impervious metal sheath.

(2) Other Locations. In other locations, any applicable cable permitted in Table 725.154(G) shall be permitted to be used.

(... Continued on next page ...)

See the following page for additional information related to this change.

Comments: 18-93, 18-94, 18-95, 18-96, 18-98, 18-99, 18-100

Proposal: 18-249a

LED Sign Illumination Systems, Secondary Wiring (continued)

Code Language

(... Continued from previous page ...)

(B) Installation. Secondary wiring shall be installed in accordance with (B)(1) and (B)(2).

(1) Support wiring shall be installed in a neat and workmanlike manner. Cables and conductors installed exposed on the surface of ceilings and sidewalls shall be supported by the building structure in such a manner that the cable is not be damaged by normal building use. Such cables shall be supported by straps, staples, hangers, cable ties, or similar fittings designed and installed so as not to damage the cable. The installation shall also comply with 300.4(D).

(2) Connections in cable and conductors shall be made with listed insulating devices and be accessible after installation. Where made in a wall, connections shall be enclosed in a listed box.

(C) Protection Against Physical Damage. Where subject to physical damage, the conductors shall be protected and installed in accordance with 300.4.

(D) Grounding and Bonding. Grounding and bonding shall be in accordance with 600.7.

Courtesy of YESCO

Significance of the Change

This new section, in Part II of Article 600, applies specifically to the secondary wiring installed for light-emitting diode (LED) sign illumination systems. Its four subdivisions provide specific installation requirements for the Class 2 secondary wiring of this equipment. It must be recognized that the driving language of 600.33 requires conformance to the manufacturer's installation instructions. Because these systems are required to be listed in accordance with 600.3, they are equipped with specific installation instructions that must be followed. The requirements in this new section address size and installation requirements for circuit conductors and cables in such systems, as well as protection from physical damage. Installation must be completed in a neat and workmanlike manner, and the grounding and bonding requirements of 600.7 apply. Note that a new exception to 600.7(B)(1) relaxes the bonding requirement for metal parts associated with Class 2 secondary circuits, a change that reflects that the *NEC* is now incorporating rules that apply specifically to sign and outline lighting systems powered by Class 2 circuits.

See the previous page for additional information related to this change.

Exception

NECA Photo - © 2010 Rob Colgan

Code Language

620.53 Car Light, Receptacle(s), and Ventilation Disconnecting Means. ***(Unchanged)***

Exception: Where an individual branch circuit supplies car lighting, receptacle(s), and a ventilation motor not exceeding 2 hp, the disconnecting means required by 620.53 shall be permitted to comply with 430.109(C). This disconnecting means shall be listed and shall be capable of being locked in the open position. The provision for locking or adding a lock to the disconnecting means shall be installed on or at the switch or circuit breaker used as the disconnecting means and shall remain in place with or without the lock installed. Portable means for adding a lock to the switch or circuit breaker shall not be permitted as the means required to be installed at the disconnecting means and shall remain with the equipment.

Significance of the Change

This new exception aligns the disconnecting means requirements with other *NEC* provisions that permit the required disconnecting means to be located out of sight from the equipment it serves. The conditions in the exception are that the disconnecting means meet the requirements in 430.109(C) addressing acceptable (listed) types of disconnects. To locate the disconnecting means remote from the equipment, it must be capable of being locked in the open position and the provisions for adding a lock must remain with the switch or circuit breaker, whether or not the lock is installed. Portable means for adding a lock to the disconnecting means are not permitted as the locking provision required for the installation. This concept is quite simple, in that the provisions for adding the lock must be part of the installed equipment which is covered by the *NEC*.

Change Summary

A new exception has been added following 620.53. The exception recognizes a lockable disconnecting means in place of the required in-sight disconnect covered by the general rule.

Comment: None
Proposal: 12-33

Definition of Electric Vehicle

Code Language

625.2 Definitions.

Electric Vehicle. An automotive-type vehicle for on-road use, such as passenger automobiles, buses, trucks, vans, neighborhood electric vehicles, electric motorcycles, and the like, primarily powered by an electric motor that draws current from a rechargeable storage battery, fuel cell, photovoltaic array, or other source of electric current. Plug-in hybrid electric vehicles (PHEV) are considered electric vehicles. For the purpose of this article, off-road, self-propelled electric vehicles, such as industrial trucks, hoists, lifts, transports, golf carts, airline ground support equipment, tractors, boats, and the like, are not included.

Change Summary

A new second sentence has been added to this definition as follows: "Plug-in hybrid electric vehicles (PHEV) are considered electric vehicles."

I-Stock Photo, Courtesy of NECA

Significance of the Change

This revision makes clear that hybrid vehicles incorporating electric power are considered electric vehicles and that all rules pertaining to electric vehicles in Article 625 are applicable. Information in the substantiation for this revision indicated that this industry has developed a new type of vehicle with many similarities to electric vehicles with respect to their connection to a supply of electricity for vehicle charging. The revision clarifies that hybrid vehicles that include onboard electric power and charging capabilities are governed by the rules of Article 625. Plug-in hybrid electric vehicles are an alternative and equivalent type of supply equipment for wiring systems. The new definition of Plug-in Hybrid Electric Vehicle (PHEV) indicates that these vehicles are intended for on-road use, with the ability to store and use off-vehicle electric energy in the rechargeable energy storage system. The PHEV also has a second source of motive power. Electric motorcycles are now included in the definition of an electric vehicle.

Comment: 12-26
Proposal: 12-43

Definition of Electric Vehicle Inlet

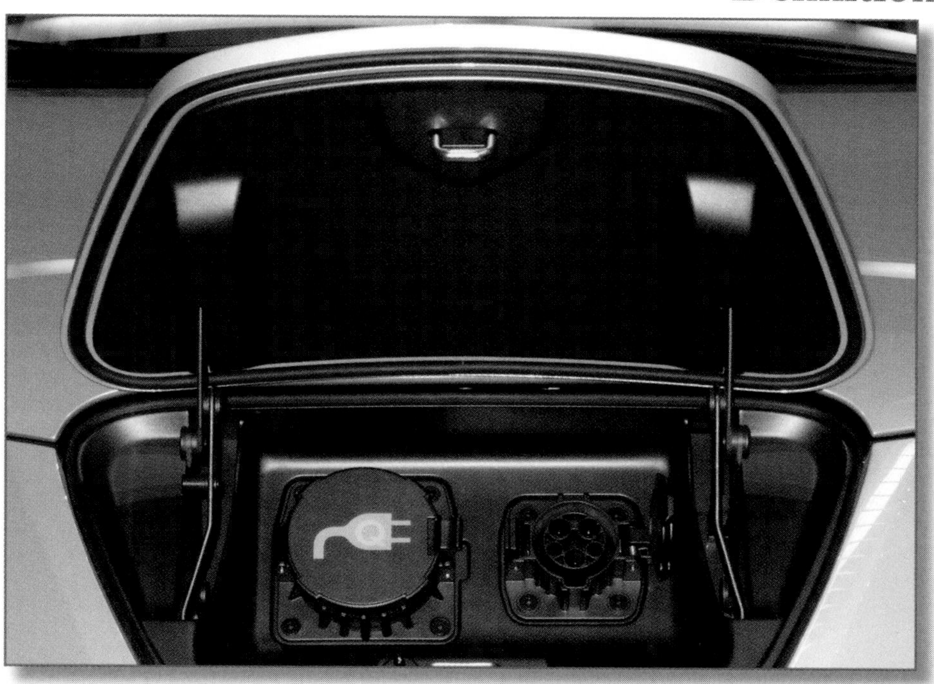

I-Stock Photo, Courtesy of NECA

Code Language

625.2 Definitions.

Electric Vehicle Inlet. The device on the electric vehicle into which the electric vehicle connector is inserted for power transfer and information exchange. This device is part of the electric vehicle coupler. For the purposes of this *Code*, the electric vehicle inlet is considered to be part of the electric vehicle and not part of the electric vehicle supply equipment.

Informational Note: For further information, see 625.26 for interactive systems.

Significance of the Change

This revision to the definition of *Electric Vehicle Inlet* acknowledges the bi-directional energy transfer as already permitted and addresses in 625.26, for interactive systems. This revision clarifies that the purpose of the electrical vehicle inlet is for uses beyond just charging capabilities. Power is permitted to be provided in both directions through the inlet connection of some plug-in hybrid electric vehicles. Note that these types of vehicles with onboard capabilities as interactive power sources are required to be listed. When in the charging mode, these vehicles consume power. When used as standby power sources, they provide power and are limited to use as optional standby systems. These special systems and equipment must also meet the applicable requirements in Article 705 covering interconnected power production sources.

Change Summary

The term *power transfer* has been added to this definition along with an informational note that refers users to 625.26.

Comment: None

Proposal: 12-46

625.2

Article 625 Electric Vehicle Charging System
Part I General

Definition of Plug-in Hybrid Electric Vehicle (PHEV)

Code Language

625.2 Definitions.

Plug-in Hybrid Electric Vehicle (PHEV). A type of electric vehicle intended for on-road use with the ability to store and use off-vehicle electrical energy in the rechargeable energy storage system, and having a second source of motive power.

Courtesy of Eaton Corporation

Change Summary

A new definition of the term *Plug-in Hybrid Electric Vehicle* (PHEV) has been added to 625.2, and a new sentence has been added to the current definition of Electric Vehicle, as follows: "Plug-in hybrid electric vehicles (PHEV) are considered electric vehicles."

Significance of the Change

The motor vehicle industry has developed a new type of vehicle with many similarities to electric vehicles with respect to their connection to a supply of electricity for vehicle charging. This revision identifies plug-in hybrid electric vehicle supply equipment as an alternative and equivalent type of supply equipment for premises wiring systems. This new definition is included in 625.2 and referenced in the scope of Article 625, to clarify that these types of vehicles are covered by the rules in this article. This definition indicates that not only does a plug-in hybrid electric vehicle include an onboard source of motive power, but the power storage capabilities can also be used as an alternative power source to supply wiring systems. These vehicles include a recharging system for the onboard battery storage units. A new definition of the term *Rechargeable Energy Storage System* has also been added to 625.2 in addition to the revision of the existing term *Electric Vehicle*. The revisions in 625.2 are a coordinated set of new definitions and changes that incorporate plug-in hybrid electric vehicles in the requirements of Article 625.

Comments: 12-24, 12-26

Proposals: 12-42a, 12-43

Definition of Rechargeable Energy Storage System

I-Stock Photo, Courtesy of NECA

Code Language

625.2 Definitions.

Rechargeable Energy Storage System. Any power source that has the capability to be charged and discharged.

Informational Note: Batteries, capacitors, and electro mechanical flywheels are examples of rechargeable energy storage systems.

Significance of the Change

The term *rechargeable energy storage system* is used within the definition of Plug-in Hybrid Electric Vehicle (PHEV), necessitating its explanatory definition in 625.2. This new definition is required for application to electric vehicles that are charged and discharged and lists batteries, capacitors, and electro mechanical flywheels as examples of other technologies that could be used as energy storage systems. This definition also aligns with SAE J1715, *Hybrid Electric Vehicle (HEV) & Electric Vehicle (EV) Terminology*.

This definition clarifies that the onboard batteries of electric vehicles can be used for more than just the motive functions of a vehicle, for example, to supply standby power systems for premises wiring. This new definition clearly indicates that this battery's purpose and usage is energy storage without limiting its use of energy to powering a vehicle. The expanded use of vehicle energy storage systems will necessitate additional *NEC* rules and conformance to current requirements that apply to interconnected electrical power production sources.

Change Summary

A new definition for the term *Rechargeable Energy Storage System* has been added to 625.2. This term appears in the new definition of the term *Plug-in Hybrid Electric Vehicle* (PHEV). A new informational note describes examples of rechargeable energy storage systems.

Comment: 12-25

Proposals: 12-50, 12-51, 12-52

Feeder and Service Load Calculations

Code Language

626.11. Feeder and Service Load Calculations.

(A) ***(Unchanged)***

(B) ***(Unchanged)***

(C) ***(Unchanged)***

(D) Conductor Rating. Truck space branch-circuit supplied loads shall be considered to be continuous.

Change Summary

Section 626.11(D) has been revised to indicate that the "branch circuit-supplied loads shall be considered to be continuous" and to remove the text that indicated the branch circuit conductors had to have an ampacity not less than the loads supplied.

Significance of the Change

This revision clarifies the sizing requirements for conductors supplying truck spaces. Not only must the branch circuit be capable of supplying the load served, as previously required, but the branch circuit-supplied loads are to be considered as continuous loads for calculation purposes. Thus the conductors are required to be calculated with an ampacity of not less than 125 percent of the load supplied. The term *Continuous Load* is defined in Article 100 as a load where the maximum current is expected for 3 hours or more, which is certainly the case for electrified truck parking spaces covered in Article 626.

Comment: None

Proposal: 12-74

Disconnecting Means

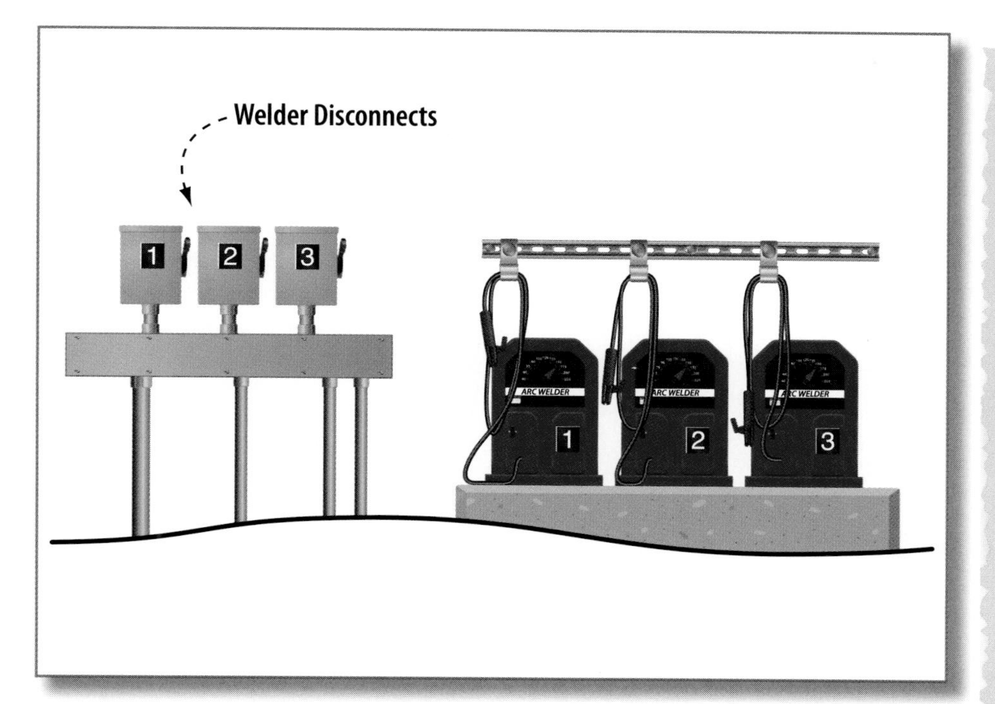

Code Language

630.13 Disconnecting Means. An identified disconnecting means shall be provided in the supply circuit for each arc welder that is not equipped with a disconnect mounted as an integral part of the welder.

The disconnecting means shall be a switch or circuit breaker, and its rating shall be not less than that necessary to accommodate overcurrent protection as specified under 630.12.

Significance of the Change

Adding the word *identified* in this section underscores the requirement that the disconnecting means for each arc welder that is not otherwise so equipped with an integral disconnect must identify which arc welder it supplies. It is not uncommon to have welders installed in groups with disconnecting means that are field-installed, not retrofitted as an integral part of the welder itself. In these types of installations, the disconnecting means in each welder supply circuit must clearly identify the welder it supplies. This revision enhances the safety of workers and service personnel who use or work on this type of equipment.

Change Summary

The word *identified* has been inserted before the term *disconnecting means* in 630.13. The identification requirement will assist personnel in identifying the appropriate means to disconnect supply power to an individual welder that may be located among a group of welders.

Comment: None

Proposal: 12-107

640.45

Article 640 Audio Signal Processing, Amplification, and Reproduction Equipment
Part III Portable and Temporary Audio System Installations

Protection of Wiring

Code Language

640.45 Protection of Wiring. Where accessible to the public, flexible cords and cables laid or run on the ground or on the floor shall be covered with approved nonconductive mats. Cables and mats shall be arranged so as not to present a tripping hazard. The cover requirements of 300.5 shall not apply to wiring protected by burial.

Change Summary

A new last sentence has been added to 640.45 as follows: The cover requirements of 300.5 shall not apply to wiring protected by burial.

Significance of the Change

The significance of this revision is that an exemption from the minimum burial depths required by 300.5 is now in place for audio circuit conductors and cables that are installed in temporary or portable conditions. Information in the substantiation for change indicated that the proposed revision would incorporate the currently accepted allowances of comparable applications at carnivals, as covered in 525.20(G). It should be understood that 640.45 is included within Part III of Article 640, which applies only to temporary and portable audio installations. These permitted uses are similar for audio cable, for which the permission for shallow burial is well suited, especially for temporary wiring applications. This revision provides consistency with the rules for installations presently covered under Article 525 for temporary underground installations of power conductors, except the allowance applies to temporary audio circuits with limited energy.

Comment: None
Proposal: 12-120

645.1

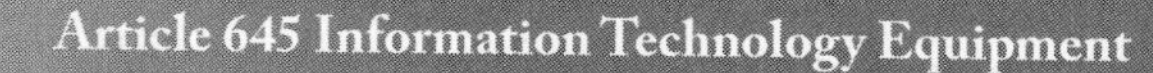

Scope

I-Stock Photo, Courtesy of NECA

Code Language

645.1 Scope. This article covers equipment, power-supply wiring, equipment interconnecting wiring, and grounding of information technology equipment and systems in an information technology equipment room.

Informational Note: For further information, see NFPA 75-2009, *Standard for the Protection of Information Technology Equipment*, which covers the requirements for the protection of information technology equipment and information technology equipment areas.

Change Summary

The words *and systems* have been added to 645.1, and an informational note following 645.1 provides a useful reference to NFPA 75, *Standard for the Protection of Information Technology Equipment*. The revision to the informational note provides a short description of what is contained in and specifically covered by NFPA 75.

Significance of the Change

The informational note provides users with a reference to NFPA 75, *Standard for the Protection of Information Technology Equipment*. This standard provides designers and installers with additional useful information that should be used in IT equipment room design and applications. Article 645 of the *NEC* can be applied to information technology rooms and installations only if all of the requirements in 645.4 have been met. The reference to NFPA 75 serves to assist those wishing to design and operate IT rooms in accordance with the provisions of Article 645, instead of requiring compliance with all the applicable requirements in the *NEC*, including those in Chapters 1 through 4.

Comment: 12-71

Proposals: 12-121, 12-122

Definitions

Code Language

645.2 Definitions.

Critical Operations Data System. An information technology equipment system that requires continuous operation for reasons of public safety, emergency management, national security, or business continuity.

Remote Disconnect Control. An electric device and circuit that controls a disconnecting means through a relay or equivalent device.

Zone. A physically identifiable area (such as barriers or separation by distance) within an information technology equipment room, with dedicated power and cooling systems for the information technology equipment or systems.

Change Summary

Three new definitions have been added to 645.2: Critical Operations Data System, Remote Disconnect Control, and Zone. Additionally, 645.10 has been expanded and restructured into two subdivisions covering remote disconnecting means controls and critical operations data systems.

Comment: None

Proposal: 12-129

Significance of the Change

Action by CMP-12 on Proposal 12-129 resulted in clarified provisions and three new definitions for 645.10. These definitions are necessary for conformance to the *NEC Style Manual* and for application of new subdivisions (A) and (B) of 645.10. The new definitions provide users with a clear description of what constitutes critical operations data systems, remote disconnect controls, and zones. While Article 708, new in the 2008 *NEC*, provided new requirements governing critical operations power systems, none of its provisions related to critical data systems. Subdivision (B) of 645.10 now includes requirements for eliminating remote control disconnecting means in critical data systems, for which the concerns are great for critical data system loss and disruption.

The new definition of *Remote Disconnect Control* distinguishes between the disconnecting means and the control circuits that cause the power disconnect to operate. The 2008 *NEC* introduced the concept of separate zones for cooling systems in an IT application, in Article 645, yet no definitive criteria defining what constitutes a zone were provided. The new definition makes clear that a zone is a separated portion of a room such that power only within that portion of the zone can be shut down and so that fire or products of combustion cannot be spread to adjacent areas or other zones within the IT room. At the same time, other separate zones in the IT room can remain operational.

645.2

Article 645 Information Technology Equipment

Definition of Information Technology Equipment

I-Stock Photo, Courtesy of NECA

Code Language

645.2 Definitions.

Information Technology Equipment (ITE). Equipment and systems rated 600V or less, normally found in offices or other business establishments and similar environments classified as ordinary locations, that are used for creation, and manipulation of data, voice, video, and similar signals that are not communications equipment, as defined in Part I of Article 100 and do not process communications circuits as defined in 800.2.

Informational Note: For information on listing requirements for both information technology equipment and communications equipment, see UL 60950-1, *Information Technology Equipment - Safety - Part 1: General Requirements*.

Significance of the Change

This new definition is essential for distinguishing between information technology equipment and communications equipment, to effectively determine which article applies to an installation. The new definition of Information Technology Equipment is important because it provides a differentiation between communications equipment and circuits that are part of the IT room installation. As a result, the communications equipment and circuits within an IT room are part of the equipment and are covered by the rules in Article 645. It is also likely that both communications equipment and circuits, along with information technology equipment and interconnecting circuits, are installed in the same IT room or data center. It is important to emphasize that all the requirements in 645.4 must be complied with before Article 645 can be used for IT installations. Otherwise, application of all other requirements in the *NEC*, including those in Chapters 1 through 4 and 5 through 8, would be required as necessary.

Change Summary

A new definition of the term *Information Technology Equipment* (ITE) has been added to 645.2. A new informational note referencing UL 60950-1, *Information Technology Equipment - Safety - Part 1: General Requirements*, follows the definition.

Comment: 12-74

Proposal: 12-126

Definition of Information Technology Equipment Room

Code Language

645.2 Definitions.

Information Technology Equipment Room. A room within the information technology equipment area that contains the information technology equipment [75:3.3.9].

I-Stock Photo, Courtesy of NECA

Change Summary

A new definition of the term *Information Technology Equipment Room* has been added to 645.2.

Significance of the Change

The term *information technology equipment room* appears several times throughout Article 645. In accordance with 2.2.2.2 of the *NEC Style Manual*, definitions of terms used within an article must be provided in the x.2 section of that article. The new definition of this term is extracted from NFPA 75-2009, *Standard for the Protection of Information Technology Equipment*. This definition provides consistent correlation with the defined term in NFPA 75 and simply describes what constitutes an information technology equipment room: a room intended for placement and operation of information technology equipment.

Comment: None

Proposal: 12-127

Other Articles

Code Language

645.3 Other Articles. Circuits and equipment shall comply with 645.3(A) through (G), as applicable.

(A) Spread of Fire or Products of Combustion

(B) Plenums

(C) Grounding

(D) Electrical Classification of Data Circuits

(E) Fire Alarm Equipment

(F) Communications Equipment

Informational Note: See Part I of Article 100, Definitions, for the definition of communications equipment.

(G) Community Antenna Television and Radio Distribution Systems Equipment

Change Summary

Section 645.3 Other Articles, composed of seven subdivisions and one informational note, has been added to Article 645. This revision clarifies which other articles apply to information technology equipment installations.

Significance of the Change

This revision creates correlation between Article 645 and other applicable requirements related to the spread of fire and products of combustion. In addition, the relationship between Article 645 and Chapters 5, 6, and 7 has been clarified by the addition of references to rules in those chapters. A specific subdivision (B) now addresses plenums and provides references to other articles that include requirements applicable to all wiring installed in plenums. Specific grounding rules apply to the non-current-carrying conductive members of fiber optic cables. Subdivision (C) provides a reference to 770.101 for such specific grounding requirements. New subdivision (D) clarifies the electrical classification of Class 2 and Class 3 circuits in the same cable assembly containing communications circuits.

Fire alarm systems and equipment in information technology rooms must comply with the applicable rules in Article 760. Subdivisions (F) and (G) provide references to Articles 800 and 820, respectively. The rules in Article 800 apply to communications equipment installed in IT rooms, while Article 645 applies to the powering of communications equipment in IT rooms. Article 820 applies to community antenna television and radio distribution systems and equipment installed in IT rooms. Article 645 applies to community antenna television and radio distribution systems and equipment powered from an IT room.

Comments: 12-77, 12-81
Proposal: 12-132

Special Requirements for Information Technology Equipment Room

Code Language

645.4 Special Requirements for Information Technology Equipment Room. This article shall be permitted to provide alternate wiring methods to the provisions of Chapters 1 through 4 for power wiring, 725.154 for signaling wiring, and 770.113(C) and Table 770.154(a) for optical fiber cabling when all of the following conditions are met:

(1) ***(Unchanged)***

(2) ***(Unchanged except editorial)***

Informational Note: For further information, see NFPA 75-2009, *Standard for the Protection of Information Technology Equipment*, Chapter 10, 10.1, 10.1.1, 10.1.2, and 10.1.3.

(3) All information technology and communications equipment installed in the room is listed.

(... Continued on next page ...)

I-Stock Photo, Courtesy of NECA

Change Summary

This section has been revised to improve clarity and usability. The existing informational notes have been revised to reflect the current edition of NFPA 75 and the applicable sections where appropriate. Additionally, a new list item (6) and associated informational note have been added to this section. The new list item (6) clarifies that only equipment and wiring related to the operation of the information technology room is installed within that room.

Special Requirements for Information Technology Equipment Room

Significance of the Change

The driving text of this section has been revised to clarify that Article 645 is permitted to apply when all the conditions provided in list items (1) through (6) have been met. The alternatives provided in Article 645 are permitted for power wiring, 725.154 for signaling wiring, and 770.113(C) and Table 770.154(a) for optical fiber cabling, when all requirements in list items (1) through (6) have been met. Additionally, the existing informational notes have been revised to reflect the current edition of NFPA 75 and the applicable sections or chapters of the standard as appropriate. List item (3) has been expanded to include communications equipment in the room requiring listing. List item (4) has been revised to restrict unauthorized persons from occupying the room, to make clear that the room is to be occupied only by personnel needed for the maintenance and functional operation of the installed information technology equipment.

New list item (6) clarifies that only electrical equipment and wiring associated with operation of the information technology room is permitted in the room. This revision emphasizes the purpose of the room and its sole dedication to information technology systems and equipment and the operation of information technology systems contained therein. The new informational note following new list item (6) provides users with a list of items associated with the operation of an information technology room, such as those systems required for security, environmental stability, and building safety systems.

Code Language

(... Continued from previous page ...)

(4) The room is occupied by, and accessible to, only those personnel needed for the maintenance and functional operation of the installed information technology equipment.

(5) ***(Unchanged)***

Informational Note: For further information on room construction requirements, see NFPA 75-2009, *Standard for the Protection of Information Technology Equipment*, Chapter 5.

(6) Only electrical equipment and wiring associated with the operation of the information technology room is installed in the room.

Informational Note: HVAC systems, communications systems, and monitoring systems such as telephone, fire alarm systems, security systems, water detection systems, and other related protective equipment are examples of equipment associated with the operation of the information technology room.

(Note: See NEC Section 645.4 for complete text.)

Comment: 12-83

Proposals: 12-134, 12-135

645.5

Article 645 Information Technology Equipment

Supply Circuits and Interconnecting Cables

Code Language

645.5 Supply Circuits and Interconnecting Cables.

(A) Branch-Circuit Conductors. The branch-circuit conductors supplying one or more units of information technology equipment shall have an ampacity not less than 125 percent of the total connected load.

(B) Power Supply Cords. Information technology equipment shall be permitted to be connected to a branch circuit by a power supply cord.

(1) Power supply cords shall not exceed 4.5 m (15 ft).

(2) Power cords shall be listed and a type permitted for use on listed information technology equipment or shall be constructed of listed flexible cord and listed attachment plugs and cord connectors of a type permitted for information technology equipment.

Informational Note: One method of determining if cords are of a type permitted for the purpose is found in UL 60950-1, *Safety of Information Technology Equipment - Safety - Part 1: General Requirements.*

(C) Interconnecting Cables. Separate information technology equipment units shall be permitted to be interconnected by means of listed cables and cable assemblies. The 4.5 m (15 ft) limitation in 645.5(B)(1) shall not apply to interconnecting cables.

(D) Physical Protection. Where exposed to physical damage, supply circuits and interconnecting cables shall be protected.

(E) Under Raised Floors. Power cables, communications cables, connecting cables, interconnecting cables, cord-and-plug connections, and receptacles associated with the information technology equipment shall be permitted under a raised floor, provided the following conditions are met:

(1) The raised floor is of approved construction, and the area under the floor is accessible.

(2) The branch-circuit supply conductors to receptacles or field-wired equipment are in rigid metal conduit, rigid nonmetallic conduit, intermediate metal conduit, electrical metallic tubing, electrical nonmetallic tubing, metal wireway, nonmetallic wireway, surface metal raceway with metal cover, nonmetallic surface raceway, flexible metal conduit, liquidtight flexible metal conduit, or liquidtight flexible nonmetallic conduit, Type MI cable, Type MC cable, or Type AC cable and associated metallic and nonmetallic boxes or enclosures. These supply conductors shall be installed in accordance with the requirements of 300.11.

(Only editorial changes in list items (3) through (6) and in Subdivision F.)

(G) Abandoned Supply Circuits and Interconnecting Cables. The accessible portion of abandoned supply circuits and interconnecting cables shall be removed unless contained in a raceway.

(See NEC for full text of this section.)

Supply Circuits and Interconnecting Cables (continued)

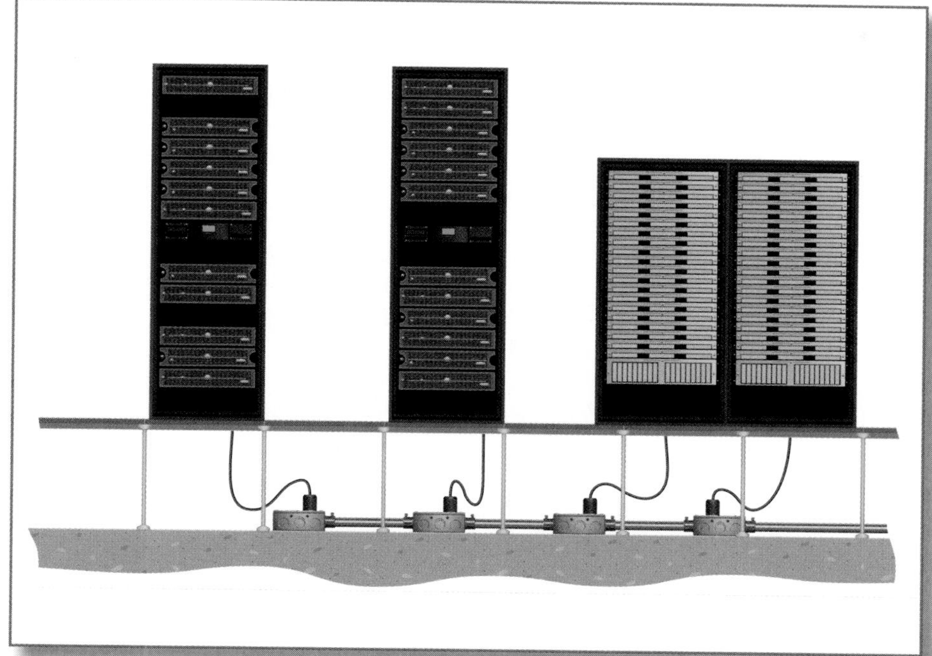

Change Summary

The term *information technology equipment* has been added to subdivisions (A), (B), and (C), and the term *flexible supply cords* has replaced *flexible cord* in subdivision (B). Additionally, subdivision (D) Physical Protection has been added, while the remaining subdivisions have been re-identified for correct alphanumeric sequence. The phrase "and associated metallic and nonmetallic boxes or enclosures" has been added to (E)(2), and the word *metal* has been deleted from subdivision (G), former subdivision (F).

Significance of the Change

This section has been revised to improve usability and clarity and to specifically address the requirements for power supply cords used with listed information technology equipment. The term *data processing system* has been replaced with *information technology equipment* in a few locations for consistency. Subdivision (B)(2) has been revised to clarify that power supply cords can be used for IT equipment but must be listed and of a type for that specific use. The new informational note following subdivision (B) provides a method of verification of suitability of flexible cords for use in supplying IT equipment. The informational note references UL 60950-1, *Safety of Information Technology Equipment - Safety - Part 1: General Requirements*. The attachment plugs and cord connectors are also required to be listed for information technology equipment.

In subdivision (C) the revision clarifies that the 4.5 ft length limitation for power supply cords does not apply to interconnecting cables of IT equipment. The revision to (E)(2) clarifies that not only are the wiring methods provided in this list, but any associated metallic or nonmetallic boxes and enclosures are also now included. This revision recognizes the outlet boxes that are installed with these wiring methods for the branch circuits installed for information technology equipment. Deleting the word *metal* from subdivision (G) allows the use of metallic or nonmetallic raceways for abandoned cables that are permitted to remain under the raised floor of an information technology room. The remaining changes to this section are editorial in nature to accommodate the numbering changes.

Comment: 12-89

Proposals: 12-139, 12-143, 12-145

Disconnecting Means

Code Language

645.10 Disconnecting Means. An approved means shall be provided to disconnect power to all… ***(the remainder unchanged except for the addition of a new last sentence).*** The disconnecting means shall be implemented by either (A) or (B).

Exception: Installations qualifying under the provisions of Article 685.

(A) Remote Disconnect Controls.

(1) Remote disconnect controls shall be located at approved locations readily accessible in case of fire to authorized personnel and emergency responders.

(2) The remote disconnect controls for the control of electronic equipment power and HVAC systems shall be grouped and identified. A single means to control both systems shall be permitted.

(3) Where multiple zones are created, each zone shall have an approved means to confine fire or products of combustion to within the zone.

(4) Additional means to prevent unintentional operation of remote disconnect controls shall be permitted.

Informational Note: For further information, see NFPA 75-2009, *Standard for the Protection of Information Technology Equipment.*

(B) Critical Operations Data Systems. Remote disconnecting controls shall not be required for critical operations data systems when all of the following conditions are met:

(1) An approved procedure has been established and maintained for removing power and air movement within the room or zone.

(2) Qualified personnel are continuously available to meet emergency responders and to advise them of disconnecting methods.

(3) A smoke-sensing fire detection system is in place.

Informational Note: For further information, see NFPA 72-2007, *National Fire Alarm Code.*

(4) An approved fire suppression system suitable for the application is in place.

(5) Cables installed under a raised floor, other than branch circuit wiring and power cords installed in compliance with 645.5(D)(2) or (D)(3), or in compliance with 300.22(C), 725.154(A), 770.113(C) and Table 770.154(a), 800.113(C) and Table 800.154(a), or 820.113(C) and Table 820.154(a).

Disconnecting Means (continued)

Courtesy of Cogburn Brothers, Inc.

Change Summary

Section 645.10 has been expanded and restructured into two subdivisions covering remote disconnecting means controls and critical operations data systems, and a new last sentence has been added to this section. This expansion of 645.10 will now provide direction for users about disconnecting means for critical operations data systems.

Significance of the Change

Action by CMP-12 on Proposal 12-129 resulted in new definitions and in a revised and expanded 645.10. The new definitions are necessary for conformance to the *NEC Style Manual* and for application to the terms used in new (A) and (B) of 645.10. These new definitions provide users with a clear description of what constitutes critical operations data systems, remote disconnect controls, and zones. Subdivision (A) addresses installation requirements and operational characteristics of remote disconnecting means controls. These are control devices such as panic buttons that actuate the disconnecting means through control circuit wiring. The revised section indicates that the remote disconnecting means location must be an approved one and that grouping and identification of the electronic equipment and HVAC system disconnecting means are required where more than a single disconnecting means are installed.

Subdivision (B) now includes conditions under which remote control disconnecting means may be eliminated in critical data systems. The concerns are for critical data system loss and disruption. Five specific conditions must be met before the provisions in (B) can be applied and the remote disconnecting means control can be eliminated. Two new informational notes provide references to NFPA 75, *Standard for the Protection of Information Technology*, and NFPA 72, *National Fire Alarm Code*, both of which contain the necessary provisions for constructing information technology equipment rooms and providing suitable protection for the building and operators of such facilities.

Comment: 12-92

Proposal: 12-129

645.25

Article 645 Information Technology Equipment

Engineering Supervision

Code Language

645.25 Engineering Supervision. As an alternative to the feeder and service load calculations required by Parts III and IV of Article 220, feeder and service load calculations for new or existing loads shall be permitted to be used if performed by qualified persons under engineering supervision.

Change Summary

A new section, 645.25 Engineering Supervision, has been added to Article 645. In addition, Table 220.3 has been revised to include the reference to Article 645, to correlate with the new calculation alternative provided in Article 645.

Significance of the Change

This revision provides the alternative of using engineering supervision in the process of calculating feeders and service loads associated with information technology room installations. This provision is applicable for both new and existing installations. The option to use engineering supervision for these calculations is conditioned on the process being performed by qualified persons under engineering supervision. This alternative relaxes the requirement of Parts III and IV of Article 220 for new and existing installations. If the alternative offered by 645.25 is not selected for calculating new or existing loads associated with IT equipment room installations, the requirements in Parts III and IV of Article 220 would be applicable as a general rule.

Comment: 12-93

Proposal: 12-148

Short-Circuit Current Rating

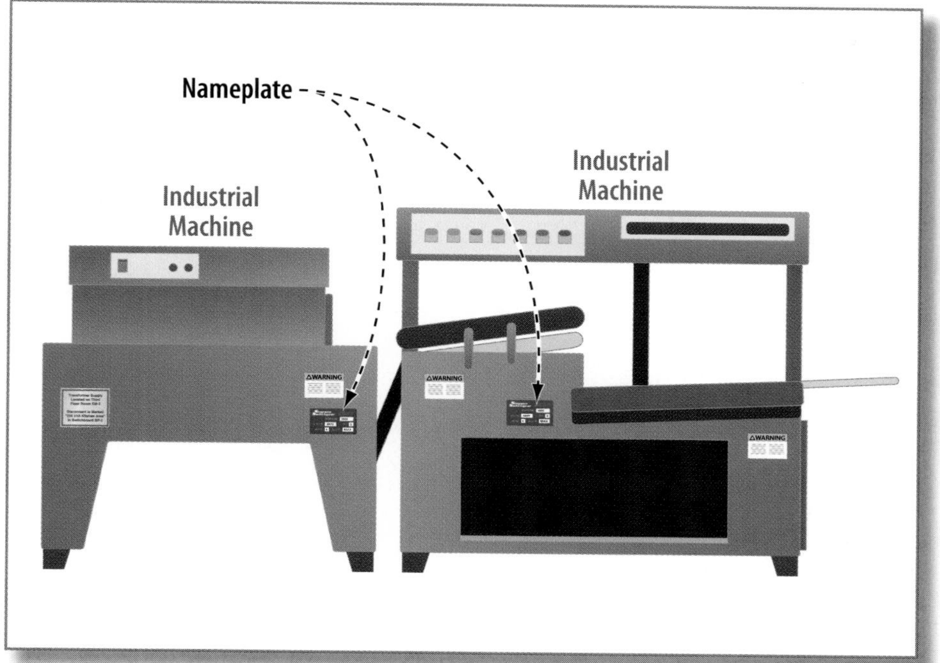

Code Language

670.5 Short-Circuit Current Rating. Industrial machinery shall not be installed where the available fault current exceeds its marked short-circuit current rating as marked in accordance with 670.3(A)(4).

Significance of the Change

This new section correlates with and clarifies the short-circuit current rating requirements of 670.3(A)(4). This requirement indicates that industrial machinery is not permitted to be installed on a system if the amount of available fault current exceeds the equipment ratings. This is also a general requirement in 110.9. Section 670.3 regulates only the label data and its required markings, whereas this new rule addresses limits of use and installation of industrial machinery on circuits that exceed the overall short-circuit current rating of the equipment. Information in the substantiation also pointed out similar requirements within the Article 285 of the *NEC* and Section 4.8 of NFPA 79-2007, *Electrical Standard for Industrial Machinery*.

Change Summary

Section 670.5, Short-Circuit Current Rating, has been added to Article 670. This new requirement correlates with the general rule in 110.9 covering equipment to be applied on systems that do not exceed the amount of available fault-current ratings of such equipment.

Comment: 12-111

Proposal: 12-177

Definition of Dry-Niche Luminaire

Code Language

680.2 Definitions.

Dry-Niche Luminaire. A luminaire intended for installation in the floor or wall of a pool, spa, or fountain in a niche that is sealed against the entry of water.

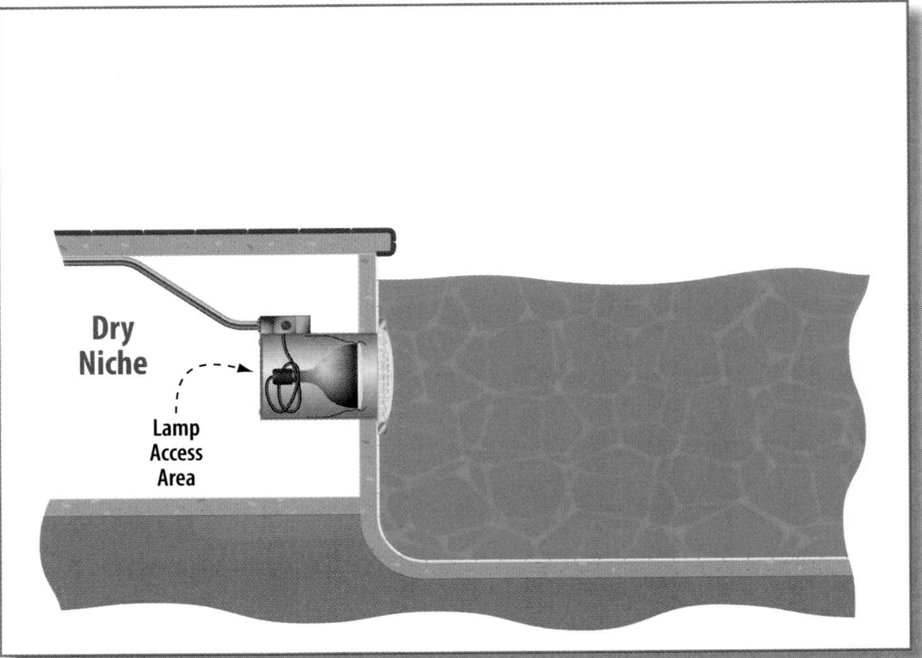

Change Summary

The words "the floor" and "spa" have been added to the definition of *Dry-Niche Luminaire* in 680.2.

Significance of the Change

The 2008 definition of *Dry-Niche Luminaire* was limited to application in pools and fountains, and to installation in the walls of such equipment or installations. This definition has been expanded to cover spas in addition to pools and fountains. Information in the substantiation pointed out that many new designs of equipment extend beyond pools to spas or other similar bodies of water covered by Article 680. Many of these new designs also include dry-niche luminaires installed in the floor of such equipment. The definition as revised provides consistency with actual applications of dry-niche luminaires in pools, spas, and fountains, and it no longer addresses those types of dry-niche luminaires installed only in the walls of such equipment.

Comment: None

Proposal: 17-90

Definition of Low Voltage Contact Limit

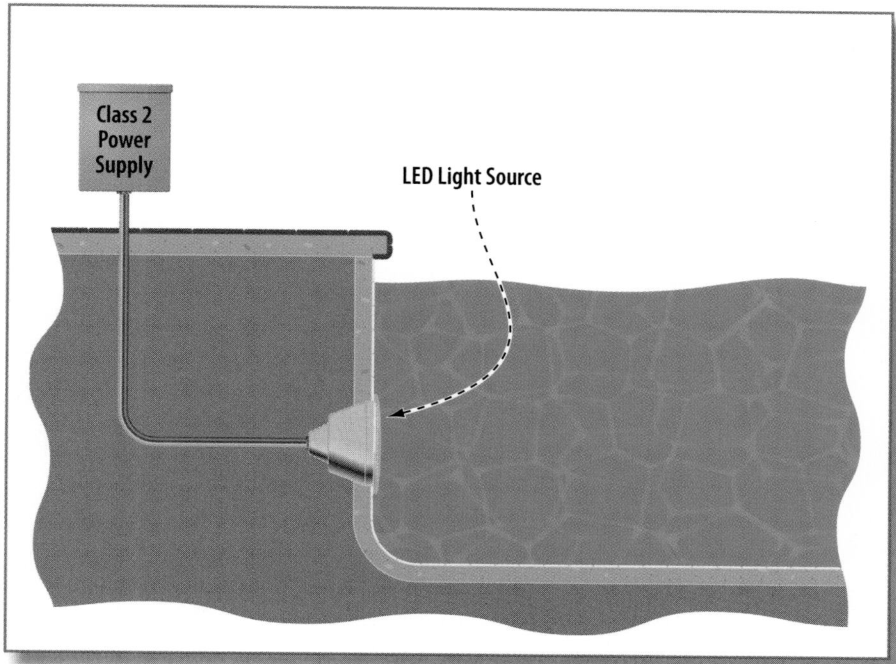

Code Language

680.2 Definitions.

Low Voltage Contact Limit. A voltage not exceeding the following values:

(1) 15 volts (RMS) for sinusoidal ac.

(2) 21.2 volts peak for nonsinusoidal ac.

(3) 30 volts for continuous dc.

(4) 12.4 volts peak for dc that is interrupted at a rate of 10 to 200 Hz.

Significance of the Change

Substantiation provided with Proposal 17-106 identified the need for this new term, associated with emerging lighting technologies that would eventually be more commonly applied in underwater designs and equipment covered by Article 680 of the *NEC*. Energy-saving technologies such as light-emitting diodes (LEDs) require power supplies that produce other than conventional sinusoidal AC, such as continuous DC power supplies and other switch mode power supplies. The voltage values in the Article 680 rules are based on sinusoidal AC voltages. The same wet contact voltage values for sinusoidal AC, nonsinusoidal AC, continuous DC, and switched mode DC formerly were provided in Article 725 but have been relocated to Table 11(A) Note 2 and Table 11(B) Note 4. These tables are used solely for listing references and are related to requirements within Article 725 for use of listed equipment (power supplies). While Article 680 does include many requirements for listed equipment, some installation rules still apply to equipment that may or may not be listed. Including these voltage values and the new definition provides useful descriptive information for new requirements within Article 680 that address low-voltage contact limits.

Change Summary

A new definition of the term *Low Voltage Contact Limit* has been added to 680.2.

Comment: 17-41

Proposal: 17-106

GFCI Protection

Code Language

(C) GFCI Protection. Outlets supplying pool pump motors connected to single-phase, 120 volt through 240 volt branch circuits, rated 15 or 20 amperes, whether by receptacle or by direct connection, shall be provided with ground-fault circuit-interrupter protection for personnel.

Change Summary

Subdivision (B) of 680.22 formerly addressed requirements for ground-fault circuit-interrupter protection for pool pump motors. This requirement has been relocated as new subdivision (C) under 680.21, which covers requirements for motors. Additionally, the phrase "120 volt or 240 volt" has been changed to "120 volt through 240 volt."

Significance of the Change

Ground-fault circuit-interrupter protection is required for certain pool pump motors in accordance with 680.21(C) [formerly 680.22(B)]. The revision to this rule expands the GFCI requirement to all motors falling within the range of 120 volts to 240 volts. As previously worded, the GFCI requirement literally did not apply to 200-, 208-, or 220-volt motors. This revision clarifies the requirement to provide ground-fault circuit-interrupter protection for all outlets supplying pool pump motors that are protected by either 15- or 20-ampere overcurrent protective devices. This GFCI requirement applies to all motors in this voltage and ampere range that are either directly connected or connected by a cord and attachment plug through a receptacle.

Comment: 17-46
Proposal: 17-126

680.23(A)(3)

Article 680 Swimming Pools, Fountains, and Similar Installations
Part II Permanently Installed Pools

GFCI Protection, Relamping

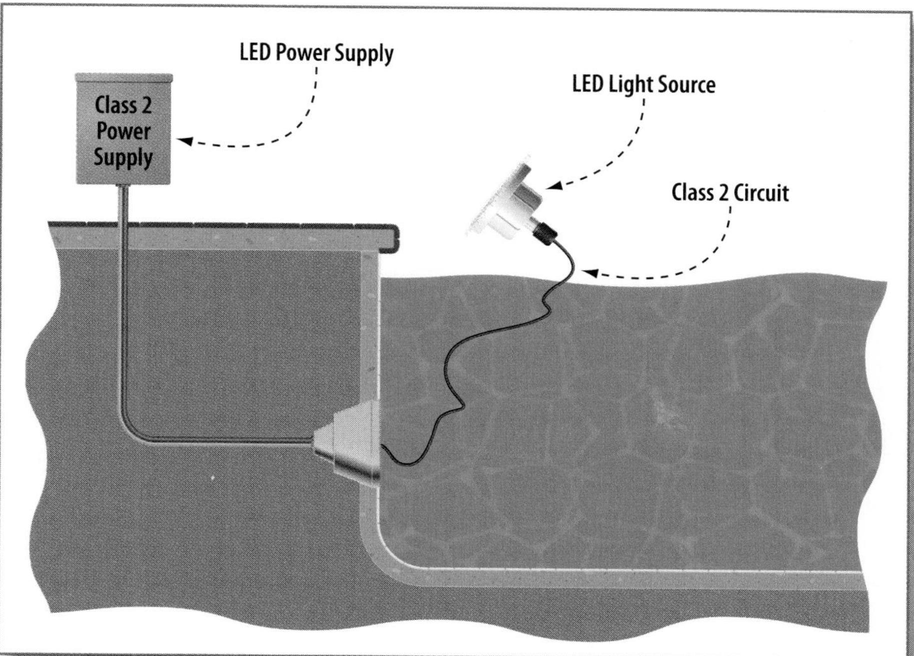

Code Language

(3) GFCI Protection, Relamping. A ground-fault circuit interrupter shall be installed in the branch circuit supplying luminaires operating at more than the low-voltage contact limit such that there is no shock hazard during relamping. The installation of the ground-fault circuit interrupter shall be such that there is no shock hazard with any likely fault-condition combination that involves a person in a conductive path from any ungrounded part of the branch circuit or the luminaire to ground.

Significance of the Change

This section continues to require ground-fault circuit-interrupter protection for personnel during underwater luminaire relamping operations for those situations where a shock hazard exists. This change removes the previous voltage threshold of 15 volts AC and provides a clear reference to the "low voltage contact limit" as defined in 680.2. The revisions to this section recognize new emerging technologies such as light-emitting diode (LED) lighting systems that operate on other than conventional sinusoidal AC voltages. By including the reference to the low-voltage contact limits as defined in 680.2, the concerns during relamping operations are addressed as not being shock hazards until these defined values are exceeded.

Change Summary

The words "15 volts" have been removed from this section and replaced by the words "the low voltage contact limit." The revisions to this section correlate with the new defined term *Low Voltage Contact Limit* in 680.2.

Comment: 17-58
Proposal: 17-139

Compliance

Code Language

(8) Compliance. Compliance with these requirements shall be obtained by the use of a listed underwater luminaire and by installation of a listed ground-fault circuit interrupter in the branch circuit or a listed transformer or power supply for luminaires operating at not more than the low voltage contact limit.

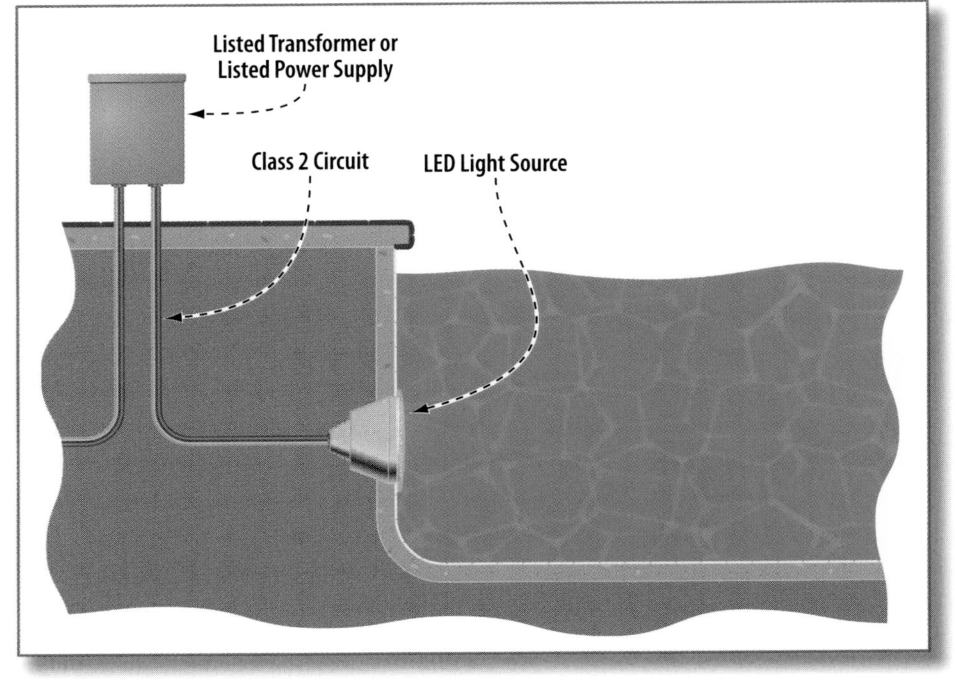

Change Summary

The words "15 volts" have been deleted from this section and replaced by the words "the low voltage contact limit." The revisions to this section correlate with the new defined term *Low Voltage Contact Limit* in 680.2. Power supplies are also addressed in this section as revised.

Significance of the Change

This section addresses compliance with shock protection hazards associated with underwater luminaires and requires the use of a listed underwater luminaire and installation of a listed ground-fault circuit interrupter in the branch circuit or a listed transformer for luminaires operating at not more than the low voltage contact limit. This revision removes the 15-volt threshold for underwater luminaires and replaces it with reference to the low voltage contact limit that is defined in 680.2. As long as the low-voltage contact limits (as defined and provided in the new definition) are not exceeded, ground-fault circuit interrupter protection is not required. These voltage sources are produced by specific transformers or power supplies manufactured for use with specific luminaires that are equipped with light-emitting diode technologies. The listing requirement in this section ensures that evaluation and certification by qualified electrical testing laboratories is required for this equipment intended for use in underwater lighting applications.

Comment: 17-60
Proposal: 17-142

680.24(A)(2)

Article 680 Swimming Pools, Fountains, and Similar Installations
Part II Permanently Installed Pools

Junction Box Installation

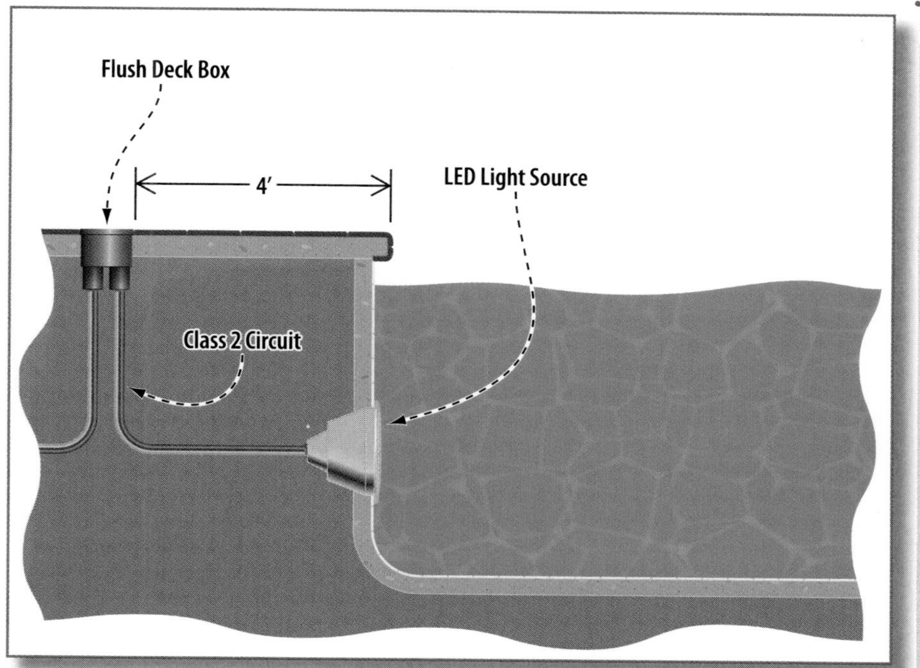

Code Language

(2) Installation. Where the luminaire operates over the low-voltage contact limit, the junction box location shall comply with (A)(2)(a) and (A)(2)(b). Where the luminaire operates at the low-voltage contact limit or less, the junction box location shall be permitted to comply with (A)(2)(c).

(a) ***(No changes)***

(b) ***(No changes)***

(c) *Flush Deck Box.* If used on a lighting system operating at the low-voltage contact limit or less, a flush deck box shall be permitted if both of the following conditions are met:

(1) An approved potting compound is used to fill the box to prevent the entrance of moisture.

(2) The flush deck box is located not less than 1.2 m (4 ft) from the inside wall of the pool.

Significance of the Change

Section 680.24(A)(2) covers installation requirements and locations for junction boxes and electrical enclosures for transformers or ground-fault circuit interrupters. The 15- volt threshold in 680.24(A)(2) and (A)(2)(c) has been replaced by the term *low-voltage contact limit* in both locations. Additionally, a new definition of *Low-Voltage Contact Limit* is now provided in 680.2 that gives maximum voltage thresholds where shock hazards exist and includes voltages produced by transformers and power supplies in sinusoidal AC, nonsinusoidal AC, constant DC, and switched DC voltage levels. These voltage values are coordinated with the maximum voltage levels provided in Tables 11(A) and 11(B) in Chapter 9 that are used for listed equipment. The result is that ground-fault circuit-interrupter protection is not required if any of the low-voltage contact limits as defined in 680.2 are not exceeded and the installation locations addressed in 680.24(A)(2) for these boxes and enclosures are permitted where the low-voltage contact limits are not exceeded.

Change Summary

The words “15 volts” have been removed from this section and replaced by the words “the low voltage contact limit” in list items (A)(2) and (A)(2)(c). The revisions to this section correlate with the newly defined term *Low Voltage Contact Limit* in 680.2.

Comment: 17-70
Proposal: 17-159

Equipotential Bonding

Code Language

680.26(B) Bonded Parts ***(No changes)***

680.26(B)(1)(b) *Copper Conductor Grid.* A copper conductor grid shall be provided and shall comply with (b)(1) through (b)(4).

680.26(B)(1)(b)(2) Conform to the contour of the pool

680.26(B)(2) Perimeter Surfaces. The perimeter surface shall extend for 1 m (3 ft) horizontally beyond the inside walls of the pool and shall include unpaved surfaces, as well as poured concrete surfaces and other types of paving. Perimeter surfaces less than 1 m (3 ft) separated by a permanent wall or building 1.5 m (5 ft) in height or more shall require equipotential bonding on the pool side of the permanent wall or building. ***(Remainder is unchanged by this revision.)***

680.26(B)(7) Fixed Metal Parts. All fixed metal parts shall be bonded including, but not limited to, metal sheathed cables and raceways, metal piping, metal awnings, metal fences, and metal door and window frames.

Exception No. 1: Those separated from the pool by a permanent barrier that prevents contact by a person shall not be required to be bonded.

(See 680.26(B) for full text.)

Change Summary

Section 680.26(B)(1)(b)(1) has been revised by adding the following sentence: "The bonding shall be in accordance with 250.8 or other approved means." Section 680.26(B)(1)(b)(2) was revised by shortening it to read as follows: "Conform to the contour of the pool." Section 680.26(B)(2) was revised by adding a second sentence as follows: "Perimeter surfaces less than 1 m (3 ft) separated by a permanent wall or building 1.5 m (5 ft) in height or more shall require an equipotential bonding grid on the pool side of the permanent wall or building." Section 680.26(B)(7) was revised by adding the words "fixed" and "parts" to the title and expanding the unlimited list of parts that should be bonded to the equipotential bonding grid. Exception No. 1 to 680.26(B)(7) has been revised to read as follows: "Exception No. 1: Those separated from the pool by a permanent barrier that prevents contact by a person shall not be required to be bonded."

Equipotential Bonding (continued)

Significance of the Change

Section 680.26 has been revised slightly in multiple locations to improve usability and clarity of the contained requirements. The new second sentence in 680.26(B)(1)(b)(1) provides users with clarification on the bonding connections meeting the requirements in 250.8 or by other approved means. The clarification in 680.26(B)(1)(b)(2) now indicates that the equipotential bonding grid must follow the contour of the pool only. The deck can be at a different contour than that of the pool. The key is to have the bonding grid extend from the waterline contour. The new second sentence in 680.26(B)(2) clarifies the equipotential bonding requirements for perimeter surfaces when the pool is separated by a permanent wall or building that is at least 5 feet in height. In these instances, equipotential bonding is required only on the pool side of the permanent wall or building. List item (7) to 680.26(B) has been revised to address all fixed metal parts as indicated in the revised title, and the noninclusive list of parts has been expanded to include some of the other metal parts falling under this requirement, for example, metal awnings, metal fences, and metal door and window frames. The phrase "that prevents contact by a person" was added to Exception No. 1 to 680.26(B)(7) to clarify that the barrier must prevent contact by persons in order to meet the provisions of the exception. Other editorial changes have been incorporated to improve clarity in the existing provisions.

Comments: 17-81, 17-83

Proposals: 17-172, 17-173, 17-174, 17-175, 17-184, 17-185

680.43 (Exc. No.2)

Article 680 Swimming Pools, Fountains, and Similar Installations
Part IV Spas and Hot Tubs

Indoor Installations

Code Language

Exception No. 2: The equipotential bonding requirements for perimeter surfaces in 680.26(B)(2) shall not apply to a listed self-contained spa or hot tub installed above a finished floor.

I-Stock Photo, Courtesy of NECA

Change Summary

A new Exception No. 2 applying to indoor installations of spas or hot tubs has been added to 680.43. This exception applies specifically to a listed self-contained spa or hot tub package assembly.

Significance of the Change

In previous edition of the *NEC*, 680.43 required that the rules in Parts I and II of Article 680 apply to indoor spa and hot tub installations except as modified by 680.43. This meant that all the equipotential bonding requirements contained in 680.26(B)(2) were applicable to these installations. This new exception relaxes the equipotential bonding requirement for perimeter surfaces associated with a listed self-contained spa or hot tub installed indoors and above the finished floor. This revision applies only to indoor installations covered in 680.43. The outdoor spa and hot tub installations addressed in 680.42 are not affected by this change. Outdoor spas and hot tubs are required to comply with the rules in Part I and II of Article 680, including the equipotential bonding requirements in 680.26, other than the bonding provisions addressed in 680.42(B), which recognizes metal-to-metal contact on a common frame and relaxes the bonding requirement for metal bands or hoops used to secure wooden staves.

Comment: 17-102

Proposal: 17-207

680.43(D)(5)

Article 680 Swimming Pools, Fountains, and Similar Installations
Part IV Spas and Hot Tubs

Bonding

Code Language

(5) Electrical devices and controls that are not associated with the spas or hot tubs and that are located less than 1.5 m (5 ft) from such units; otherwise, they shall be bonded to the spa or hot tub system.

Significance of the Change

Section 680.43(D) provides the requirements for bonding conductive parts in the vicinity of indoor hot tub and spa installations. List item (5) has been revised by removing the word "not," thus clarifying that any electrical devices and controls that are not associated with the spas or hot tubs and are located less than 5 feet from such units are required to be bonded to the spa or hot tub electrical supply system. The objective of this bonding requirement is to equalize potential differences between all conductive surfaces less than 5 feet from the inside walls of the spa or hot tub and to connect these conductive surfaces to the same potential of the grounding and bonding system for the spa or hot tub unit.

Change Summary

The word "not" that previously appeared in the phrase "...located not less than..." has been removed from this section.

Comment: None

Proposal: 17-211

680.62(B) (Exc.)

Article 680 Swimming Pools, Fountains, and Similar Installations
Part VI Pools and Tubs for Therapeutic Use

Bonding

Code Language

Exception: Small conductive surfaces not likely to become energized, such as air and water jets and drain fittings not connected to metallic piping, and towel bars, mirror frames, and similar nonelectrical equipment not connected to metal framing, shall not be required to be bonded.

Change Summary

A new exception has been added to 680.62(B). This new exception exempts bonding requirements for small metal parts that are not likely to become energized.

Significance of the Change

Section 680.62(B) provides bonding requirements for metal parts associated with pools and tubs for therapeutic use. This new exception provides relief from the bonding requirement for small conductive surfaces that are not connected electrically to any metal framing. The new exception is applicable to small conductive surfaces that are not likely to become energized, a decision that typically requires judgment by the enforcing authority. Typical small conductive surfaces covered by this exception include but are not limited to such items as air and water jets and drain fittings not connected to metal piping, towel bars, mirror frames, and other nonelectrical equipment. One of the key factors in properly applying this exception is that these small conductive surfaces are not electrically connected to metal structural framing that could become energized. The new exception provides practical relaxation from a bonding requirement that would otherwise be applied to small conductive surfaces that are very unlikely to become energized and present shock hazards in the vicinity of this type of equipment.

Comment: 17-108
Proposal: 17-223

Article 680 Swimming Pools, Fountains, and Similar Installations
Part VII Hydromassage Bathtubs

Bonding

Code Language

680.74 Bonding. All metal piping systems and all grounded metal parts... The bonding jumper shall not be required to be connected to a double insulated circulating pump motor. The 8 AWG or larger solid copper bonding jumper shall be required for equipotential bonding in the area of the hydromassage bathtub and shall not be required to be extended or attached to any remote panelboard, service equipment, or any electrode. The 8 AWG or larger solid copper bonding jumper shall be long enough to terminate on a replacement non-double-insulated pump motor and shall be terminated to the equipment grounding conductor of the branch circuit of the motor when a double-insulated circulating pump motor is used.

(See NEC 680.74 for complete text of this section.)

Significance of the Change

This revision requires that the 8 AWG copper bonding conductor be long enough to terminate on a replacement non-double-insulated pump motor in the hydromassage bath tub. This new requirement is similar to that in 680.26(B)(6)(a) requiring an 8 AWG or larger solid copper bonding jumper long enough to terminate on a replacement non-double insulated pool pump motor. This bonding conductor is required to be connected to the equipment grounding conductor of the pump motor branch circuit if a double-insulated circulating pump motor is installed. This connection ensures required continuity between the equipment grounding conductor of the supply circuit for the equipment and the equipotential bonding circuit of the hydromassage bathtub. The 8 AWG solid copper bonding conductor is not required to be run to the supplying panelboard or service equipment.

Change Summary

A new last sentence has been added to 680.74 as follows: "The 8 AWG or larger solid copper bonding jumper shall be long enough to terminate on a replacement non-double-insulated pump motor and shall be terminated to the equipment grounding conductor of the branch circuit of the motor when a double-insulated circulating pump motor is used."

Comment: None

Proposal: 17-230

690.2

Article 690 Solar Photovoltaic (PV) Systems
Part I General

Definitions of Monopole Subarray and Subarray

Code Language

690.2 Definitions.

Monopole Subarray. A PV subarray that has two conductors in the output circuit, one positive (+) and one negative (−). Two monopole PV subarrays are used to form a bipolar PV array.

Subarray. An electrical subset of a PV array.

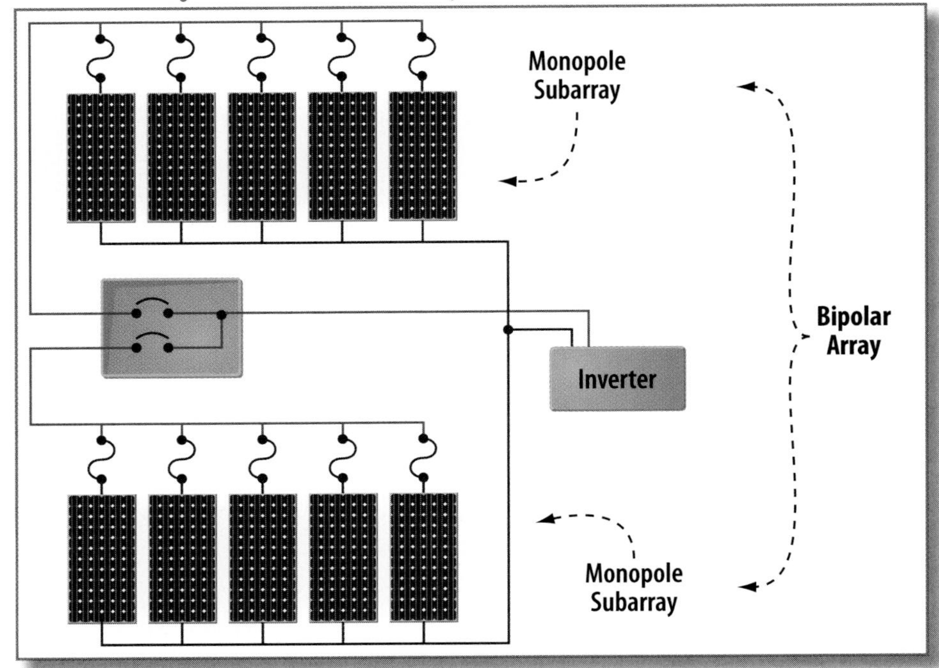

Change Summary

Two new definitions have been added to 690.2 for clarity and usability: *Monopole Subarray* and *Subarray*. These terms are are now defined for clarity and usability.

Significance of the Change

Subarray is now defined as an "electrical subset" of a PV array. The term *subset* is defined as "a set contained within a set." Therefore, any discernible portion of a PV array can be considered as a "subarray." A specific type of subarray is also newly defined in the *NEC*. *Monopole Subarray* is defined as having two conductors (one positive and one negative) in the output circuit. The definition further clarifies that two "monopole subarrays" are used to form a bipolar PV array. The term Bipolar PV Array is defined in 690.2 as follows:

Bipolar Photovoltaic Array. A photovoltaic array that has two outputs, each having opposite polarity to a common reference point or center tap.

The addition of these definitions is necessary to aid the *Code* user in applying the rules of Article 690, where these terms are found in 690.4(B)(1) and (B)(2), 690.4(F), and 690.7(E).

Comment: None
Proposals: 4-174, 4-179

690.4(B)

Article 90 Solar Photovoltaic (PV) Systems
Part I General

Identification and Grouping

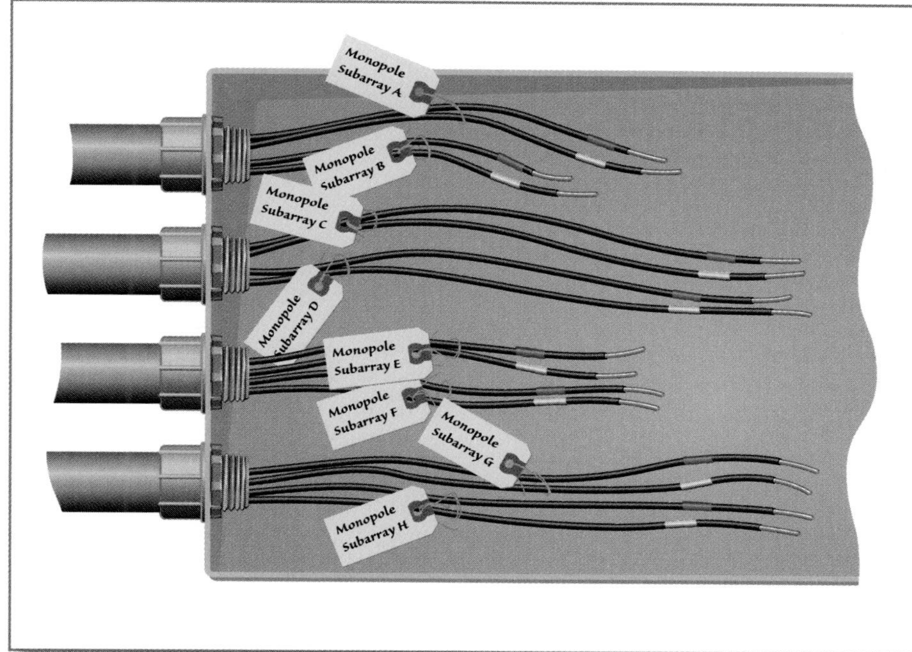

Change Summary

Section 690.4(B) prohibits PV source and output circuits from occupying the same raceway, cable tray, cable, outlet box, or junction box as other "non-PV" systems. New requirements for identification and grouping of all PV conductors are provided in 690.4(B)(1) through (B)(4). This includes PV source circuits, PV output/inverter circuits, and where more than one PV system exists.

Significance of the Change:

The parent text of 690.4(B) has been revised to clarify that PV source and output circuits may not occupy the same raceway, cable tray, cable, outlet box, or junction box as other "non-PV" systems, unless the conductors of the different systems are separated by a partition. Additionally, new requirements for identification and grouping have been added for the following: for PV source circuits in 690.4(B)(1), for PV output/inverter circuits in 690.4(B)(2), and for conductors of multiple PV systems in 690.4(B)(3). The required identification must be at all termination, connection, and splice points, and the means of identification must be by separate color coding, marking tape, tagging, or other approved means.

Section 690.4(B)(4) now requires grouping of conductors where conductors of more than one PV system (PV source circuits, PV output/inverter circuits) will occupy the same junction box or raceway with removable cover(s). This grouping requires that AC and DC conductors of each system be grouped separately by wire ties or similar means at least once, then grouped at intervals not to exceed 6 feet.

See the following page for additional information related to this change.

Comments: 4-64, 4-66

Proposal: 4-184

Identification and Grouping (continued)

Code Language

690.4 Installation

(B) Identification and Grouping. Photovoltaic source circuits and PV output circuits shall not be contained in the same raceway, cable tray, cable, outlet box, junction box, or similar fitting as conductors, feeders, or branch circuits of other non-PV systems, unless the conductors of the different systems are separated by a partition. PV system conductors shall be identified and grouped as required by 690.4(B)(1) through (4). The means of identification shall be permitted by separate color coding, marking tape, tagging, or other approved means.

(1) Photovoltaic Source Circuits. Photovoltaic source circuits shall be identified at all points of termination, connection, and splices.

(2) Photovoltaic Output and Inverter Circuits. The conductors of PV output circuits and inverter input and output circuits shall be identified at all points of termination, connection, and splices.

(3) Conductors of Multiple Systems. Where the conductors of more than one PV system occupy the same junction box, raceway, or equipment, the conductors of each system shall be identified at all termination, connection, and splice points.

Exception: Where the identification of the conductors is evident by spacing or arrangement, further identification shall not be required.

(4) Grouping. Where the conductors of more than one PV system occupy the same junction box or raceway with a removable cover(s), the ac and dc conductors of each system shall be grouped separately by wire ties or similar means at least once, and then shall be grouped at intervals not to exceed 1.8 m (6 ft).

Exception: The requirement for grouping shall not apply if the circuit enters from a cable or raceway unique to the circuit that makes the grouping obvious.

See the previous page for additional information related to this change.

690.4(E)

Article 690 Solar Photovoltaic (PV) Systems
Part I General

Qualified Persons

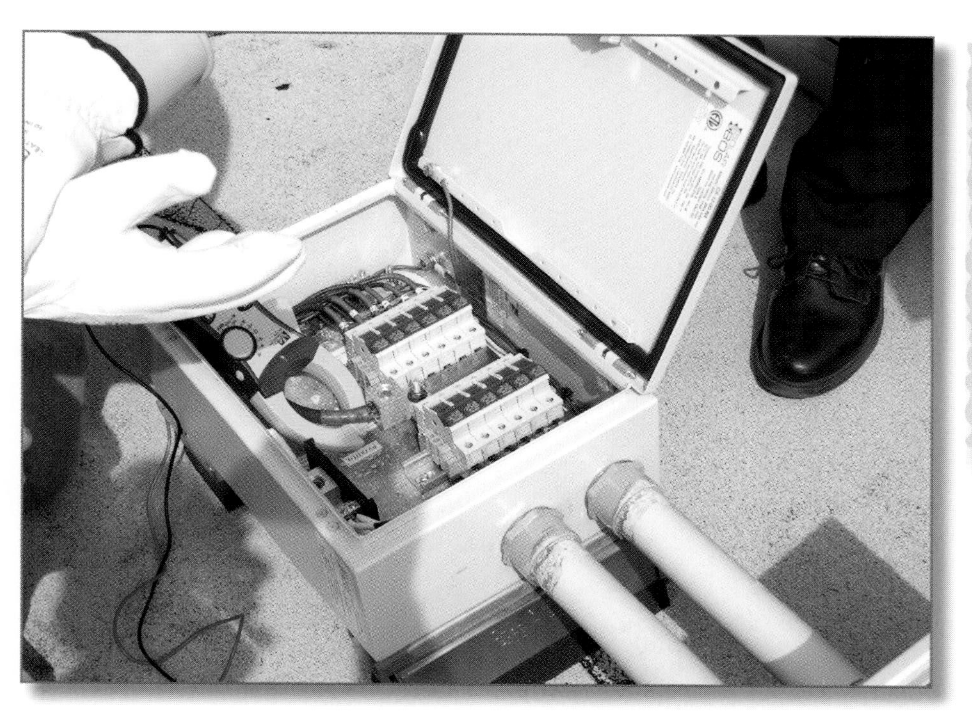

Code Language

690.4 Installation

(E) Wiring and Connections. The equipment and systems in 690.4(A) through (D) and all associated wiring and interconnections shall be installed only by qualified persons.

Informational Note: See Article 100 for the definition of *qualified person*.

Significance of the Change

This new requirement in 690.4 mandates that only qualified persons install photovoltaic systems. It is commonly understood that electrical installations covered by the *Code* should always be performed by properly trained workers and contractors. This is what the public expects: safe electrical systems for persons and property. Unfortunately, fast-emerging PV technologies and systems of all sizes have drawn many less than qualified entrepreneurs to the electrical field. Amidst all the allure of this growing new technology, increasing numbers of PV installations are being performed by workers not fully trained and qualified in the electrical field. PV systems are not plug-and-play, and this is not a job for a handyman. This is electrical work that should be performed only by qualified electricians and contractors.

Knowledge of electrical systems, general requirements in the *NEC*, and the rules in Article 690, is essential in addition to any specialty certification for installers of this type of equipment. Many state and local jurisdictions are tuned into this growing concern and are passing laws that require specific credentials for PV installers and contractors. It should be recognized that in many cases even qualified electrical workers and contractors must attain credentials as certified electrical workers and contractors for PV systems.

Change Summary

A new first-level subdivision, 690.4(E) Wiring and Connections, has been added to require that all equipment, associated wiring, and interconnections of PV systems be installed only by qualified persons. This mandates that the entire PV system be installed only by a qualified person. Article 100 contains the definition of *Qualified Person*.

Comment: 4-69

Proposal: 4-186

690.4(F)

Article 690 Solar Photovoltaic (PV) Systems
Part I General

Circuit Routing

Code Language

690.4 Installation

(F) Circuit Routing. Photovoltaic source and PV output conductors, in and out of conduit, and inside of a building or structure, shall be routed along building structural members such as beams, rafters, trusses, and columns where the location of those structural members can be determined by observation. Where circuits are imbedded in built-up, laminate, or membrane roofing materials in roof areas not covered by PV modules and associated equipment, the location of circuits shall be clearly marked.

Change Summary

Section 690.4(F) provides new requirements for the physical routing of photovoltaic source and PV output conductors. Routing must be along building structural members such as beams, rafters, trusses, and columns where possible. Where circuits are imbedded in the roof, the location of circuits must be clearly marked.

Comment: None

Proposal: 4-187

Significance of the Change

This new requirement is intended to provide firefighters with the ability to ventilate roofs during a fire without fear of cutting into energized conductors of a rooftop PV system. This requirement mandates that photovoltaic source and PV output conductors be physically routed along building structural members such as beams, rafters, trusses, and columns where the location of these structural members can be made through observation. While the location of these members may be difficult to observe from the rooftop, access to the roof area from inside the building or structure will reveal their location. Physically routing these conductors in this manner lowers the probability they will be contacted by firefighters in the event of a fire or by tradespeople installing new fans or HVAC equipment.

This requirement also mandates that where PV module systems are integrated into the roof and the associated circuits are imbedded in the roof, the location of these circuits must be clearly marked. This would require a physical marking on the rooftop to warn firefighters and tradespeople of the circuits contained in the roof. While no prescriptive means for applying this marking is included in the new requirement, this marking should be permanent and easily seen in any weather condition. For example, in an area where snow is prevalent, the marking may need to be raised at intervals to warn firefighters.

Bipolar Photovoltaic Systems

Code Language

690.4 Installation

(G) Bipolar Photovoltaic Systems. Where the sum, without consideration of polarity, of the PV system voltages of the two monopole subarrays exceeds the rating of the conductors and connected equipment, monopole subarrays in a bipolar PV system shall be physically separated, and the electrical output circuits from each monopole subarray shall be installed in separate raceways until connected to the inverter. The disconnecting means and overcurrent protective devices for each monopole subarray output shall be in separate enclosures. All conductors from each separate monopole subarray shall be routed in the same raceway.

Exception: Listed switchgear rated for the maximum voltage between circuits and containing a physical barrier separating the disconnecting means for each monopole subarray shall be permitted to be used instead of disconnecting means in separate enclosures.

Significance of the Change

This new requirement recognizes the potential danger of combining the conductors of two monopole subarrays where the sum of the voltages exceeds the rating of the conductors and connected equipment. The sum of the voltages of monopole subarrays is added without consideration of polarity. This new first-level subdivision also requires that the disconnecting means and overcurrent protective devices for each monopole subarray output be in separate enclosures and that all conductors from each separate monopole subarray be routed in the same raceway.

An exception has been added to permit the disconnecting means and overcurrent protective devices for each monopole subarray to be installed in the same equipment provided that (1) it is listed switchgear rated for the maximum voltage between circuits and (2) the equipment contains a physical barrier separating the disconnecting means for each monopole subarray.

If individual monopole subarray conductors come into contact with each other, the sum of the open-circuit monopole subarray voltages will in most cases be between 800 and 1200 volts. This overvoltage may be applied to switchgear, conductors, PV modules, and other equipment listed for only 600 volts. Separation of these system conductors will ensure that such a scenario does not occur.

Change Summary

A new requirement in 690.4(G) mandates that where two monopole subarrays are to be connected as a bipolar system and the sum of the individual voltages of the monopole subarrays exceeds conductor insulation or equipment voltage ratings, all conductors must be physically separated until connected to the inverter.

Comment: 4-71

Proposal: 4-188

690.4(H)

Article 690 Solar Photovoltaic (PV) Systems
Part I General

Multiple Inverters

Code Language

690.4 Installation

(H) Multiple Inverters. A PV system shall be permitted to have multiple utility-interactive inverters installed in or on a single building or structure. Where the inverters are remotely located from each other, a directory in accordance with 705.10 shall be installed at each dc PV system disconnecting means, at each ac disconnecting means, and at the main service disconnecting means showing the location of all ac and dc PV system disconnecting means in the building.

Exception: A directory shall not be required where all inverters and PV dc disconnecting means are grouped at the main service disconnecting means.

Change Summary

A new subdivision has been added to prescriptively permit the installation of multiple utility-interactive inverters in or on a single building or structure. Where the inverters are located remotely from each other, a directory in accordance with 705.10 is required.

Comment: None

Proposal: 4-189

Significance of the Change

This new first-level subdivision provides prescriptive requirements for the number of utility-interactive inverters installed in or on a single building or structure, and marking requirements where they are installed remotely from one another. Typical PV installations may consist of a number of small utility-interactive inverters installed in or on a building or structure and 690.4(H) now clearly permits the installation. This new requirement mandates that where more than one utility-interactive inverter is installed and they are remote from one another, a directory must be developed and installed in accordance with 705.10, which reads as follows:

> A permanent plaque or directory, denoting all electric power sources on or in the premises, shall be installed at each service equipment location and at locations of all electric power production sources capable of being interconnected.

The directory is required to be installed at each DC PV system disconnecting means, at each AC disconnecting means, and at the main service disconnecting means, and to show the location of all AC and DC PV system disconnecting means in the building. An exception follows to exempt the directory where all the inverters and PV DC disconnecting means are grouped together with the main service disconnecting means.

Back-fed Circuit Breakers

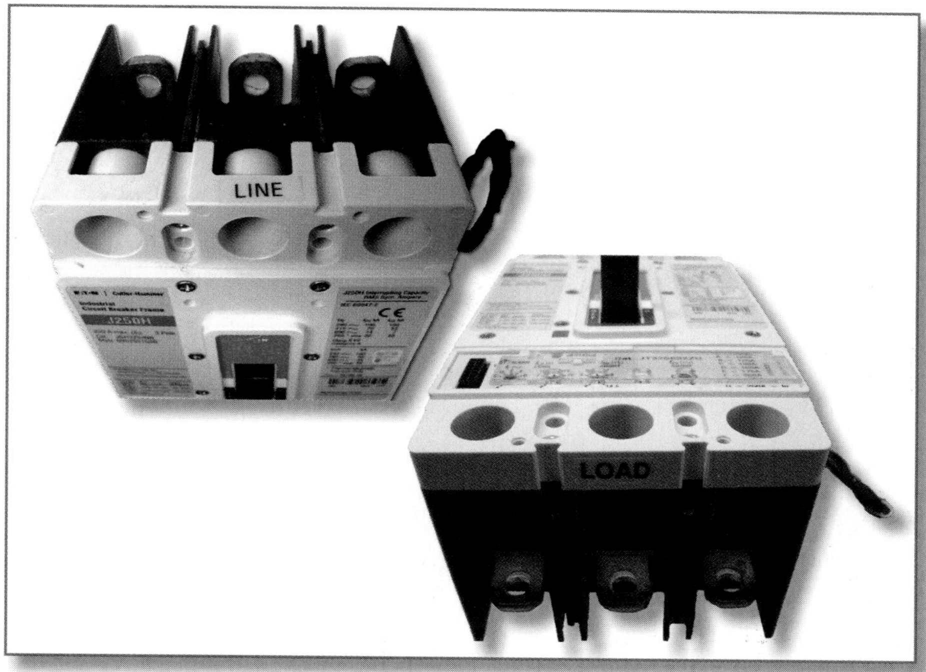

Courtesy of Eaton Corporation

Code Language

690.10 Stand-Alone Systems

(E) Back-fed Circuit Breakers. Plug-in type back-fed circuit breakers connected to a stand-alone inverter output in either stand-alone or utility-interactive systems shall be secured in accordance with 408.36(D). Circuit breakers that are marked "line" and "load" shall not be backfed.

Significance of the Change

The vast majority of PV installations are accomplished with utility-interactive inverters supplying power from the inverter output to the grid through a back-fed circuit breaker in a panelboard. Back-fed circuit breakers are exempt from the additional securement requirement of 408.36(D). This exemption, previously found in 690.64(B)(6) in the 2008 *NEC*, is now located in 705.12 via a revised 690.64.

However, inverters in stand-alone systems and the stand-alone outputs of utility-interactive inverters act as voltage sources without anti-islanding circuits. This means that the stand-alone output circuit does not shut down when the utility-interactive output circuits are disconnected. Installers familiar with utility-interactive systems may fail to clamp the back-fed circuit breaker connected to the output of a stand-alone inverter or the stand-alone output of an inverter in a utility-interactive system. This application of a back-fed breaker represents a safety hazard if the circuit breaker is inadvertently unplugged from a panelboard while energized.

The last sentence of this new requirement states that "A circuit breaker marked line and load may not be used in this manner." It should be noted that circuit breakers are not "identified" for back-feed use. The only identification to determine whether a circuit breaker is not permitted to be back-fed is where the device is marked Line and Load.

Change Summary

This new requirement mandates that where a "plug-in" type circuit breaker is used to control/protect a stand alone inverter output in either stand-alone or utility-interactive systems, the circuit breaker must have an additional means of securement to the panelboard. A circuit breaker marked line and load may not be used in this manner.

Comment: None

Proposal: 4-203

690.11

Article 690 Solar Photovoltaic (PV) Systems
Part II Circuit Requirements

AFCI (Direct Current)

Code Language

690.11 Arc-Fault Circuit Protection (Direct Current). Photovoltaic systems with dc source circuits, dc output circuits, or both, on or penetrating a building operating at a PV system maximum system voltage of 80 volts or greater shall be protected by a listed (dc) arc-fault circuit interrupter, PV type, or other system components listed to provide equivalent protection. The PV arc-fault protection means shall comply with the following requirements:

(1) The system shall detect and interrupt arcing faults resulting from a failure in the intended continuity of a conductor, connection, module, or other system component in the dc PV source and output circuits.

(2) The system shall disable or disconnect one of the following:

a. Inverters or charge controllers connected to the fault circuit when the fault is detected

b. System components within the arcing circuit

(3) The system shall require that the disabled or disconnected equipment be manually restarted.

(4) The system shall have an annunciator that provides a visual indication that the circuit interrupter has operated. This indication shall not reset automatically.

Change Summary

Direct current (DC) AFCI protection is now required for all PV systems with DC source and/or output circuits on or penetrating a building operating at a PV system maximum system voltage of 80 volts or greater.

Comment: 4-83

Proposals: 4-204, 4-205

Significance of the Change

Photovoltaic (PV) systems are installed outdoors on rooftops of dwellings and commercial structures, exposing them to extreme environmental conditions, including wind, rain, snow, ice, dirt, and temperature extremes. System components may deteriorate over time. Insulation and component failure can occur, causing a DC arcing fault that can easily start a fire on the dwelling or commercial occupancy. After a PV system is installed and continues to produce energy year after year, it is unlikely that routine inspections or maintenance will be performed by qualified people.

A DC fault, including a DC arcing fault, is far more difficult to interrupt than an AC fault because of the non-time varying (non-zero crossing) nature of direct current. The most common DC arcing fault in a PV system is a series arcing fault, and it can occur anywhere in the DC portion of the system.

This new section requires devices to detect and interrupt these destructive DC arcing faults. This requires the disabling or disconnecting of the inverters or charge controllers connected to the faulted circuit, or of system components within the arcing circuit. The AFCI protective device or system must have an annunciator, with visual indication that must be manually disabled to ensure that the homeowner or commercial operator is aware a problem exists.

Opening Grounded Conductor

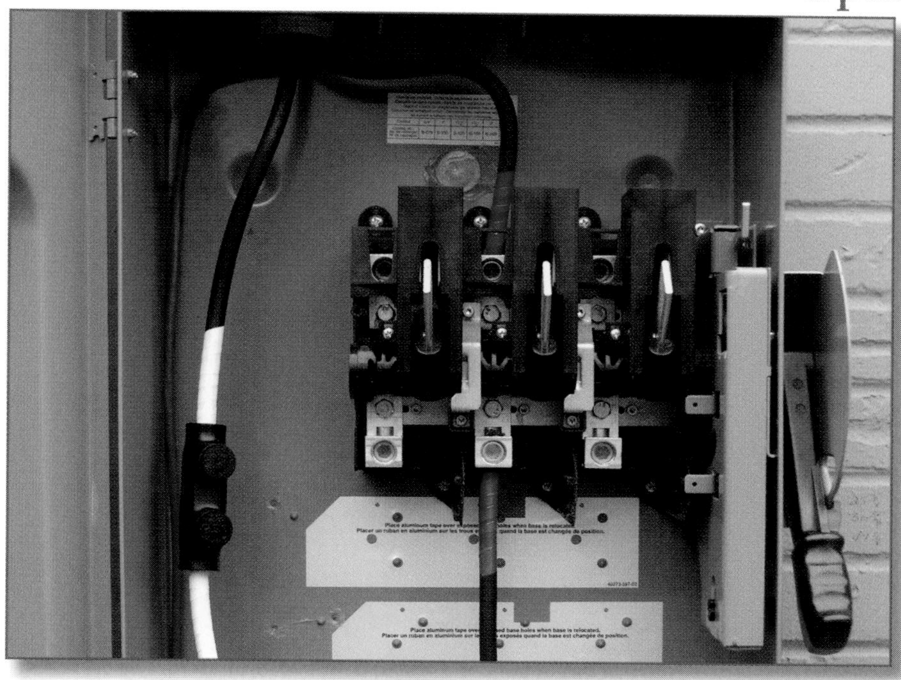

Code Language

690.13 All Conductors.

(Editorial revisions only in section)

Exception No. 1: A switch or circuit breaker that is part of a ground-fault detection system required by 690.5, or that is part of an arc-fault detection/interruption system required by 690.11, shall be permitted to open the grounded conductor when that switch or circuit breaker is automatically opened as a normal function of the device in responding to ground faults.

Exception No. 2: A disconnecting switch shall be permitted in a grounded conductor providing the following conditions are met:

(1) The switch is used only for PV array maintenance.

(2) The switch is accessible only by qualified persons.

(3) The switch is rated for the maximum dc voltage and current that could be present during any operation, including ground fault conditions.

Significance of the Change

The parent text of 690.13 has been revised to clarify that the intent of the *NEC* is that the DC conductors of a PV system, as an energy source, have provisions to allow them to be disconnected from all other (non-PV) conductors in a building or structure. While the general rule of 690.13 prohibits the opening of a grounded conductor, two exceptions are allowed. Exception No.1 permits the opening of the grounded conductor by a switch or circuit breaker for a ground-fault detection system required in 690.5 and is revised to include the new AFCI requirement in 690.11. The provisions of Exception No. 1 permit the grounded conductor to be opened only under a fault condition. When a fault does occur or during maintenance, the location and correction of ground faults in PV arrays may require that the ungrounded conductor be disconnected from the system.

A new Exception No. 2 has been added to permit a maintenance switch. A maintenance switch is permitted only for PV array maintenance, with access only to qualified persons, and must be rated for the maximum DC voltage and current that could be present during any operation, including ground-fault conditions.

Change Summary

Exception No.1 has been expanded to permit both ground-fault devices and arc-fault devices operating in a fault or service condition to open (unground) the grounded conductor. A new exception permits a maintenance switch to open the grounded conductor.

Comment: 4-86

Proposals: 4-206, 4-208

690.16(B)

Article 690 Solar Photovoltaic (PV) Systems
Part III Disconnecting Means

Fuse Servicing

Code Language

690.16 Fuses

(B) Fuse Servicing. Disconnecting means shall be installed on PV output circuits where overcurrent devices (fuses) must be serviced that cannot be isolated from energized circuits. The disconnecting means shall be within sight of and accessible to, the location of the fuse or integral with fuseholder and shall comply with 690.17. Where the disconnecting means are located more than 1.8 m (6 feet) from the overcurrent device, a directory showing the location of each disconnect shall be installed at the overcurrent device location.

Non-load-break rated disconnecting means shall be marked "Do not open under load."

Change Summary

A new requirement has been added in 690.16 to require that a disconnecting means be provided to isolate a fuseholder during the servicing of a fuse.

Courtesy of Cooper Bussmann

Significance of the Change

This revision requires that inverters manufactured with internal PV source and output circuit combining fuses on the input circuits connected directly to the inverter input terminals be provided with an external disconnecting means. Without an external disconnecting means, these fuses cannot be safely serviced when the PV array is illuminated. Typical fuses in the inverters used in these larger systems may be 100 amps or larger. Fuses must be serviced in a safe manner, and this usually means that they should be disconnected from all sources of voltage. Most PV combiners use "finger safe" fuseholders for this purpose.

Comment: None

Proposal: 4-218

690.31(B)

Article 690 Solar Photovoltaic (PV) Systems
Part IV Wiring Methods

Conduit Fill

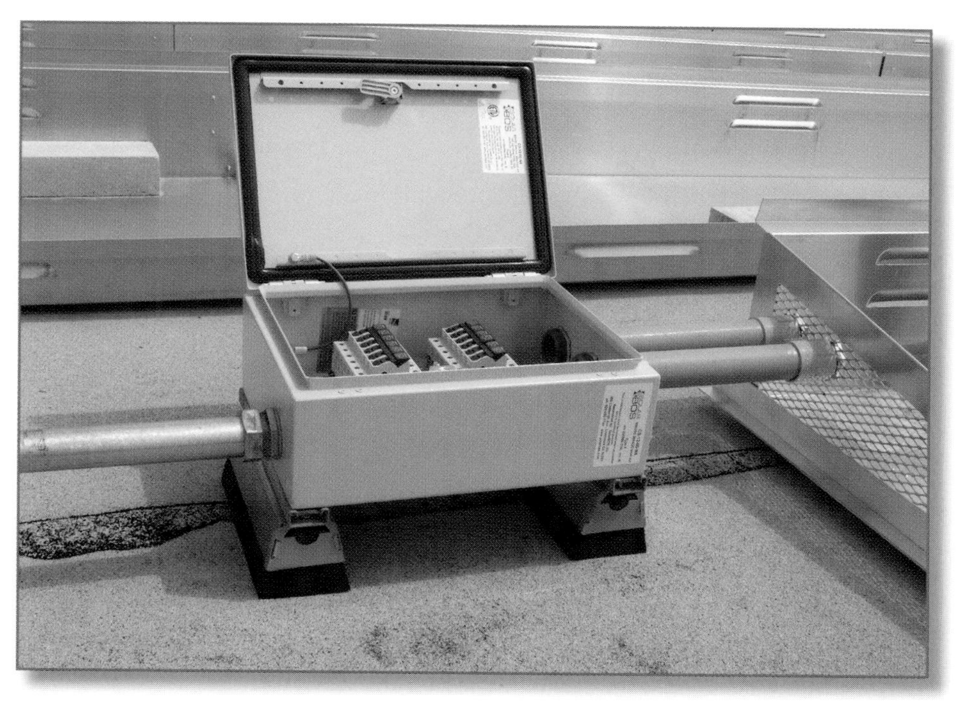

Code Language

690.31 Methods Permitted.

(B) Single-Conductor Cable. Single-conductor cable Type USE-2, and single-conductor cable listed and labeled as photovoltaic (PV) wire shall be permitted in exposed outdoor locations in photovoltaic source circuits for photovoltaic module interconnections within the photovoltaic array.

Exception: Raceways shall be used when required by 690.31(A).

Informational Note: Photovoltaic (PV) Wire (also Photovoltaic (PV) Cable) has a non-standard outer diameter. Conduit fill may be calculated using Table 1, Chapter 9.

Significance of the Change

This revision occurs in Part IV Wiring Methods. Section 690.31 is titled Methods Permitted and addresses permitted wiring methods in six first-level subdivisions. Section 690.31(B) addresses single-conductor cable and permits single-conductor cable Type USE-2 and single-conductor cable listed and labeled as photovoltaic (PV) wire in exposed outdoor locations in photovoltaic source circuits for photovoltaic module interconnections within the photovoltaic array. Additionally, 690.31(A) requires that raceways must be used as follows:"Where photovoltaic source and output circuits operating at maximum system voltages greater than 30 volts are installed in readily accessible locations, circuit conductors shall be installed in a raceway."

The new informational note informs the *Code* user that PV wire/cable has a non-standard outer diameter. This means that Annex C cannot be used and that the conduit fill must be calculated according to Table 1, Chapter 9. The outside diameter of the cable must be used to calculate the conduit fill. Note that the outside diameter may be referenced to a larger size conductor referenced in Annex C.

Change Summary

A new informational note has been added to inform the *Code* user that PV wire/cable has a non-standard outer diameter. Conduit fill must then be calculated because the conduit fill tables in Annex C cannot be used as PV cable has a non-standard diameter.

Comment: 4-94

Proposal: 4-224

690.31(E)

Article 690 Solar Photovoltaic (PV) Systems
Part IV Wiring Methods

Type MC Cable

Code Language

690.31 Methods Permitted

(E) Direct-Current Photovoltaic Source and Output Circuits Inside a Building. Where dc photovoltaic source or output circuits from a building-integrated or other photovoltaic system are run inside a building or structure, they shall be contained in metal raceways, Type MC metal clad cable that complies with 250.118(10), or metal enclosures from the point of penetration of the surface of the building or structure to the first readily accessible disconnecting means. The disconnecting means shall comply with 690.14(A), (B), and (D). The wiring methods shall comply with the additional installation requirements in (1) through (4).

Change Summary

Type MC cable is now a permitted wiring method where DC photovoltaic source or output circuits from a building-integrated or other photovoltaic system are run inside a building or structure.

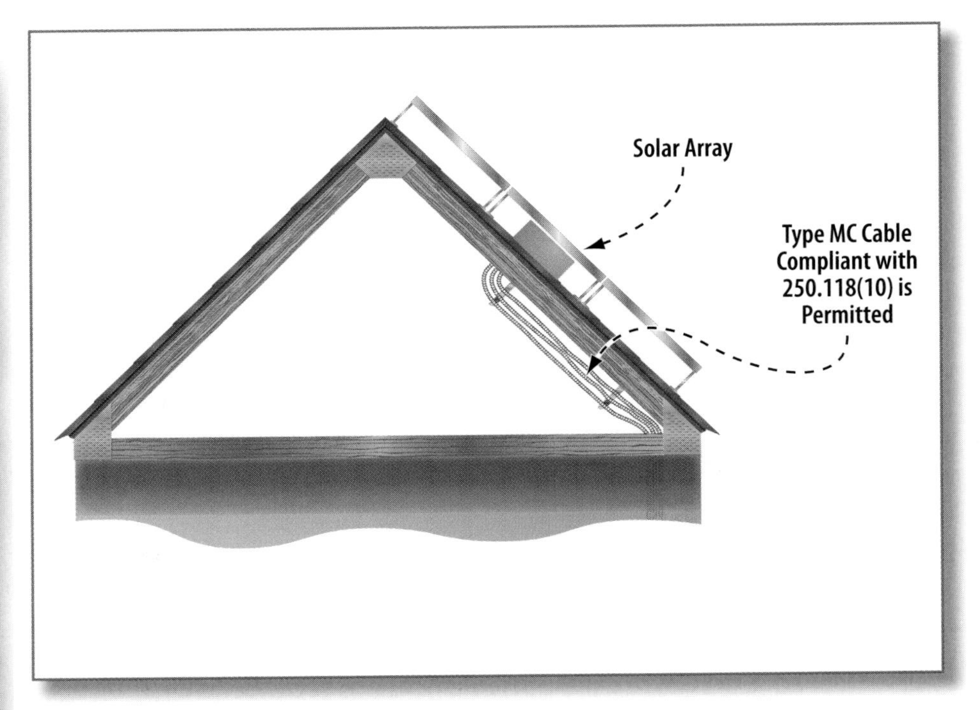

Significance of the Change

The previous requirements of 690.31(E) required that DC PV source and output circuits run inside a building or structure be contained in metal raceways, or metal enclosures, from the point of penetration of the surface of the building or structure to the first readily accessible disconnecting means. This requirement exists to protect the DC PV source and output circuits from damage within the building or structure. A DC fault, including a DC arcing fault, is far more difficult to interrupt than an AC fault because of the non-time varying (non-zero crossing) nature of DC.

This revision expands the permitted wiring methods to include Type MC metal clad cable that complies with 250.118(10). The physical construction of Type MC cable provides adequate protection for the DC PV source and output circuits. The requirement that the MC cable comply with 250.118(10) ensures that an effective ground-fault current path exists.

Comment: 4-97

Proposal: 4-228

Wiring Methods

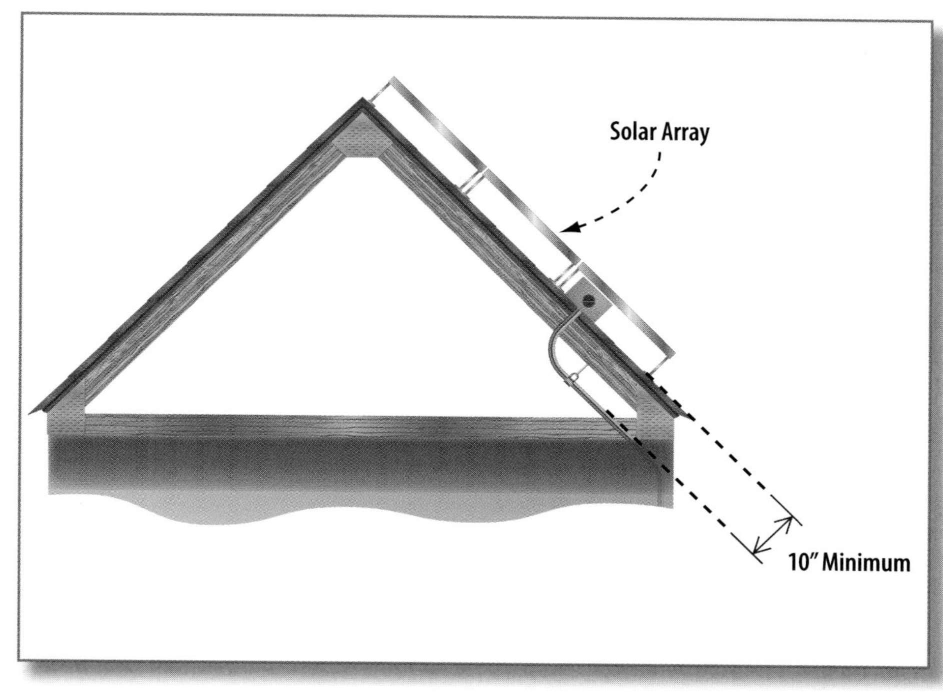

Code Language

690.31 Methods Permitted

(E) Direct-Current Photovoltaic Source and Output Circuits Inside a Building.

(1) Beneath Roofs. Wiring methods shall not be installed within 25 cm (10 in.) of the roof decking or sheathing except where directly below the roof surface covered by PV modules and associated equipment. Circuits shall be run perpendicular to the roof penetration point to supports a minimum of 25 cm (10 in.) below the roof decking.

Informational Note: The 25 cm (10 in.) requirement is to prevent accidental damage from saws used by firefighters for roof ventilation during a structure fire.

Significance of the Change

A new requirement has been added to protect PV wiring methods from damage where holes or other penetrations may be made in the roof of a building or structure. In the event of fire, firefighters may cut holes in a roof to ventilate the structure while fighting the fire. During building maintenance or renovation, trades may cut holes in the roof for many reasons. This new requirement now prohibits all DC PV source and output raceways and MC cable from being installed within 10 inches of the roof decking or sheathing except where directly below the roof surface covered by PV modules and associated equipment. Where these circuits penetrate the roof, they are required to be run perpendicular to the roof penetration point to supports a minimum of 25 cm (10 in.) below the roof decking. This means that on a sloped roof the raceway or MC cable must be run perpendicular to the roof to a support that is a minimum of 25 cm (10 in.) below the roof decking.

A new informational note informs the *Code* user that the primary purpose of this requirement is to prevent accidental damage from saws used by firefighters for roof ventilation during a structure fire. It is understood that in a fire or other emergency, there is no time to research what is below the roof.

Change Summary

A new requirement prohibits all DC PV source and output raceways and MC cable from being installed within 10 inches of the roof decking or sheathing except where directly below the roof surface covered by PV modules and associated equipment. Additionally, where circuits penetrate the roof, they must be run perpendicular to the roof penetration point to supports a minimum of 10 inches below the roof.

Comment: 4-97

Proposal: 4-228

690.31(E)(2)

Article 690 Solar Photovoltaic (PV) Systems
Part IV Wiring Methods

Wiring Methods

Code Language

690.31 Methods Permitted

(E) Direct-Current Photovoltaic Source and Output Circuits Inside a Building.

(2) Flexible Wiring Methods. Where flexible metal conduit (FMC) smaller than metric designator 21 (trade size ¾) or Type MC cable smaller than 25 mm (1 in.) in diameter containing PV power circuit conductors is installed across ceilings or floor joists, the raceway or cable shall be protected by substantial guard strips that are at least as high as the raceway or cable. Where run exposed, other than within 1.8 m (6 ft) of their connection to equipment, these wiring methods shall closely follow the building surface or be protected from physical damage by an approved means.

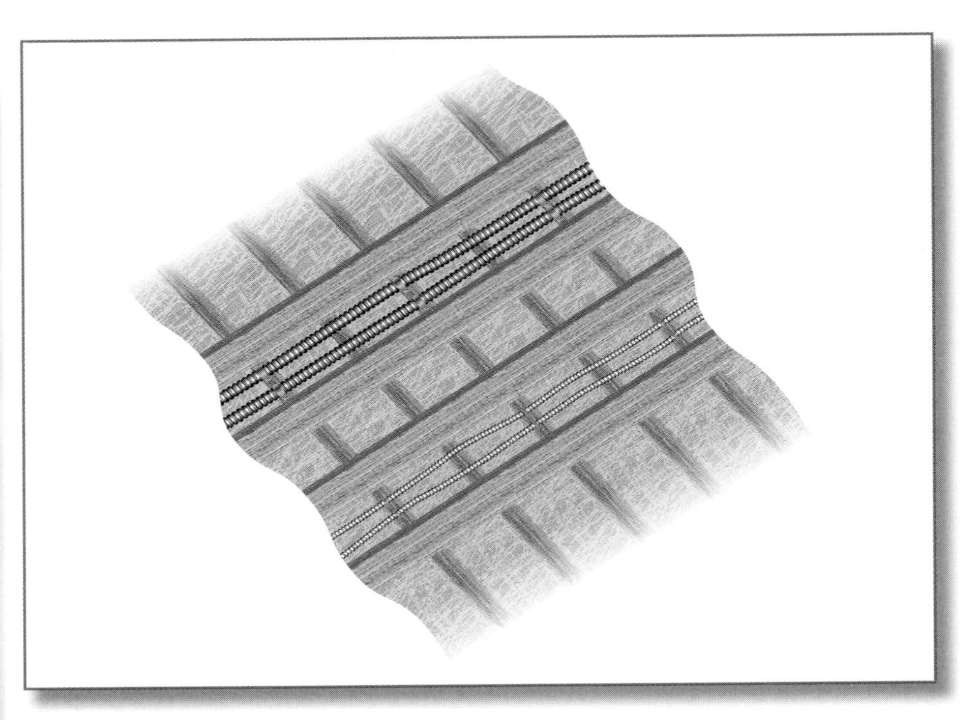

Change Summary

For protection against physical damage, flexible metal conduits smaller than ¾ inch and MC cable smaller than 1 inch in diameter are now required to be run along the building surface. Where installed across ceilings or floor joists, they must be protected by substantial guard strips that are at least as high as the raceway or cable.

Significance of the Change

This new requirement mandates that where flexible metal conduit smaller ¾ trade size and Type MC cable smaller than 1 inch in diameter and containing PV power circuit conductors are installed across ceilings or floor joists, they must be protected by substantial guard strips. Additionally, where run exposed in a building or structure, other than within 6 feet of their connection to equipment, flexible metal conduit smaller ¾ trade size and Type MC cable smaller than 1 inch in diameter and containing PV power circuit conductors must closely follow the building surface or be protected from physical damage by an approved means.

Where FMC and MC cable are used in these areas, they are subject to damage from homeowners and occupants. Attic spaces, for example, are accessed for maintenance and storage by homeowners and building occupants, exposing these flexible wiring methods to physical damage.

Comment: 4-97

Proposal: 4-228

690.31(E)(3) & (4)

Article 690 Solar Photovoltaic (PV) Systems
Part IV Wiring Methods

Marking

Code Language

690.31 Methods Permitted

(E) Direct-Current Photovoltaic Source and Output Circuits Inside a Building.

(3) Marking or Labeling Required. The following wiring methods and enclosures that contain PV power source conductors shall be marked with the wording "Photovoltaic Power Source" by means of permanently affixed labels or other approved permanent marking:

(1) Exposed raceways, cable trays, and other wiring methods.

(2) Covers or enclosures of pull boxes and junction boxes.

(3) Conduit bodies in which any of the available conduit openings are unused.

(4) Marking and Labeling Methods and Locations. The labels or markings shall be visible after installation. Photovoltaic power circuit labels shall appear on every section of the wiring system that is separated by enclosures, walls, partitions, ceilings, or floors. Spacing between labels or markings, or between a label and a marking, shall not be more than 3 m (10 ft). Labels required by this section shall be suitable for the environment where they are installed.

Significance of the Change

Where photovoltaic power source conductors enter a building, 690.31(E) requires that they be installed in metal raceways, Type MC metal-clad cable that complies with 250.118(10), or metal enclosures, from the point of penetration of the surface of the building or structure to the first readily accessible disconnecting means. Two new second-level subdivisions have been added for marking of these permitted wiring methods within the building.

Section 690.31(E)(3) requires all permitted wiring methods, including raceways, cable tray, MC cable, covers of enclosures/boxes, and conduit bodies, be marked as a "Photovoltaic Power Source."

Section 690.31(E)(4) requires that the markings be visible after installation and that all wiring methods be marked on every section of the wiring system that is separated by enclosures, walls, partitions, ceilings, or floors. This means that every accessible portion of the wiring method must be marked. Such marking is required to be placed on the wiring method at intervals not over 10 feet and to be suitable for the environment in which it is installed.

Change Summary

PV power source conductors run inside a building are now required to be marked as "Photovoltaic Power Source." This marking is required for all wiring methods, boxes, and conduit bodies.

Comment: 4-97

Proposal: 4-228

Equipment Grounding

Code Language

690.43 Equipment Grounding. Equipment grounding conductors and devices shall comply with 690.43(A) through (F).

(A) Equipment Grounding Required…

(B) Equipment Grounding Conductor Required…

(C) Structure as Equipment Grounding Conductor… Metallic mounting structures, other than building steel, used for grounding purposes shall be identified as equipment grounding conductors or shall have identified bonding jumpers or devices connected between the separate metallic sections and shall be bonded to the grounding system.

(D) Photovoltaic Mounting Systems and Devices. Devices and systems used for mounting PV modules that are also used to provide grounding of the module frames shall be identified for the purpose of grounding PV modules.

(E) Adjacent Modules…

(F) All Conductors Together…

Change Summary

The equipment grounding requirements of 690.43 have been logically separated into first-level subdivisions for clarity and usability. New requirements have been added for metallic mounting structures and for mounting devices/systems for PV modules.

Comment: None

Proposal: 4-232

Significance of the Change

The requirements for equipment grounding of PV systems have been revised and separated into six new first-level subdivisions to improve usability and clarity. Subdivision 690.43(C) has been expanded to allow listed and identified devices to bond both the PV modules and other equipment to the mounting structure. These devices are required to be listed and identified for the purpose, because the mounting structures for PV modules are generally aluminum, creating corrosion and deterioration issues where a copper connection must be made. New text now requires that metallic mounting structures, other than building steel, used for grounding purposes be identified as equipment grounding conductors or that they have identified bonding jumpers or devices connected between the separate metallic sections and be bonded to the grounding system. This requirement is necessary because in most cases aluminum support structures are designed only for structural support, not electrical continuity.

New 690.43(D) now permits devices and systems used for mounting PV modules to also provide grounding of the module frames, provided they are identified for the purpose of grounding PV modules. Again, these devices must be identified for the purpose.

690.47(B)

Article 690 Solar Photovoltaic (PV) Systems
Part V Grounding

Direct Current Grounding Electrode System

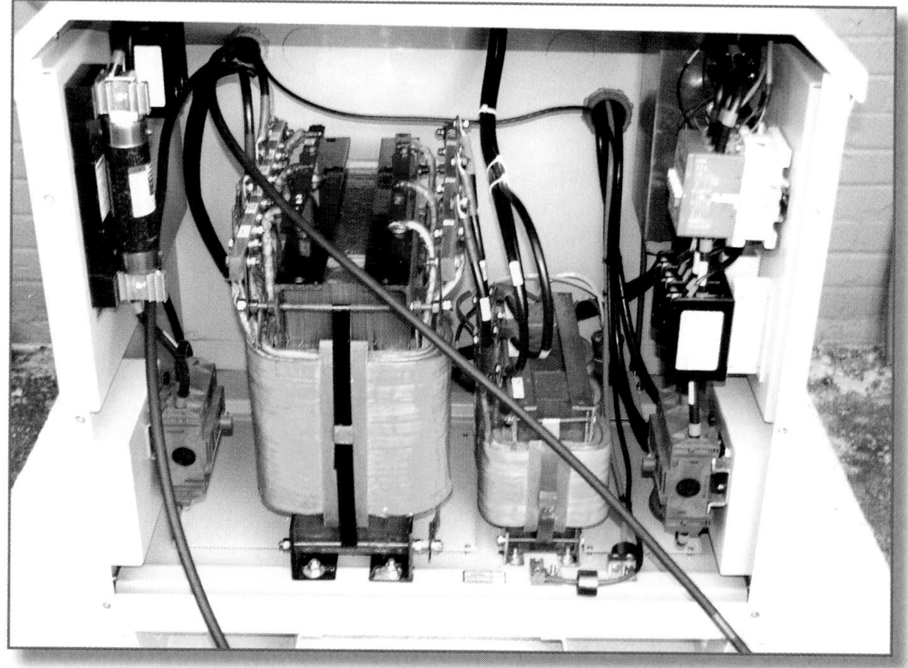

Code Language

690.47 Grounding Electrode System

(B) Direct-Current Systems. If installing a dc system, a grounding electrode system shall be provided in accordance with 250.166 for grounded systems or 250.169 for ungrounded systems. The grounding electrode conductor shall be installed in accordance with 250.64.

A common dc grounding electrode conductor shall be permitted to serve multiple inverters. The size of the common grounding electrode and the tap conductors shall be in accordance with 250.166. The tap conductors shall be connected to the common grounding electrode conductor by exothermic welding or with connectors listed as grounding and bonding equipment in such a manner that the common grounding electrode conductor remains without a splice or joint.

Change Summary

A new second paragraph has been added to 690.43(B) to permit a single grounding electrode conductor to serve multiple inverters.

Significance of the Change

This revision permits a single grounding electrode conductor to serve more than one inverter. The size of the grounding electrode conductor for multiple inverters need not be larger than the largest grounding electrode conductor for any of the inverters. Grounding electrode conductor taps are sized as individual conductors. The grounding electrode conductor tap conductors are required to be connected to the common grounding electrode conductor by (1) exothermic welding or (2) with connectors listed as grounding and bonding equipment. The termination of the tap conductors must be in such a manner that the common grounding electrode conductor remains without a splice or joint.

This revision is necessary due to the high number of systems using multiple small inverters versus one large inverter for economical reasons. Each of these smaller inverters along with their connected modules is an individual DC system. This revision now permits a common DC grounding electrode to provide the necessary connection to earth for all inverters. The existing text continues to require that the grounding electrode conductor be sized according to 250.166 for grounded systems, or according to 250.169 for ungrounded systems. The grounding electrode conductor must be installed in accordance with 250.64.

Comment: 4-99

Proposal: 4-234

690.47(C)

Article 690 Solar Photovoltaic (PV) Systems
Part V Grounding

System Grounding, AC, DC

Code Language

690.47 Grounding Electrode System

(C) Systems with Alternating-Current and Direct-Current Grounding Requirements. Photovoltaic systems having dc circuits and ac circuits with no direct connection between the dc grounded conductor and ac grounded conductor shall have a dc grounding system. The dc grounding system shall be bonded to the ac grounding system by one of the methods listed in (1), (2), or (3).

This section shall not apply to ac PV modules.

When using the methods of (C)(2) or (C)(3), the existing ac grounding electrode system shall meet the applicable requirements of Article 250, Part III.

Informational Note No. 1: ANSI/UL 1741, *Standard for Inverters, Converters, and Controllers for Use in Independent Power Systems*, requires that any inverter or charge controller that has a bonding jumper between the grounded dc conductor and the grounding system connection point have that point marked as a grounding electrode conductor (GEC) connection point. In PV inverters, the terminals for the dc equipment grounding conductors and the terminals for ac equipment grounding conductors are generally connected to, or electrically in common with, a grounding busbar that has a marked dc GEC terminal.

Informational Note No. 2: For utility-interactive systems, the existing premises grounding system serves as the ac grounding system.

(1) Separate Direct-Current Grounding Electrode System Bonded to the Alternating-Current Grounding Electrode System. A separate dc grounding electrode or system shall be installed, and it shall be bonded directly to the ac grounding-electrode system. The size of any bonding jumper(s) between the ac and dc systems shall be based on the larger size of the existing ac grounding electrode conductor or the size of the dc grounding electrode conductor specified by 250.166. The dc grounding electrode system conductor(s) or the bonding jumpers to the ac grounding electrode system shall not be used as a substitute for any required ac equipment grounding conductors.

(2) Common Direct-Current and Alternating-Current Grounding Electrode. A dc grounding electrode conductor of the size specified by 250.166 shall be run from the marked dc grounding electrode connection point to the ac grounding electrode. Where an ac grounding electrode is not accessible, the dc grounding electrode conductor shall be connected to the ac grounding electrode conductor in accordance with 250.64(C)(1). This dc grounding electrode conductor shall not be used as a substitute for any required ac equipment grounding conductors.

(3) Combined Direct-Current Grounding Electrode Conductor and Alternating Current Equipment Grounding Conductor. An unspliced, or irreversibly spliced, combined grounding conductor shall be run from the marked dc grounding electrode conductor connection point along with the ac circuit conductors to the grounding busbar in the associated ac equipment. This combined grounding conductor shall be the larger of the sizes specified by 250.122 or 250.166 and shall be installed in accordance with 250.64(E).

System Grounding, AC, DC (continued)

Change Summary

The grounding electrode system requirements for PV systems with AC and DC grounding requirements have been clarified and expanded in a user-friendly format.

Significance of the Change

This parent text of 690.47(C) now clearly requires that where a PV system has DC circuits and AC circuits with no direct connection between the DC grounded conductor and AC grounded conductor, a DC grounding system is required. The parent text also mandates that the DC grounding system be bonded to the AC grounding system by one of the methods illustrated in three new second-level subdivisions, all titled to provide clarity and usability for the *Code* user and the specific installation.

The three bonding methods provided include one that establishes a new DC grounding electrode system and two that utilize an existing AC grounding electrode system. The parent text requires that where a PV system is installed and utilizes an existing AC grounding electrode system, the installer and inspector must perform a visual inspection of the existing AC grounding electrode system to ensure that it meets the applicable requirements of Article 250, Part III.

Two new Informational Notes are provided to inform the *Code* user and the enforcer with information necessary to understand and apply these rules. Informational Note No. 1 explains that listed inverters, converters, and controllers designed for use in independent power systems require a bonding jumper between the grounded DC conductor and the grounding system connection point and that the point be marked as the GEC connection point. This note also informs the *Code* user that in PV inverters, the terminals for the DC equipment grounding conductors and the terminals for AC equipment grounding conductors are generally connected to or electrically in common with a grounding busbar that has a marked DC GEC terminal. Informational Note No. 2 informs the *Code* user that in a utility-interactive system, the existing premises grounding system serves as the AC grounding system.

Comment: 4-100

Proposal: 4-235

690.62(A)

Article 690 Solar Photovoltaic (PV) Systems
Part VII Connection to Other Sources

Neutral Ampacity

Code Language

690.62 Ampacity of Neutral Conductor

(A) Inverter Outputs Connected Between Neutral Conductor and Ungrounded Conductor(s). Where the outputs of single or multiple single-phase inverter(s) are connected between the neutral conductor and one or more of the ungrounded conductors of a 3-phase 4-wire, wye-connected system or a 120/240V single-phase system, the ampacity of the neutral conductor shall be no less than the greater of (1) or (2):

(1) 125 percent of the continuous load plus 100% of the noncontinuous load on that neutral conductor or

(2) 125 percent of the sum of the rated output current of all inverters considering worst-case imbalance.

Change Summary

The ampacity calculation for the neutral conductor of inverter outputs that are connected between a system neutral conductor and an ungrounded conductor has been revised to improve clarity and accuracy in a new 690.62(A). The existing requirement for a neutral used solely for instrumentation,voltage detection or phase detection is reloacted in 690.62(B).

Comment: 4-103

Proposal: 4-242

Example:

Consider a 480/277V, 3-phase, 4-wire, wye system. The existing maximum, connected, unbalanced load current in the neutral is 40 amps.

40 x 1.25 = 50 amps

Consider two 7 kW inverters are connected between each phase and neutral. A total of six inverters are connected. Rated output current of each inverter is 27.3 amps. When all six inverters are producing rated current, the neutral currents from the inverters are near zero. In a worst-case situation, only two inverters connected on one phase are working at rated output and the others are shut off or have failed. The currents in the neutral from these two inverters would total 2 x 27.3 amps or 54.6 amps, and this should be used at 125% for the required ampacity of the neutral, since it is larger than the 40 amps of load current.

54.6 x 1.25 = 68.25 amps

Significance of the Change

This revision clearly separates the ampacity calculations into the two scenarios that could create a maximum amount of current. The previous requirement mandated that these values be added, when in reality these two current values are not additive. The first value considers the load supplied and requires sizing at 125 percent of the continuous load plus 100 percent of the noncontinuous load on that neutral conductor. If the inverters are not operating, the neutral must be able to carry any connected load currents. Operation of the inverters in the presence of load currents tends to decrease currents in the neutral, so sizing in this manner is appropriate.

The second value considers the inverter output and requires sizing at 125 percent of the sum of the rated output current of all inverters considering worst-case imbalance. If there are no loads, the circuit must carry the full rated output of the inverter(s). Where multiple inverters are installed and connected phase-to-neutral, consideration must be given to situations where one or more inverters could fail or be turned off, or where the connected array is shaded, thus eliminating any balance between the phases and increasing the neutral currents. Sizing at 125 percent of rated output is needed to ensure that the neutral conductor ampacity is consistent with the ampacity calculated elsewhere in the *Code*.

692.4(C)

Article 692 Fuel Cell Systems
Part I General

Qualified Persons

Code Language

692.4 Installation

(C) System Installation. Fuel cell systems including all associated wiring and interconnections shall be installed by only qualified persons.

Informational Note: See Article 100 for the definition of *qualified person.*

Significance of the Change

The new requirement in 692.4(C) mandates that only qualified persons install fuel cell systems, equipment, and associated wiring. It is commonly understood that electrical installations covered by the *Code* should always performed by properly trained workers and contractors. This is what the public expects: safe electrical systems for persons and property. Unfortunately, fast emerging fuel cell technologies and systems of all sizes have drawn many less than qualified entrepreneurs into the electrical field. Amidst all the allure of this growing new technology, increasingly more fuel cell installations are being performed by workers not fully trained and qualified in the electrical field. Fuel cell systems are not plug-and-play, and this is not a job for a handyman. This is electrical work that should be performed by qualified electricians and contractors.

Knowledge of the electrical systems, general requirements of the *NEC*, in addition to the rules in Article 692, is essential in addition to any specialty certification for installers of this type of equipment. Many state and local jurisdictions are tuned into this growing concern and are passing laws that require specific credentials and certification for installers and contractors.

Change Summary

New subdivision (C) System Installation has been added to 692.4, followed by a new informational note to refer users to the *NEC* definition of the term *Qualified Person*.

Comment: 4-109

Proposal: 4-252

Article 694

Article 694 Small Wind Electric Systems

Small Wind Systems

Code Language

694.1 Scope. The provisions of this article apply to small wind (turbine) electric systems that consist of one or more wind electric generators with individual generators having a rated power up to and including 100 kW. These systems can include generators, alternators, inverters, and controllers.

Informational Note: Small wind electric systems can be interactive with other electrical power production sources or might be stand-alone systems. Small wind electric systems can have ac or dc output, with or without electrical energy storage, such as batteries. See Informational Note Figures 694.1 and 694.2.

694.2 Definitions ***(There are 14 new definitions)***

Parts

I. General
II. Circuit Requirements
III. Disconnecting Means
IV. Wiring Methods
V. Grounding
VI. Marking
VII. Connection to Other Sources
VIII. Storage Batteries
IX. Systems over 600 Volts

Change Summary

A new article has been added to the *NEC* to address wind (turbine) electric systems that consist of one or more wind electric generators with individual generators having a rated power up to and including 100 kW.

Comments: 4-116, 118, 119, 120, 121, 122, 123

Proposal: 4-263

Significance of the Change

This new article is designed in a similar structure and format as Articles 690 and 692, and in many cases the text is extremely similar. It is separated into 14 parts to logically separate installation requirements now necessary to address the large number of small wind turbine systems being installed in the United States and internationally. The 2008 *NEC* included no specific installation requirements to address the unique operating characteristics of these small wind electric systems. While many installations are stand-alone applications, most are utility interactive, and so requirements similar to Article 690 apply. Article 694 applies to small wind systems consisting of a single wind generator(s) up to and including 100 kW, so its requirements are limited only by the size of the individual generator(s). Systems may include hundreds of wind generators and be covered by this article, provided the individual generators are not larger than 100 kW.

These small wind systems have become increasingly popular in rural and urban areas as an energy source to offset utility costs and in an effort to move toward environmentally friendly energy independence. The push toward renewable energy sources is creating a huge market for small wind systems, and federal and local incentives are further stimulating the growth of wind technology.

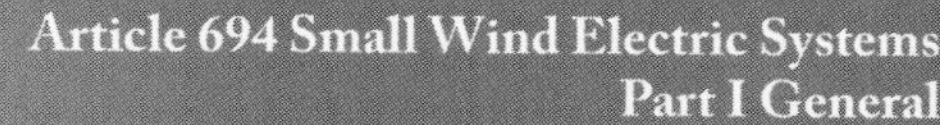

Article 694 Small Wind Electric Systems
Part I General

Qualified Persons

NECA Photo - © 2010 Rob Colgan

Code Language

694.7 Installation. Systems covered by this article shall be installed only by qualified persons.

Informational note: See Article 100 for the definition of qualified person.

Significance of the Change

This new requirement in 694.7 mandates that only qualified persons may install small wind electrical systems and equipment covered by Article 694. It is commonly understood that electrical installations covered by the *Code* should always performed by trained, qualified workers and contractors. This is what the unsuspecting public expects: safe electrical systems for persons and property. Unfortunately, fast emerging wind power generation technologies and systems of all sizes have drawn many less than qualified entrepreneurs to the electrical field. Amidst all the allure of this growing new technology, increasing numbers of small wind power system installations are being performed by workers not fully trained and qualified in the electrical field. This is electrical work that should be performed only by qualified electricians and contractors.

Knowledge of electrical systems, general requirements in the *NEC*, in addition to the new requirements in Article 694 is essential in addition to any specialty certification for installers of this type of equipment. Many state and local jurisdictions are tuned into this growing concern and are passing laws that require specific credentials for installers and contractors of small wind electric systems.

Change Summary

Section 694.7 is titled "Installation" in a new Article 694. The parent text of this section requires that small wind systems be installed only by qualified persons. A new informational note is included to refer *Code* users to the definition of the term *Qualified Person* in Article 100.

Comment: 4-110

Proposal: 4-121

695.3(A)

Article 695 Fire Pumps

Individual Sources

Code Language

695.3 Power Source(s) for Electric Motor-Driven Fire Pumps. Electric motor-driven fire pumps shall have a reliable source of power.

(A) Individual Sources. Where reliable, and where capable of carrying indefinitely the sum of the locked-rotor current of the fire pump motor(s) and the pressure maintenance pump motor(s) and the full-load current of the associated fire pump accessory equipment when connected to this power supply, the power source for an electric motor driven fire pump shall be one or more of the following.

(1) Electric Utility Service Connection. *(No change)*

(2) On-Site Power Production Facility. *(No change)*

(3) Dedicated Feeder. A dedicated feeder shall be permitted where it is derived from a service connection as described in 695.3(A)(1). [20:9.2.2(3)]

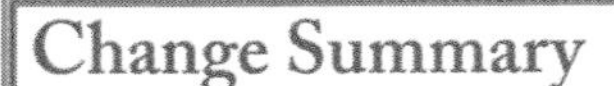

Change Summary

A new second-level subdivision has been added to 695.3(A) to recognize that a dedicated feeder derived directly from a service connection is permitted as an individual source.

Significance of the Change

Section 695.3(A) contains prescriptive requirements for individual sources that must be capable of carrying indefinitely the sum of the locked-rotor current of the fire pump motor(s) and the pressure maintenance pump motor(s) and the full-load current of the associated fire pump accessory equipment. The 2008 edition of the *NEC* limited the individual sources to 695.3(A)(1) Electric Utility Service Connection and to (A)(2) On-Site Power Production Facility. However, the service connection permitted in 695.3(A)(1) was permitted to have a single disconnecting means and overcurrent protection in accordance with 695.4. Once this service-supplied connection is provided with a disconnecting means and overcurrent protection, the conductors are by definition no longer service conductors — they are feeders.

This revision provides additional clarity for the *Code* user and correlates with the requirements of 9.2.2(3) of NFPA 20, *Standard for the Installation of Stationary Pumps for Fire Protection*, which permits a "dedicated feeder connection derived directly from the dedicated service to the fire pump installation."

Comment: None

Proposal: 13-60a

Multiple Sources

Code Language

695.3 Power Source(s) for Electric Motor-Driven Fire Pumps.

(B) Multiple Sources. If reliable power cannot be obtained from a source described in 695.3(A), power shall be supplied by one of the following: [20:9.3.2]

(1) Individual Sources. An approved combination of two or more of the sources from 695.3(A).

(2) Individual Source and On-site Standby Generator. An approved combination of one or more of the sources in 695.3(A) and an on-site standby generator complying with 695.3(D). [20:9.3.4]

Exception to (B)(1) and (B)(2): An alternate source of power shall not be required where a back-up engine-driven or back-up steam turbine-driven fire pump is installed. [20:9.3.3]

Change Summary

This revision of 695.3(B) provides clarity for the *Code* user. Only an approved combination of individual sources in 695.3(B)(1) or an approved combination of an individual source and an on-site standby generator in 695.3(B)(2) are permitted. The remainder of 695.3(B) has been editorially relocated for clarity. An exception has been added to address installations where a backup engine-driven or backup steam turbine-driven fire pump is installed.

Significance of the Change

This revision clarifies the requirement for the most common type of electrical fire pump installation. The fact that the parent text of 695.3 requires that electric motor-driven fire pumps have a *reliable source of power* translates into the need for a second or alternate source if the primary source should fail for some reason. A typical electrical fire pump installation consists of a service-supplied connection and an on-site standby generator. An exception to the alternate source requirement addresses installations where a backup engine-driven or backup steam turbine-driven fire pump is installed.

This revision into first-level subdivisions provides clarity and usability for power source requirements for the majority of electric fire pump installations. The remainder of 695.3(B) as written in the 2008 *NEC* has been editorially relocated for clarity: the generator requirements into a new 695.3(D), the feeder source requirements into a new 695.3(C), and the arrangement requirements into a new (E). Note that the *NEC* does not have primary jurisdiction over requirements for electric fire pumps. NFPA 20, *Standard for the Installation of Stationary Pumps for Fire Protection*, has primary jurisdiction of all stationary fire pumps. Also note that NFPA 20, in 9.3.1, requires at least one alternate source of power only where the height of the structure is beyond the pumping capacity of the fire department apparatus.

Comment: 13-81

Proposal: 13-60a

Multibuilding Campus-Style Complexes

Code Language

695.3 Power Source(s) for Electric Motor-Driven Fire Pumps.

(C) Multibuilding Campus-Style Complexes. If the sources in 695.3(A) are not practicable and the installation is part of a multibuilding campus style complex, feeder sources shall be permitted if approved by the authority having jurisdiction and installed in accordance with either (C)(1) and (C)(3) or (C)(2) and (C)(3).

(1) Feeder Sources. Two or more feeders shall be permitted as more than one power source if such feeders are connected to, or derived from, separate utility services. The connection(s), overcurrent protective device(s), and disconnecting means for such feeders shall meet the requirements of 695.4(B).

(2) Feeder and Alternate Source. A feeder shall be permitted as a normal source of power if an alternate source of power independent from the feeder is provided. The connection(s), overcurrent protective device(s), and disconnecting means for such feeders shall meet the requirements of 695.4(B).

(3) Selective Coordination. The overcurrent protection device(s) in each disconnecting means shall be selectively coordinated with any other supply-side overcurrent protective device(s).

Change Summary

Clarity is provided for multibuilding campus-style installations. New prescriptive requirements address a combination feeder and alternate source installation and selective coordination.

Comment: 13-83

Proposal: 13-60a

Significance of the Change

The previous text of 695.3(B)(2) Feeder Sources has been relocated to new first-level subdivision 695.3(C) and renamed for clarity as Multibuilding Campus-Style Complexes. Revision of the title was necessary because the feeder sources within are permitted only in multibuilding campus-style complexes. Section 695.3(C)(1) reflects the previous requirement and permits two or more feeders as more than one power source if connected to or derived from separate utility services. Section 695.3(C)(2) permits a feeder as a normal source of power if an alternate source of power independent from the feeder is provided. Note that both 695.3(B)(1) and (B)(2) require compliance with 695.4(B).

It is important to note that the parent text of 695.3(C) requires compliance with "either (C)(1) and (C)(3) or (C)(2) and (C)(3)." Regardless of whether 695(C)(1) or (C)(2) is applied, 695.3(C)(3) now requires that the overcurrent protective device in each feeder disconnecting means be selectively coordinated with any other supply side overcurrent protective device. This new requirement for selective coordination is added to correlate with 9.2.2(4)(e) of NFPA 20.

This revision of requirements for permitted use of feeder sources in multibuilding campus-style complexes provides clarity and usability for the *Code* user.

Phase Converters

Courtesy of PHASE-A-MATIC Incorporated

Code Language

695.3 Power Source(s) for Electric Motor-Driven Fire Pumps.

(F) Phase Converters. Phase converters shall not be permitted to be used for fire pump service. [20:9.1.7]

Significance of the Change

This new requirement clearly prohibits the use of a "phase converter" to supply an electric fire pump. This change is necessary to correlate the *NEC* requirements for electrical installation with NFPA 20, *Standard for the Installation of Stationary Pumps for Fire Protection*, which has primary jurisdiction over fire pumps. NFPA 20 prohibits the use of phase converters in the following section:

"9.1.7 Phase converters shall not be used to supply power to a fire pump."

It is imperative that the user of the *NEC* understand that the installation of a "phase converter" falls under the scope of Article 455 in the *NEC*. A "phase converter" is defined along with an informational note in Article 455 as follows:

455.2 Definitions.

Phase Converter. An electrical device that converts single-phase power to 3-phase electric power.

Informational Note: Phase converters have characteristics that modify the starting torque and locked-rotor current of motors served, and consideration is required in selecting a phase converter for a specific load.

This first-level subdivision will not prohibit transformers arranged in a scott connection to convert 3-phase to 2-phase or vice versa.

Change Summary

A new first-level subdivision, 695.3(F), clearly prohibits the use of phase converters to supply an electric fire pump. This prohibition is within the purview of NFPA 20 and is extracted into the *NEC*.

Comment: None

Proposal: 13-60a

695.4(B)

Article 695 Fire Pumps

Disconnect and OCPD

Code Language

695.4 Continuity of Power

(B) Connection Through Disconnecting Means and Overcurrent Device

(1) Number of Disconnecting Means

(a) *General.* … ***(Editorial revisions)***

(b) *Feeder Sources.* … ***(Editorial revisions)***

(c) *On-Site Standby Generator*. Where an on-site generator is used to supply a fire pump, an additional disconnecting means and associated overcurrent protective device(s) shall be permitted.

(2) Overcurrent Device Selection

(a) *Individual Sources.* … ***(Editorial modifications)***

(b) *On-Site Standby Generators.* Overcurrent protective devices between an on-site standby generator and a fire pump controller shall be selected and sized to allow for instantaneous pickup of the full pump room load, but shall not be larger than the value selected to comply with 430.62 to provide short-circuit protection only. [20:9.6.1.1]

(3) Disconnecting Means. … ***(Editorial revisions)***

Change Summary

The text of 695.4(B) is revised for clarity. An additional disconnecting means and associated OCPDs are now clearly permitted where an on-site generator is used to supply a fire pump. The sizing of OCPDs is clarified.

Comments: 13-88, 13-92

Proposal: 13-77a

Significance of the Change

Section 695.4(B) has been editorially revised for clarity. Its parent text has been separated into new second-level subdivision 695.4(B)(1) Number of Disconnecting Means, followed by three second-level subdivisions. Section 695.4(B)(1)(a) *General* contains the existing requirement for a single disconnecting means and associated overcurrent protective device(s). Section 695.4(B)(1)(b) *Feeder Sources* contains the existing permission for an additional disconnecting means and associated overcurrent protective device(s) in a multibuilding campus-style installation. Section 695.4(B)(1)(c) *On-Site Standby Generator* contains new text permitting an additional disconnecting means and associated overcurrent protective device(s) where an on-site generator is used to supply a fire pump.

Two third-level subdivisions have been added to 695.4(B)(2) for clarity. Section 695.4(B)(2)(a) *Individual Sources* contains the existing requirement and now clarifies that where the locked rotor current value does not correspond to a standard overcurrent device size, the next standard overcurrent device size in accordance with 240.6, must be used. Section 695.4(B)(2)(b) *On-Site Standby Generators* contains the existing requirement with new text that clarifies that the OCPD from the generator is not required to be sized to handle locked rotor current. The existing disconnect requirements have been moved to new second-level subdivision 695.4(B)(3) Disconnecting Means.

695.6(D)

Pump Wiring

Code Language

695.6 Power Wiring

(D) Pump Wiring. All wiring from the controllers to the pump motors shall be in rigid metal conduit, intermediate metal conduit, electrical metallic tubing, liquidtight flexible metal conduit, or liquidtight flexible nonmetallic conduit Type LFNC-B, listed Type MC cable with an impervious covering, or Type MI cable.

Significance of the Change

The purpose of this requirement, formerly 695.6(E) in the 2008 *NEC*, is the physical protection of conductors between the fire pump controller and the fire pump. Electrical metallic tubing (EMT) has been added to the list of permitted wiring methods on the basis of its recognition in Article 358 for use where subject to physical damage. Type MC cable, conversely, is permitted to be installed between the fire pump controller and the fire pump, provided it is not subject to physical damage. See 330.12(1). Although EMT is recognized for use where subject to physical damage, it is not permitted where subject to severe physical damage. See 358.12(1).

This first-level subdivision also recognizes rigid metal conduit, intermediate metal conduit, liquidtight flexible metal conduit, or liquidtight flexible nonmetallic conduit Type LFNC-B, listed Type MC cable with an impervious covering, or Type MI cable.

Change Summary

The permitted wiring methods from the fire pump controller to the pump motors have been expanded to include electrical metallic tubing (EMT.)

Comment: None

Proposals: 13-114, 13-115, 13-116

695.6(H)

Article 695 Fire Pumps

Power Controller

Code Language

695.6 Power Wiring

(H) Listed Electrical Circuit Protective System to Controller Wiring. Electrical circuit protective system installation shall comply with any restrictions provided in the listing of the electrical circuit protective system used and the following shall apply:

(1) A junction box shall be installed ahead of the fire pump controller a minimum of 300 mm (12 in.) beyond the fire-rated wall or floor bounding the fire zone.

(2) Where required by the manufacturer of a listed electrical circuit protective system or by the listing, or as required elsewhere in this *Code*, the raceway between a junction box and the fire pump controller shall be sealed at the junction box end as required and in accordance with the instructions of the manufacturer. [**20**:9.8.2]

(3) Standard wiring between the junction box and the controller shall be permitted. [**20**:9.8.3]

Change Summary

These new requirements have been added to the *NEC* to correlate with NFPA 20, *Standard for the Installation of Stationary Pumps for Fire Protection*. The installation of electrical circuit protective systems such as Type MI cable is now prohibited from terminating directly into a fire pump controller.

Comment: None

Proposal: 13-97

Significance of the Change

This new first-level subdivision provides correlation between the *NEC* and NFPA 20, *Standard for the Installation of Stationary Pumps for Fire Protection*. That standard has primary jurisdiction over stationary fire pump installations, while the *NEC* has jurisdiction over the electrical installation. However, the *NEC* must follow the performance-based requirements of NFPA 20.

This new subdivision prohibits the installation of electrical circuit protective systems such as Type MI cable from terminating directly into a fire pump controller. Instead, a junction box is required by 695.6(H)(1) to be installed ahead of the fire pump controller, with a raceway exiting the junction box and terminating in the controller. The junction box must be located a minimum of 12 in. beyond the fire-rated wall or floor bounding the fire zone in which the fire pump controller is located. Section 695.6(H)(2) requires that the raceway between the junction box and the fire pump controller be sealed at the junction box end as required and in accordance with the instructions of the manufacturer. Standard wiring between the junction box and the controller is permitted.

695.6(I)

Junction Boxes

Code Language

695.6 Power Wiring

(I) Junction Boxes. Where fire pump wiring to or from a fire pump controller is routed through a junction box, the following requirements shall be met:

(1) The junction box shall be securely mounted. [**20**:9.7(1)]

(2) Mounting and installing of a junction box shall not violate the enclosure type rating of the fire pump controller(s). [**20**:9.7(2)]

(3) Mounting and installing of a junction box shall not violate the integrity of the fire pump controller(s) and shall not affect the short circuit rating of the controller(s). [**20**:9.7(3)]

(4) As a minimum, a Type 2, drip-proof enclosure (junction box) shall be used where installed in the fire pump room. The enclosure shall be listed to match the fire pump controller enclosure type rating. [**20**:9.7(4)]

(5) Terminals, junction blocks, wire connectors, and splices, where used, shall be listed. [**20**:9.7(5)]

(6) A fire pump controller or fire pump power transfer switch, where provided, shall not be used as a junction box to supply other equipment, including a pressure maintenance (jockey) pump(s).

Significance of the Change

This new first-level subdivision provides correlation between NFPA 20 and the *NEC*. NFPA 20 has primary jurisdiction over stationary fire pump installations, while the *NEC* has jurisdiction over electrical installations. The *NEC*, however, must follow the requirements of NFPA 20.

This new first-level subdivision mandates that where fire pump wiring to or from a fire pump controller is routed through a junction box, the box must be securely mounted. The mounting of the junction box must not violate the enclosure rating of the controller in any manner. Junction boxes used in the fire pump room are now required to be a minimum of a Type-2 drip proof enclosure and must not be rated less than the controller. Any terminals, junction blocks, wire connectors, and splices, used, must be listed.

An additional requirement prohibiting the use of the fire pump controller or transfer switch from being used as a junction box is also included. These requirements are now incorporated into the *NEC* and will be enforced by the AHJ. It is imperative that installers review the installation requirements extracted from NFPA 20 throughout Article 695 to ensure compliance with the *NEC*.

Change Summary

These new requirements for junction boxes have been added to the *NEC* to correlate with NFPA 20, *Standard for the Installation of Stationary Pumps for Fire Protection*.

Comment: None

Proposal: 13-97

Raceway Terminations

Code Language

695.6 Power Wiring

(J) Raceway Terminations. Where raceways are terminated at a fire pump controller, the following requirements shall be met: [20:9.9]

(1) Listed conduit hubs shall be used. [**20**:9.9.1]

(2) The type rating of the conduit hub(s) shall be at least equal to that of the fire pump controller. [**20**:9.9.2]

(3) The installation instructions of the manufacturer of the fire pump controller shall be followed. [**20**:9.9.3]

(4) Alterations to the fire pump controller, other than conduit entry as allowed elsewhere in this *Code*, shall be approved by the authority having jurisdiction. [**20**:9.9.4]

Change Summary

These new requirements have been added to the *NEC* to correlate with NFPA 20, *Standard for the Installation of Stationary Pumps for Fire Protection*. Raceway terminations into a fire pump controller are now required to be made in a listed conduit hub.

Significance of the Change

This new first-level subdivision provides correlation between NFPA 20 and the *NEC*. NFPA 20 has primary jurisdiction over stationary fire pump installations, while the *NEC* has jurisdiction over electrical installations. However, the *NEC* must follow the requirements of NFPA 20.

This new first-level subdivision requires the following:

(1) All raceway terminations must be made in listed conduit hubs. The use of sealing-type locknuts is prohibited—instead, a hub is required.

(2) The manufacturer's installation instructions for the fire pump controller must be consulted and all instructions specific to the controller used must be followed.

(3) Any alterations to the fire pump controller, other than conduit entry, must be approved by the authority having jurisdiction. These new requirements are intended to protect the integrity of the fire pump controller with respect to the entry of water.

Comment: None

Proposal: 13-97

Control Wiring

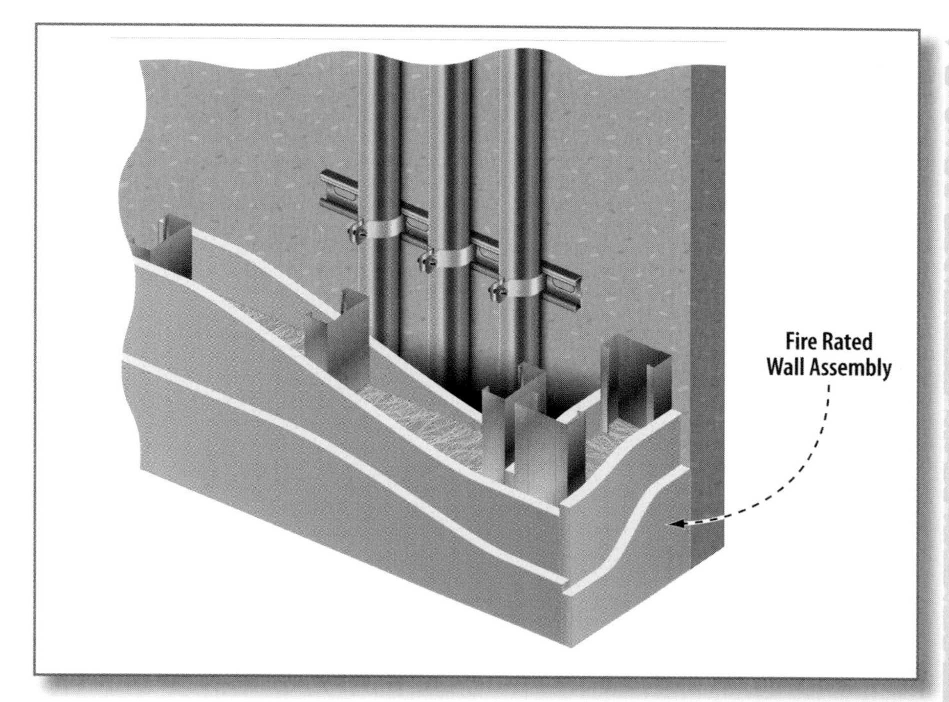

Code Language

695.14 Control Wiring

(F) Generator Control Wiring Methods. Control conductors installed between the fire pump power transfer switch and the standby generator supplying the fire pump during normal power loss shall be kept entirely independent of all other wiring. They shall be protected to resist potential damage by fire or structural failure. They shall be permitted to be routed through a building(s) using one of the following methods:

(1) Be encased in a minimum of 50 mm (2 in.) of concrete

(2) Be protected by a fire-rated assembly listed to achieve a minimum fire rating of 2 hours and dedicated to the fire pump circuits.

(3) Be a listed electrical circuit protective system with a minimum 2-hour fire rating. The installation shall comply with any restrictions provided in the listing of the electrical circuit protective system used.

Informational Note: UL guide information for electrical circuit protective systems (FHIT) contains information on proper installation requirements to maintain the fire rating.

Significance of the Change

Section 695.14(F) has been revised and a list format developed for clarity. The 2008 *NEC* required that the control conductors be installed between the fire pump power transfer switch and that the standby generator supplying the fire pump be routed through a building(s) encased in 50 mm (2 in.) of concrete or within enclosed construction dedicated to the fire pump circuits with a minimum 1-hour fire resistance rating, or circuit protective systems with a minimum of 1-hour fire resistance. This revision increases the fire rating from 1 hour to 2 hours.

These conductors must be kept entirely independent of all other wiring and must be protected to resist potential damage by fire or structural failure. The conductors are permitted either to (1) be encased in 2 inches of concrete; (2) be protected by a fire-rated assembly, such as structural elements, listed to achieve a minimum fire rating of 2 hours; or (3) be a listed electrical circuit protective system with a minimum of 2-hour fire rating. Where a 2-hour listed electrical circuit protective system is used, the installation must comply with any restrictions provided in the listing of the electrical circuit protective system. For example, 2-hour-rated conductors that are installed in steel conduit may require supports every 5 feet instead of every 10 feet.

Change Summary

A 2-hour fire rating is now clearly required for fire-rated assemblies and electrical circuit protective systems protecting control conductors installed between the fire pump power transfer switch and the standby generator supplying the fire pump. Electrical circuit protective systems are now required to be listed.

Comment: None

Proposal: 13-131

Chapter 7

Articles 700-770

Special Conditions

Significant Changes
TO THE NEC® 2011

Automatic Load Control Relay

Code Language

700.2 Definitions

Relay, Automatic Load Control. A device used to energize switched or normally-off lighting equipment from an emergency supply in the event of loss of the normal supply, and to de-energize or return the equipment to normal status when the normal supply is restored.

Informational Note: For requirements covering automatic load control relays, see ANSI/UL 924, *Emergency Lighting and Power Equipment.*

700.24 Automatic Load Control Relay. If an emergency lighting load is automatically energized upon loss of the normal supply, a listed automatic load control relay shall be permitted to energize the load. The load control relay shall not be used as transfer equipment.

Change Summary

Section 700.2 has been added to define the term *Automatic Load Control Relay* along with a new requirement in 700.24. This new requirement permits an emergency lighting load to be automatically energized upon loss of the normal supply through a listed automatic load control relay.

Comment: 13-116

Proposals: 13-142, 145, 146, 188

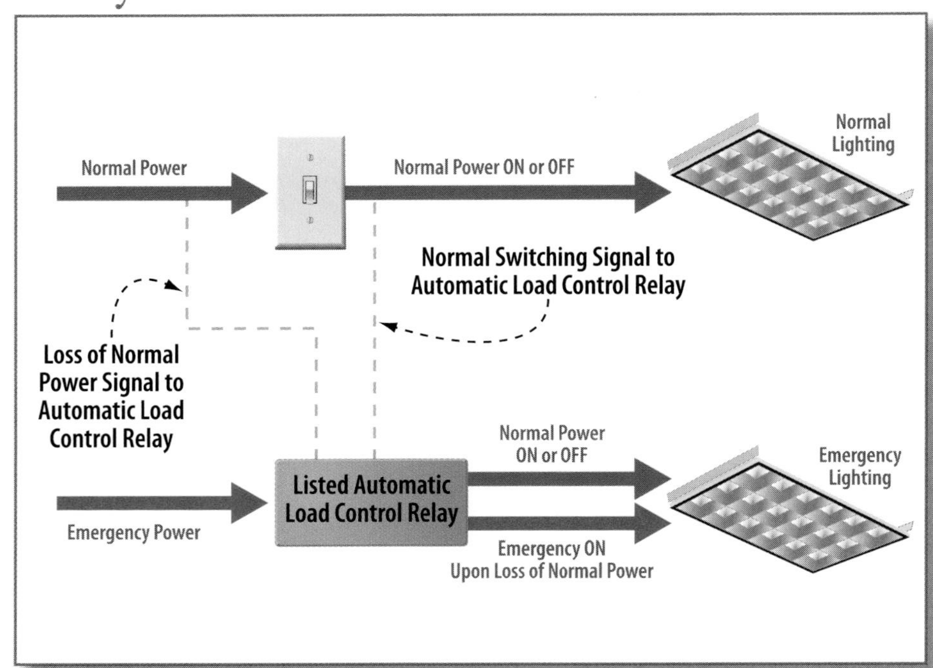

Significance of the Change

Stand-alone automatic load control relays are now readily available for use to automatically energize emergency luminaires upon loss of the normal supply. This new class of device, listed to the UL 924 standard, is clearly permitted by this new requirement to use these devices for automatic energization of emergency lighting. These relays allow the luminaire to be switched via a snap switch under normal conditions. Upon loss of normal power, the automatic load control relay will energize the load and, upon the return of normal power, will return the load to its previous normal supply condition. These devices are required to be listed and are not permitted to be used as a transfer switch.

The new definition of Automatic Load Control Relay, as well as the new requirement in 700.24, limits the use of this product to emergency lighting. The definition provides the *Code* user with a clear, concise description limiting it to emergency system use.

Section 700.10(B) has also been modified to clearly permit the use of these devices as follows:

700.10 Wiring, Emergency System. (B) Wiring.

(3) Wiring from two sources in a listed load control relay supplying exit or emergency luminaires, or in a common junction box, attached to exit or emergency luminaires.

Required Signage

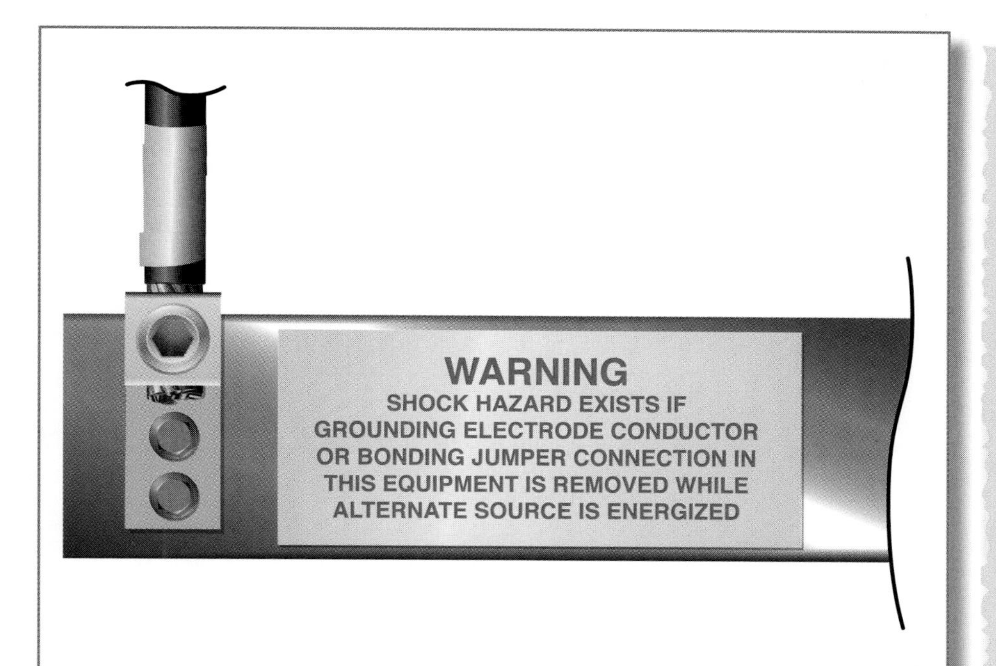

Code Language

700.7 Signs.

(B) Grounding. Where removal of a grounding or bonding connection in normal power source equipment interrupts the grounding electrode conductor connection to the alternate power source(s) grounded conductor, a warning sign shall be installed at the normal power source equipment stating:

WARNING

SHOCK HAZARD EXISTS IF GROUNDING ELECTRODE CONDUCTOR OR BONDING JUMPER CONNECTION IN THIS EQUIPMENT IS REMOVED WHILE ALTERNATE SOURCE(S) IS ENERGIZED.

Significance of the Change

This revision provides significant clarification of an existing requirement. Many users of the *NEC* incorrectly think that this requirement is simply about a connection between a grounding electrode and a grounding electrode conductor. It includes much more. This requirement addresses every termination of a normal grounded conductor between the connection to the grounding electrode in the normal system and the point at which the normal grounded conductor is bonded to the alternate power source grounded conductor.

This requirement resulted from an actual case where the emergency source was supplying power, and during that period maintenance personnel disconnected the normal source grounded conductor for testing purposes. The personnel did not realize that they were also disconnecting the grounding connection for the emergency source at the same time, since the grounded system conductor was only connected to the grounding electrode conductor in the main switchboard.

Prescriptive text is now provided for these signs. When the normal source is deenergized and the alternate source is operating, the grounded conductor of the normal system is in essence a bonding jumper connection to the grounding electrode for the alternate source.

Similar revisions have been made in 701.7(B) and 702.7(B).

Change Summary

The required signage to prohibit the removal of a grounding or bonding connection in a normal power source that would interrupt the grounding electrode conductor connection to the alternate power source has been clarified. Prescriptive text for the required sign has been provided.

Comment: 13-119

Proposal: 13-156

700.10(D)(1)

Article 700 Emergency Systems
Part II Circuit Wiring

Feeder-Circuit Wiring

Code Language

700.10 Wiring, Emergency System

(D) Fire Protection

(1) Feeder-Circuit Wiring. Feeder-circuit wiring shall meet one of the following conditions:

(1) Be installed in spaces or areas that are fully protected by an approved automatic fire suppression system

(2) Be a listed electrical circuit protective system with a minimum 2-hour fire rating

Informational Note: UL guide information for electrical circuit protection systems (FHIT) contains information on proper installation requirements to maintain the fire rating.

(3) Be protected by a listed thermal barrier system for electrical system components with a minimum 2-hour fire rating

(4) Be protected by a listed fire-rated assembly that has a minimum fire rating of 2 hours and contains only emergency wiring circuits.

(5) Be encased in a minimum of 50 mm (2 in.) of concrete

Change Summary

The minimum fire rating in 700.10(D)(1)(2), (D)(1)(3) and (D)(1)(4) has been increased to a 2-hour fire rating. The remainder of the requirement in 700.10(D) has been editorially revised. The requirement for the fire rating of spaces containing equipment in 700.10(D)(2) is also increased to a 2-hour rating.

Comment: None

Proposals: 13-172, 173, 174, 175a

Significance of the Change

This revision increases the minimum fire rating time from 1 hour to 2 hours. In 695.6(B) of the 2008 *NEC*, the minimum fire rating time was also increased from 1 hour to 2 hours. The additional hour is necessary to provide occupants enough time to safely exit a building as well as give firefighters additional time to bring the fire under control without losing power to the fire pump, the fire service elevator, and emergency lighting. The increase in fire rating applies to list items (1) through (5) as follows:

(1) An example of a space that is fully protected by an approved automatic fire suppression system would be vertical risers in a shaft that is protected with sprinklers.

(2) An example of a listed 2-hour electrical circuit protection system is a fire-rated cable type such as RHH installed in EMT.

(3) An example of a listed 2-hour thermal barrier system is a conduit wrap system to achieve the fire rating.

(4) An example of a listed 2-hour fire-rated assembly is construction building materials such as concrete board or drywall in accordance with the *UL Fire Resistance Directory*.

(5) Encasement in 2 inches of concrete is permitted to achieve the 2-hour rating.

701.6(D)

Ground Fault Indication

Code Language

701.6 Signals

(D) Ground Fault. To indicate a ground fault in solidly grounded wye, legally required standby systems of more than 150 volts to ground and circuit-protective devices rated 1000 amperes or more. The sensor for the ground fault signal devices shall be located at, or ahead of, the main system disconnecting means for the legally required standby source, and the maximum setting of the signal devices shall be for a ground-fault current of 1200 amperes. Instructions on the course of action to be taken in event of indicated ground fault shall be located at or near the sensor location.

Informational Note: For signals for generator sets, see NFPA 110-2005, *Standard for Emergency and Standby Power Systems.*

Change Summary

A new first-level subdivision has been added to 701.6 to require that audible and visual signal devices be installed to indicate a ground fault in solidly grounded wye legally required standby systems of more than 150 volts to ground and circuit-protective devices rated 1000 amperes or more.

Significance of the Change

This revision correlates the signal requirements of Articles 700 and 701 where a solidly grounded wye system of more than 150 volts to ground with circuit-protective devices rated 1000 amperes or more supply a legally required standby system. Sections 215.10 and 230.95 require that ground-fault protection for feeder and service disconnects rated 1000 amperes or more and installed on solidly grounded wye electrical systems of more than 150 volts to ground, but not exceeding 600 volts phase-to phase, be provided with ground-fault protection of equipment. However, this requirement has been modified by 700.26 and 701.26 for emergency and legally required systems. This revision now mandates that audible and visual signals be installed to signal a ground-fault condition in a legally required system where a solidly grounded wye system of more than 150 volts to ground with circuit-protective devices rated 1000 amperes or more is installed.

A change has also been made in 701.26 to refer *Code* users to the new signaling requirement.

Comment: None

Proposal: 13-218

701.12(B)(2)

Article 701 Legally Required Standby Systems
Part III Sources of Power

Fuel Transfer Pumps

Code Language

701.12 General Requirements

(B) Generator Set.

(2) Internal Combustion Engines as Prime Mover. Where internal combustion engines are used as the prime mover, an on-site fuel supply shall be provided with an on-premises fuel supply sufficient for not less than 2 hours' of full-demand operation of the system. Where power is needed for the operation of the fuel transfer pumps to deliver fuel to a generator set day tank, the pumps shall be connected to the legally required standby power system.

Change Summary

A new last sentence has been added to 701.12(B)(2) to require that fuel transfer pumps be supplied from the legally required system.

Significance of the Change

This revision correlates the power supply requirements for fuel transfer pumps in Articles 700 and 701. The previous requirement mandated only that where an internal combustion engine is used as the prime mover for a legally required standby system, an on-site fuel supply be provided with an on-premises fuel supply sufficient for not less than 2 hours' full-demand operation of the system. If a situation should occur that requires more than 2 hours' full-demand operation of the system, the generator would run out of fuel and fail.

This revision now mirrors the existing rule in 700.12(B)(2) for emergency systems. It clarifies that the power source for the fuel transfer pumps is supplied from the legally required system to ensure that the generator is supplied with fuel. A generator without fuel is no standby source at all.

Comment: None

Proposal: 13-224

705.6

Article 705 Interconnected Electric Power Production Sources
Part I General

System Installation

Code Language

705.6 System Installation. Installation of one or more electrical power production sources operating in parallel with a primary source(s) of electricity shall be installed only by qualified persons.

Informational Note: *See Article 100 for the definition of qualified person.*

Significance of the Change

Article 705 covers systems and equipment of one or more power production sources that operate in parallel with a primary source of electricity, typically supplied by a utility. These systems are often very complex, requiring specific skills and training to install interconnected power sources that will operate safely. This new requirement in 705.6 indicates that only qualified persons are permitted to install interconnected power production sources, equipment, and associated wiring. Typical interconnected power sources are photovoltaic systems, generators, wind generators, fuel cells, batteries, and so forth. It is commonly understood that electrical installations covered by the *Code* should always be performed by trained workers and contractors. This is what the unsuspecting public expects: safe electrical systems for persons and property. Amidst all the allure of growing new technologies and focus on smart grid systems, increasing numbers of interconnected power production sources are being installed. These systems are complicated, they are not plug-and-play electrical equipment, and this is definitely not a job for a handyman. This is electrical work that should be performed by qualified electricians and contractors.

Change Summary

New Section 705.6 System Installation has been added to Part I of Article 705. A new informational note now follows this section to refer users to the *NEC* definition of the term *Qualified Person*.

Comment: 4-125

Proposal: 4-266

705.12(A)

Article 705 Interconnected Electric Power Production Sources
Part I General

Supply Side

Code Language

705.12 Point of Connection

(A) Supply Side. An electric power production source shall be permitted to be connected to the supply side of the service disconnecting means as permitted in 230.82(6). The sum of the ratings of all overcurrent devices connected to power production sources shall not exceed the rating of the service.

Change Summary

A new last sentence has been added to 705.12(A) to limit the capacity of power production sources connected to the supply side of the service disconnecting means relative to the size of the service overcurrent protective device.

Significance of the Change

Section 705.12(A) permits an electric power production source to be connected to the supply side of the service disconnecting means as permitted by 230.82(6). This revision adds a new last sentence that requires that where power production sources are connected to the supply side of the service disconnecting means, the total ampere rating of all electric power production sources not exceed the ampere rating of the service. This new requirement prevents the potential output of electric power production sources connected to the supply side of the service disconnecting means from exceeding the capacity of the service. If this were to occur, the output of the electric power production sources could exceed the rating of the service overcurrent protective device, equipment, and conductors.

Comment: None

Proposal: 4-267

708.10(A)(2)

Article 708 Critical Operations Power Systems (COPS)
Part II Circuit Wiring and Equipment

Receptacle Identification

Code Language

708.10 Feeder and Branch Circuit Wiring.

(A) Identification

(2) Receptacle Identification. In a building in which COPS are present with other types of power systems described in other sections in this article, the cover plates for the receptacles or the receptacles themselves supplied from the COPS shall have a distinctive color or marking so as to be readily identifiable.

Exception: If the COPS supplies power to a DCOA that is a stand-alone building, receptacle cover plates or the receptacles themselves shall not be required to have distinctive marking.

Significance of the Change

The previous *Code* text required that all receptacles or the cover plates for the receptacles that are supplied from a COPS have a distinctive color or marking so as to readily identify the receptacle as being supplied from a COPS. The new text clarifies that this distinctive color or marking of COPS-supplied receptacles is required only where other types of power systems are present. For example, equipment that is considered critical could be fed from a COPS-supplied receptacle, and equipment not considered critical could be supplied in the same area from another power system.

A new exception has been added to completely exempt a COPS-supplied DCOA that is a stand-alone building from the marking requirement. A DCOA is defined in 708.2 as follows:

Designated Critical Operations Areas (DCOA). Areas within a facility or site designated as requiring critical operations power.

This new exception recognizes that in a stand-alone building that is designated as a DCOA, it is not necessary to mark receptacles because everything is COPS-supplied.

Change Summary

New text has been added to 708.10(A)(2) to clarify that receptacle identification is required only where other power systems are present. A new exception for a stand-alone COPS-supplied building has also been added to clearly exempt a designated critical operations area that is a stand-alone building from the marking requirement.

Comment: None

Proposal: 13-270

708.10(C)(2)

Article 708 Critical Operations Power Systems (COPS)
Part II Circuit Wiring and Equipment

Fire Protection for Feeders

Code Language

708.10 Feeder and Branch Circuit Wiring.

(C) COPS Feeder Wiring Requirements

(2) Fire Protection for Feeders. Feeders shall meet one of the following conditions:

(1) Be a listed electrical circuit protective system with a minimum 2-hour fire rating

Informational Note: UL guide information for electrical circuit protection systems (FHIT) contains information on proper installation requirements to maintain the fire rating.

(2) Be protected by a listed fire-rated assembly that has a minimum fire rating of 2 hours

(3) Be encased in a minimum of 50 mm (2 in.) of concrete

Change Summary

The minimum fire rating for COPS feeders that are a listed electrical circuit protective system or are installed in a listed fire-rated assembly has increased from a minimum fire rating of 1 hour to 2 hours.

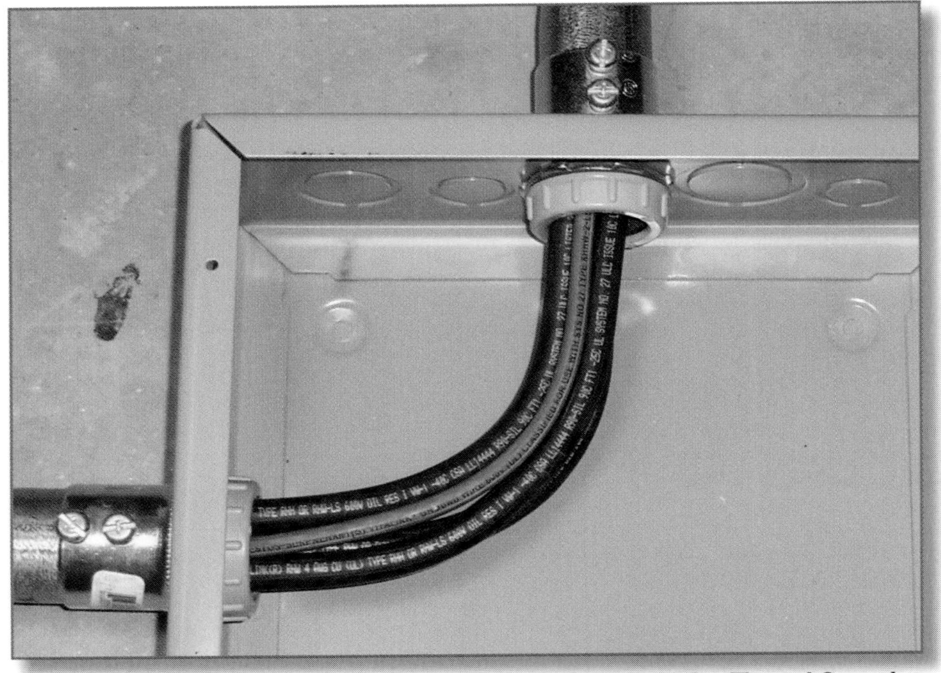

Courtesy of Tyco Thermal Controls

Significance of the Change

This revision increases the minimum fire rating for feeders supplying a critical operations power system time from 1 hour to 2 hours. In 695.6(B) of the 2008 *NEC*, the minimum fire rating time for feeders was increased from 1 hour to 2 hours for fire pumps, and a similar change has occurred this cycle in Article 700 for emergency systems. The additional hour of fire rating is necessary to ensure the continuous provision of power to the critical operations power systems. The increase in fire rating applies to list items (1) and (2) as follows:

(1) An example of a listed 2-hour electrical circuit protection system is fire-rated cable such as Type RHW installed in EMT.

(2) An example of a listed 2-hour fire-rated assembly is construction building materials such as concrete board or drywall in accordance with the *UL Fire Resistance Directory*.

(3) Encasement in a minimum of 2 inches of concrete is also permitted.

Note: A Rachem 2-hour Fire-Rated RHW product is shown in the photo.

Comment: None

Proposal: 13-273

708.14

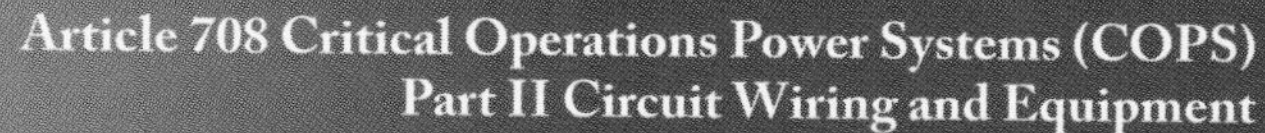

Systems

Code Language

708.14 Wiring of HVAC, Fire Alarm, Security, Emergency Communications, and Signaling Systems. All conductors or cables shall be installed using any of the metal wiring methods permitted by 708.10(C)(1) and in addition shall comply with 708.14(1) through (8), as applicable.

(1) All cables for fire alarm, security, signal systems, and emergency communications shall be shielded twisted pair cables.

(2) Shields of cables for fire alarm, security, signal systems, and emergency communications shall be continuous.

(3) ***(Editorially revised)***

(4) A listed primary protector shall be provided on all communications circuits. Listed secondary protectors shall be provided at the terminals of the communication circuits.

(5) ***(Editorially revised)***

(6) ***(No changes)***

(7) All cables for fire alarm, security, and signaling systems shall be riser-rated and shall be a listed 2-hour electrical circuit protective system. Riser emergency communication cables shall be Type CMR-CI or a listed 2-hour electrical circuit protective system.

(8) ***(Editorially revised)***

Significance of the Change

This section requires that wiring of HVAC, fire alarm, security, emergency communications, and signaling systems be installed using any of the metal wiring methods permitted by 708.10(C)(1) for physical protection. The following revisions apply as applicable:

(1) Sections 708.14(1) & (2) have been expanded to include fire alarm and security cables, and the qualifier "emergency" is new to communications circuits. These cables must now be shielded twisted pair cables, and the shields must be continuous.

(2) Section 708.14(4) now requires that a listed primary protector be provided on all communications circuits. A primary protector provides protection against overvoltage and overcurrent conditions that can occur if contact is made between outside communication conductors and power sources (electric utility) outside of a building or structure.

(3) Section 708.14(7) has been expanded to require a listed 2-hour electrical circuit protective system for all fire alarm, security, and signal systems and for emergency communications cables. The previous requirement for a 2-hour rating applied only to communications cables. Type CMR-CI or a 2-hour electrical circuit protective system is required for riser emergency communications cables.

Change Summary

Primary protectors are now required for all communications circuits. All cables must now be shielded twisted pair cables, the shields must be continuous, and all cables must be a 2-hour electrical circuit protective system.

Comment: 13-181

Proposals: 13-277, 13-281, 13-282

725.3(I) & 760.3(I)

Article 725 Class 1, Class 2, and Class 3 Remote-Control, Signaling, and Power-Limited Circuits & Article 760 Fire Alarm Systems
Part I General

Vertical Support

Code Language

725.3 Other Articles.

(I) Vertical Support for Fire-Rated Cables and Conductors. Vertical installations of circuit integrity (CI) cables and conductors installed in a raceway or conductors and cables of electrical circuit protective systems shall be installed in accordance with 300.19(B).

760.3 Other Articles.

(I) Vertical Support for Fire-Rated Cables and Conductors. Vertical installations of circuit integrity (CI) cables and conductors installed in a raceway or conductors and cables of electrical circuit protective systems shall be installed in accordance with 300.19(B).

Change Summary

A new first-level subdivision has been added to 725.3 and 760.3 to mandate that the vertical support requirements of 300.19(B) apply to all circuit integrity (CI) cables and conductors installed in raceways and cables of electrical circuit protective systems used in a Class 1, Class 2, and Class 3 remote-control, signaling, power-limited circuit and fire alarm conductors.

Courtesy of Tyco Thermal Controls

Significance of the Change

The parent text of 725.3 and 760.3 clearly states that the only sections of Article 300 that apply to the Class 1, Class 2, and Class 3 remote-control, signaling, and power-limited circuits of Article 725 and fire alarm systems in Article 760 are those sections referenced in 725.3 and 760.3.

These new first-level subdivisions now require that where circuit integrity (CI) cables and conductors installed in raceways and cables of electrical circuit protective systems are installed vertically, they must be supported in accordance with 300.19(B). These support requirements for fire-rated cables/conductors and systems are critical to ensuring that they survive in a fire situation, for the strength of copper decreases with heat and cables may fail if not properly supported.

Note: A Rachem 2-hour Fire-Rated RHW product is shown in the photo.

Comments: 3-105, 3-164

Proposals: 3-159, 3-242

725.3(J) & 760.3(K)

Bushings

Code Language

725.3 Other Articles.

(J) Bushing. A bushing shall be installed where cables emerge from raceway used for mechanical support or protection in accordance with 300.15(C).

760.3 Other Articles.

(K) Bushing. A bushing shall be installed where cables emerge from raceway used for mechanical support or protection in accordance with 300.15(C).

Change Summary

A new first-level subdivision has been added to 725.3 and 760.3 to mandate that, where a raceway is used for mechanical protection or support, a bushing be installed to protect the conductors.

Significance of the Change

The parent text of 725.3 and 760.3 clearly states that the only sections of Article 300 that apply to the Class 1, Class 2 and Class 3 remote-control, signaling, and power-limited circuits of Article 725 and the fire alarm systems in Article 760 are those sections referenced in 725.3 and 760.3. These new first-level subdivisions require that, where a raceway is used for mechanical protection or support, a bushing be installed to protect the conductors. It is typical to support and protect Class 1, Class 2, and Class 3 remote-control, signaling, and power-limited circuits and fire alarm circuits with raceways. A bushing is now required wherever this method is employed.

Note that a new first-level subdivision has been added to both 725.3 and 760.3 that applies the provisions of 300.7(A) to both articles. Where raceways or sleeves are exposed to different temperatures, they must now be sealed.

Comments: 3-115, 3-181

Proposals: 3-191, 3-272

760.3(J)

Article 760 Fire Alarm Systems
Part I General

Raceway Fill

Code Language

760.3 Other Articles

(J) Number and Size of Cables and Conductors in Raceway. Installations shall comply with 300.17.

Change Summary

The installation of fire alarm conductors in raceways is now subject to the conduit fill requirements in Table 1, Chapter 9.

Significance of the Change

The parent text of 760.3 clearly states that the only sections of Article 300 that apply to fire alarm systems are those sections referenced in this section. Fire alarm systems are typically installed in raceways. This new requirement is not intended to address a buildup of heat from fire alarm conductors installed in raceways, but to limit potential damage to conductors where raceways are overfilled. This new first-level subdivision limits the number of conductors permitted in a given raceway and refers the *Code* user to 300.17, which follows:

> **300.17 Number and Size of Conductors in Raceway.** The number and size of conductors in any raceway shall not be more than will permit dissipation of the heat and ready installation or withdrawal of the conductors without damage to the conductors or to their insulation.

The associated informational note refers the *Code* user to the fill requirements for raceways. For example, where EMT is used, 358.22 Number of Conductors requires that the fill not exceed that permitted by the percentage specified in Table 1, Chapter 9. It is important to note that the diameter of fire alarm cables is not listed in Chapter 9. Multiconductor cables are considered as a single conductor when calculating the percentage of fill.

Comment: None
Proposal: 3-244

760.41 & 760.121(B)

Article 760 Fire Alarm Systems
Part II Non–Power-Limited Fire Alarm (NPLFA) Circuits & III Power-Limited Fire Alarm (PLFA) Circuits

Fire Alarm Branch Circuits

Code Language

760.41 NPLFA Circuit Power Source Requirements & 760.121 Power Sources for PLFA Circuits

(B) Branch Circuit. The branch circuit supplying the fire alarm equipment(s) shall supply no other loads. The location of the branch-circuit overcurrent protective device shall be permanently identified at the fire alarm control unit. The circuit disconnecting means shall have red identification, shall be accessible only to qualified personnel, and shall be identified as "FIRE ALARM CIRCUIT." The red identification shall not damage the overcurrent protective devices or obscure the manufacturer's markings. This branch circuit shall not be supplied through ground-fault circuit interrupters or arc-fault circuit-interrupters.

Change Summary

The requirements for branch circuits supplying non-power-limited and power-limited fire alarm circuits have been revised. The location of the overcurrent protective device supplying the fire alarm equipment must be identified at the fire alarm equipment. The OCPD must be identified using the color red, be accessible only to qualified personnel and marked to identify it as supplying the fire alarm system.

Significance of the Change

The previous edition of the *NEC* required that an individual branch circuit supply fire alarm equipment. The revised text requires that the branch circuit supply no loads other than fire alarm equipment, meaning that a single circuit could supply two fire alarm panels.

The new marking provision requires that the location of the branch-circuit overcurrent protective device be identified at the fire alarm control unit. This marking must be "permanent" and most likely requires a plaque or permanent label of some type. Access to the OCPD is now limited to "qualified personnel" to limit the possibility of inadvertent disconnection. The circuit disconnecting means (most often a circuit breaker) is now permitted by 760.41(A) to be secured in the "on" position to prevent the inadvertent disconnection of the fire alarm system. For example, a device could be used that requires that a screw be backed out to remove the device and open the breaker.

This fire alarm circuit disconnecting means (most often a circuit breaker) is also required to have a red identification and must be labeled as "FIRE ALARM CIRCUIT." The red marking is not permitted to damage or obscure markings on the disconnecting means.

The prohibition of AFCI or GFCI protection of circuits supplying fire alarm system equipment remains.

Comments: 3-176, 3-189

Proposals: 3-259, 3-280

770.2

Article 770 Optical Fiber Cables and Raceways
Part I General

Definition of Cable Routing Assembly

Code Language

770.2 Definitions.

Cable Routing Assembly. A single channel or connected multiple channels, as well as associated fittings, forming a structural system that is used to support, route and protect high densities of wires and cables, typically communications wires and cables, optical fiber and data (Class 2 and Class 3) cables associated with information technology and communications equipment.

Change Summary

A new definition of *Cable Routing Assembly* has been added to Article 770 to address a new method for supporting, routing, and protecting fiber optical and other wires and cables associated with information technology and communications equipment. This definition is also referenced in and used throughout Articles 800 and 820.

Comment: 16-8

Proposal: 16-12

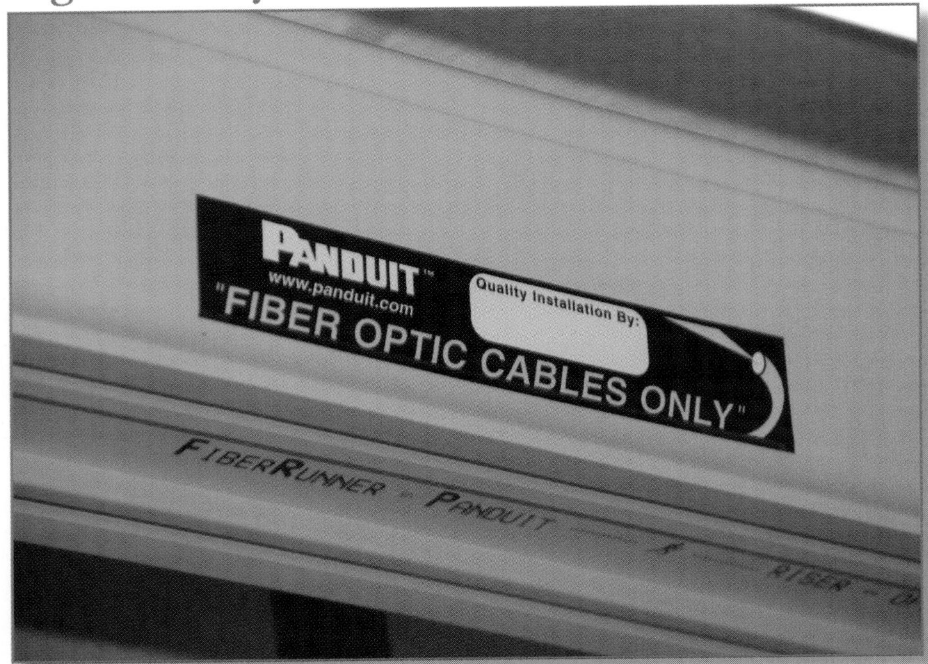

Significance of the Change

A cable routing system is a single channel or collection of channels, fittings, and mounting brackets that can be assembled to create a structure that supports, routes, and protects fiber optic cabling, CATV cabling, and high-performance copper data cabling from physical damage that can disrupt or cut off signal transmission. A cable routing system provides a method to neatly maintain existing network operations and implement new services. Cable routing systems are used in data centers and/or central offices as a cable management system to ensure optimum network performance.

This new term is referenced in and used throughout Articles 770, 800, and 820 in sections that address permitted types of raceways, enclosures, and the like. This definition applies in both Article 800 Communications Circuits and Article 820 Community Antenna Television and Radio Distribution Systems. Additionally, a new first-level subdivision has been added to 800.3(G) and 820.3(H) as follows:

Cable Routing Assemblies. The definition in 770.2, the applications in 770.154, and installation rules in 770.113 shall apply to Article 800/820.

770.113

Article 770 Optical Fiber Cables and Raceways
Part V Installation Methods Within Buildings

Installation

Significance of the Change

This revision properly relocates all installation criteria from the cable and raceway application sections to 770.113, where they logically belong. Additionally, this section has been expanded to include cable routing assemblies as defined in 770.2. This reorganization enhances usability and increases clarity and is tied to the revision of 770.154 addressing applications of listed optical fiber cables and raceways, and cable routing assemblies. The installation criteria of 770.154 were relocated to 770.113, and the hazardous locations installation requirements previously in 770.154 were relocated to 770.3. A new subdivision addresses optical fiber cable used in optical cross-connect applications. All optical fiber cables and raceways, and cable routing assemblies installed in buildings, are required to be listed.

This revision greatly increases usability by providing prescriptive list items to clearly illustrate the types of cable and raceways permitted.

Similar changes occurred in 800.113 and 800.154.

Code Language

770.113 Installation of Optical Fiber Cables and Raceways, and Cable Routing Assemblies. Installation of optical fiber cables and raceways, and cable routing assemblies shall comply with 770.113(A) through (J). Installation of raceways shall also comply with 770.12 and 770.110.

(A) Listing.

(B) Fabricated Ducts Used for Environmental Air.

(C) Other Spaces Used For Environmental Air (Plenums).

(D) Risers — Cables, Raceways and Cable Routing Assemblies in Vertical Runs.

(E) Risers — Cables and Raceways in Metal Raceways.

(F) Risers — Cables, Raceways and Cable Routing Assemblies in Fireproof Shafts.

(G) Risers — One- and Two-Family Dwellings.

(H) Cable Trays.

(I) Distributing Frames and Cross-Connect Arrays.

(J) Other Building Locations.

Change Summary

The installation requirements for optical fiber cables and raceways, and cable outing assemblies have been organized and clarified in a single section.

Comments: 16-34, 16-37

Proposal: 16-48

770.154

Article 770 Optical Fiber Cables and Raceways
Part V Installation Methods Within Buildings

Application

Code Language

770.154 Applications of Listed Optical Fiber Cables and Raceways, and Cable Routing Assemblies. Permitted and nonpermitted applications of listed optical fiber cables and raceways, and cable routing assembly types shall be as indicated in Table 770.154(a). The permitted applications shall be subject to the installation rules of 770.110 and 770.113. The substitutions for optical fiber cables requirements in Table 770.154(b) and illustrated in Figure 770.154 shall be permitted.

Change Summary

Section 770.154 has been expanded to include cable routing assemblies. A new comprehensive Table 770.154(a) has been added to provide clarity and usability in the determination of permitted optical fiber cables, raceways, and cable routing assemblies.

Significance of the Change

The application of all listed optical fiber cables, raceways, and cable routing assemblies has been comprehensively revised and reformatted in new Table 770.154(a). The parent text of 770.154 has been similarly revised to send the *Code* user to the new table when determining permitted use. This revision is tied to a rewrite of the installation requirements in 770.113 and includes a redundant reference to 770.110 and 770.113, to ensure that the *Code* user applies raceways correctly and that the installation requirements are met.

The new user-friendly table is separated into columns, allowing the *Code* user to choose an application and then determine permitted cable types, cable routing assembly types, and raceway types. The table listing cable substitutions, previously numbered 770.154(E), has been editorially changed to Table 770.154(b).

Comment: 16-44

Proposal: 16-56

Article 770 and Chapter 8

Articles 770, 800, 810, 820, 830, and 840
Parts I, III, and IV

Grounding and Bonding Terminology

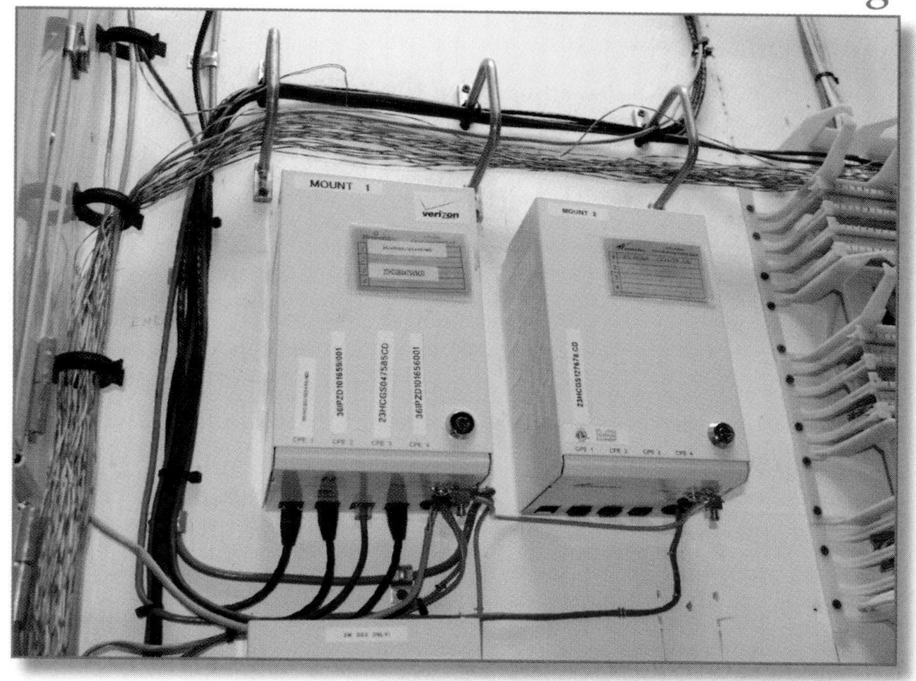

Code Language

Article 770 Fiber Optic Cables and Raceways

Part III Protection. ***(See actual text for locations where the term "grounding conductor" has been replaced" by either "grounding electrode conductor" or "bonding conductor or jumper.")***

Part IV Grounding Methods. ***(See actual text for locations where the term "grounding conductor" has been replaced by either "grounding electrode conductor" or "bonding conductor or jumper.")***

(... Continued on next page ...)

Change Summary

The term *grounding conductor* has been deleted from Articles 770, 800, 820, and 830 and has been replaced by the term *grounding electrode conductor* or the term *bonding conductor*, depending on the function covered by that specific rule. Article 840 is new, and the same revisions in grounding and bonding terminology have been incorporated into new Article 840 covering Premises-Powered Broadband Communications Systems.

See the following page for additional information related to this change.

Grounding and Bonding Terminology (continued)

Code Language

(... Continued from previous page ...)

Article 800 Communications Circuits.

Part III Protection. *(See actual text for locations where the term "grounding conductor" has been replaced" by either "grounding electrode conductor" or "bonding conductor or jumper.")*

Part IV Grounding Methods. *(See actual text for locations where the term "grounding conductor" has been replaced" by either "grounding electrode conductor" or "bonding conductor or jumper.")*

See Parts I, III, and IV of Articles 770, 800, 810, 820, 830, and 840 for actual text that reflects where the term "grounding conductor" has been replaced by the appropriately defined term.

Comments:16-84a, 16-2, 16-30a, 16-32a, 16-35, 16-37, 16-38, 16-103, 16-116, 16-122, 16-125, 16-183, 16-190, 16-197, 16-201, 16-211, 16-224, 16-266, 16-277, 16-282, 16-318, 16-334

Proposals: 16-91, 16-212, 16-225, 16-293, 16-349

Significance of the Change

CMP-5 acted favorably to deleting the term *grounding conductor* from Article 100 and Article 250 where it was previously used. This term essentially means the same thing as the defined term *Grounding Electrode Conductor*, and both defined terms address a conductor that has established a connection to ground (earth). Using two terms and definitions that refer to the same component of the grounding and bonding system is unnecessary and can lead to confusion in the field. Grounding electrode conductors are the components that establish the connection to the electrode system, whereas bonding conductors connect equipment and systems together. Since *grounding conductor* was used multiple times in Article 770 and in the Chapter 8 articles, a group of coordinated proposals were submitted to address each instance of the term *grounding conductor* and replace it with either the term *grounding electrode conductor* or the term *bonding conductor or jumper*, depending on which component the rule addresses.

The result is more consistency throughout the *NEC* regarding the use of defined grounding and bonding words or terms. CMP-16 acted favorably to the proposals and comments and incorporated a useful informational note within each of the communications articles to inform users of the revision. It should be understood that no technical changes to current requirements within these articles resulted from this series of revisions, just changes in terminology to ensure that the appropriately defined grounding or bonding term is used. The problem with the previous term is that it was very broad in meaning and was used in rules related to bonding functions and in other rules related to grounding functions.

The changes align the grounding and bonding terms used in Articles 770 through 840 with defined grounding and bonding terms provided in Article 100. The revisions should assist users in understanding which function is intended to be accomplished from a performance perspective while at the same time promoting application of the correct rule to the grounding or bonding component required for communications systems and equipment. Additionally, the new figures in Article 800 provide users with a graphic representation of grounding functions and bonding functions in simple, single-line format.

See the previous page for additional information related to this change.

The Scope of 2011 *NEC* Code-Making Panels

2011 *NEC* CODE–MAKING PANELS	
***NEC* CODE–MAKING PANEL**	**ARTICLES, ANNEX AND CHAPTER 9 MATERIAL WITHIN THE SCOPE OF THE CODE–MAKING PANEL**
1	90, 100, 110, Annex A, Annex G
2	210, 215, 220, Annex D Examples 1 – 6
3	300, 509, 720, 725, 727, 760, Chapter 9 Tables 11(A) and (B), Tables 12(A) and (B)
4	225 & 230
5	200, 250, 280, 285
6	310, 400, 402, Chapter 9 Tables 5 – 9, Annex B
7	320, 322, 324, 326, 328, 330, 332, 334, 336, 338, 340, 382, 394, 396, 398
8	342, 344, 348, 350, 352, 353, 354, 355, 356, 358, 360, 360, 362, 366, 368, 370, 372, 374, 376, 378, 380, 384, 386, 388, 390, 392, Chapter 9 Tabels 1 – 4, Annex C
9	312, 314, 404, 408, 450, 490
10	240
11	409, 430, 440, 460, 470, Annex D, Example D8
12	610, 620, 625, 626, 630, 640, 645, 647, 650, 660, 665,668, 669, 670, 685, Annex D, Examples D9 and D10
13	445, 455, 480, 690, 692, 695, 700, 701, 702, 705
14	500, 501, 502, 503, 504, 505, 506, 510, 511, 513, 514, 515, 516
15	517, 518, 520, 522, 525, 530, 540
16	770, 800, 810, 820, 830
17	422, 424, 426, 427, 680, 682
18	406, 410, 411, 600, 605
19	545, 547, 550, 551, 552, 553, 555, 604, 675, Annex D Examples D11 and D12

Chapter 8

Articles 800-840

Communications Systems

Significant Changes
TO THE NEC® 2011

800

Article 800 Communications Circuits

Bonding/Grounding

Code Language

Article 800 Communications Circuits

Informational Note: The general term *grounding conductor* as previously used in this article is replaced by either the term *bonding conductor* or the term *grounding electrode conductor* (GEC), where applicable, to more accurately reflect the application and function of the conductor. See Informational Note Figure 800(a) and Figure 800(b).

Informational Note Figure 800(a) Example of the Use of the Term *Bonding Conductor* Used in a Communications Installation.

Informational Note Figure 800(b) Example of the Use of the Term *Grounding Electrode Conductor* Used in a Communications Installation.

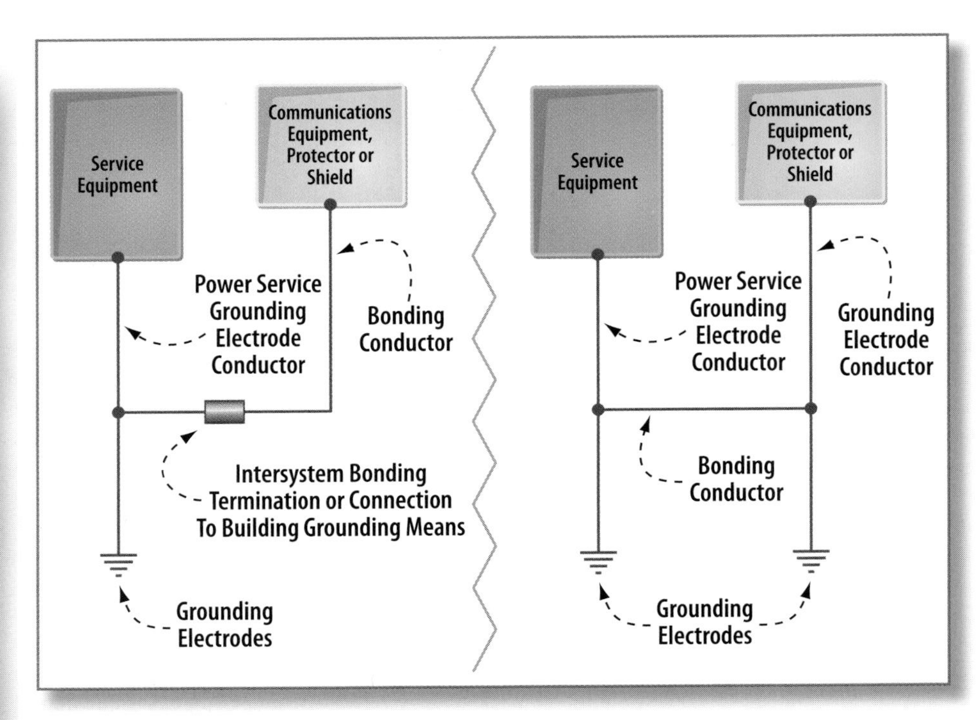

Change Summary

Throughout Chapter 8 the term *grounding conductor* has been replaced with a new proper term. A new informational note explains that the use of the terms *bonding conductor* and *grounding electrode conductor* are more accurate, and two new diagrams clearly illustrate how these terms are applied in Article 800.

Comment: 16-84a

Proposal: 16-91

Significance of the Change

A new informational note now follows the Article 800 title to inform the *Code* user that the term *grounding conductor* has been deleted and replaced for clarity throughout the article. In fact, throughout Article 770 and Chapter 8 the term *grounding conductor* has been replaced by the term *grounding electrode conductor* or the term *bonding conductor*, depending upon the individual requirement. These changes are intended to better align Chapter 8 and Article 770 with the rest of the *NEC*. This is a significant change within the communications industry, which has used the term *grounding conductor* exclusively for decades.

The two new figures provide the *Code* user with additional clarity by illustrating the use of these terms as they apply in an Article 800 installation. Informational Note Figure 800(a) provides an example for *bonding conductor* as applied in a communications installation and clearly shows that the communications system is grounded through a "bonding conductor" that is installed from the communications equipment to the intersystem bonding termination. Informational Note Figure 800(b) illustrates a communications system that has a grounding electrode system that is separate from the service grounding electrode system and a bonding conductor that is utilized for intersystem bonding.

800.154

Article 800 Communications Circuits
Part V Installation Methods Within Buildings

Application

Code Language

800.154 Applications of Listed Communications Wires, Cables, and Raceways. Permitted and nonpermitted applications of listed communications wires, cables, and raceways shall be as indicated in Table 800.154(a). The permitted applications shall be subject to the installation requirements of 800.110 and 800.113. The substitutions for communications cables listed in Table 800.154(b) and illustrated in Figure 800.154 shall be permitted.

Significance of the Change

The application of all listed communications wires, cables, and raceways has been comprehensively revised and formatted in new Table 800.154(a). The prior subdivisions have been deleted, and the parent text of 800.154 has been revised to send the *Code* user to the new table to determine permitted use. This revision is tied to a rewrite of the installation requirements in 800.113. Revision of the permitted applications includes a redundant reference in Table 800.154(a) to 800.110 and 800.113, to ensure that the *Code* user applies raceways correctly and that all installation requirements are met.

Table 800.154(a) is user friendly and is structured in columns that allow the *Code* user to choose an application, then determine permitted cable types and permitted types of raceways. Previous 2008 *NEC* Table 800.154(E) Cable Substitutions has been editorially renumbered to Table 800.154(b).

Change Summary

Section 800.154 has been revised by deleting multiple first-level subdivisions and adding a new comprehensive table. This revision provides the *Code* user with clarity and usability through new Table 800.154(a), which outlines all permitted as well as nonpermitted applications of listed communications wires, cables, and raceways.

Comment: 16-172

Proposal: 16-148

Dwelling Unit Outlet

Code Language

800.156 Dwelling Unit Communications Outlet. For new construction, a minimum of one communications outlet shall be installed within the dwelling in a readily accessible area and cabled to the service provider demarcation point.

Change Summary

The requirement for a hard-wired communications outlet in dwelling units has been clarified by mandating that the outlet be in a readily accessible area to facilitate its use by the occupant.

Significance of the Change

This revision provides clarity by requiring that the communications outlet in new dwelling unit construction be located in a readily accessible area. Without this qualifier, an installer could simply place an outlet in an unfinished basement next to the service provider demarcation point, rendering the outlet useless to the occupant. The requirement for a dwelling unit communications outlet was added to the 2008 *NEC* by mandating that all new dwelling units provide the occupant with an outlet cabled to the service provider demarcation point. However, the provision did not mandate that a service provider exist, only that a communications outlet be wired to a location at which the service provider would enter the dwelling unit.

This new requirement was added to the *NEC* for many reasons, including but not limited to installation problems, the ability to contact emergency services, and the Americans with Disabilities Act (ADA) requirements. Moreover, the use of mobile technology and cell phones has moved many homeowners to rely strictly on their mobile/cell phones and ignore their hard-wired or landline connections. This requirement recognizes that the owner of a dwelling unit will eventually move and that the next occupant may require the use of a readily accessible communications outlet.

Comment: None

Proposal: 16-184

Article 840

Article 840 Premises-Powered Broadband Communications Systems

Premises-Powered Broadband Communications Systems

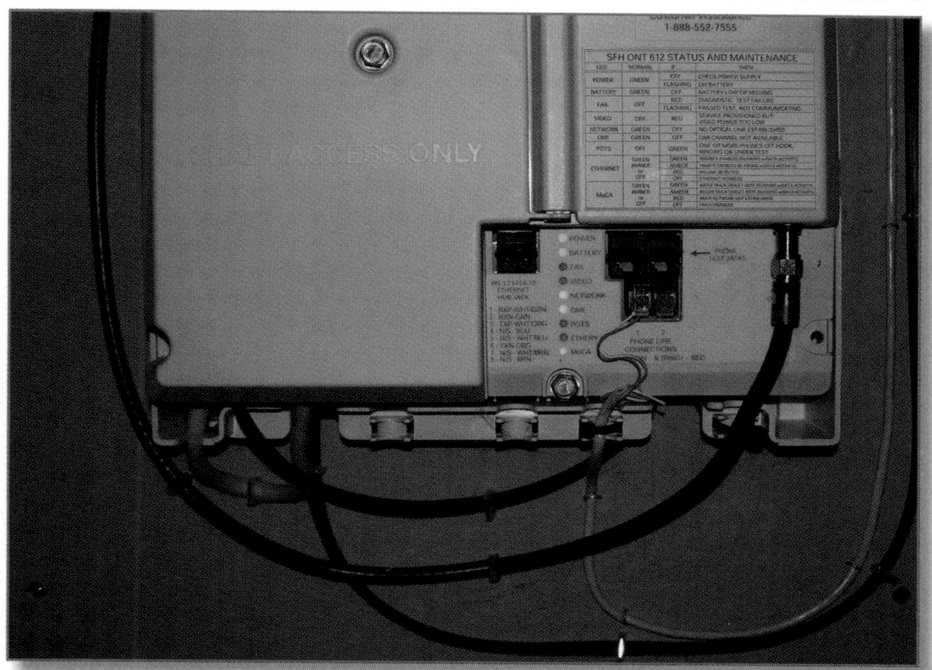

Code Language

840.1 Scope. This article covers premises-powered optical fiber-based broadband communications systems that provide any combination of voice, video, data, and interactive services through an optical network terminal (ONT).

Significance of the Change

The 2008 *NEC* did not specifically address premises-powered optical fiber-based broadband communications systems, a void that this new article will fill. Two key definitions exist in 840.2:

Fiber-to-the-Premises (FTTP). Conductive or nonconductive optical cable that is either aerial, buried, or through a raceway and is terminated at an optical network terminal (ONT) and establishing a communications network.

Optical Network Terminal (ONT). A device that converts an optical signal into component signals, including voice, audio, video, data, wireless, and interactive service electrical, and is considered to be network interface, equipment.

An informational note follows the scope of the article to explain for the *Code* user the type of system covered:

Informational Note No. 1: A typical basic system configuration consists of an optical fiber cable to the premises (FTTP) supplying a broadband signal to an ONT that converts the broadband optical signal into component electrical signals, such as traditional telephone, video, high-speed internet, and interactive services. Powering of the ONT is typically accomplished through an ONT power supply unit (OPSU) and battery backup unit (BBU) that derive their power input from the available AC at the premises. The optical fiber cable is unpowered and may be nonconductive or conductive.

Change Summary

A new Article 840 has been added to Chapter 8 to address installations of optical fiber-based communications systems that are powered, not by the service provider, but from the premises served. These systems may supply a full range of voice, video, data, and interactive services through an optical network terminal (ONT).

Comments: Multiple

Proposal: 16-349

Appendix

National Electrical Code
Additional Resources

Significant Changes
TO THE NEC® 2011

Appendix A

The *NEC* Process

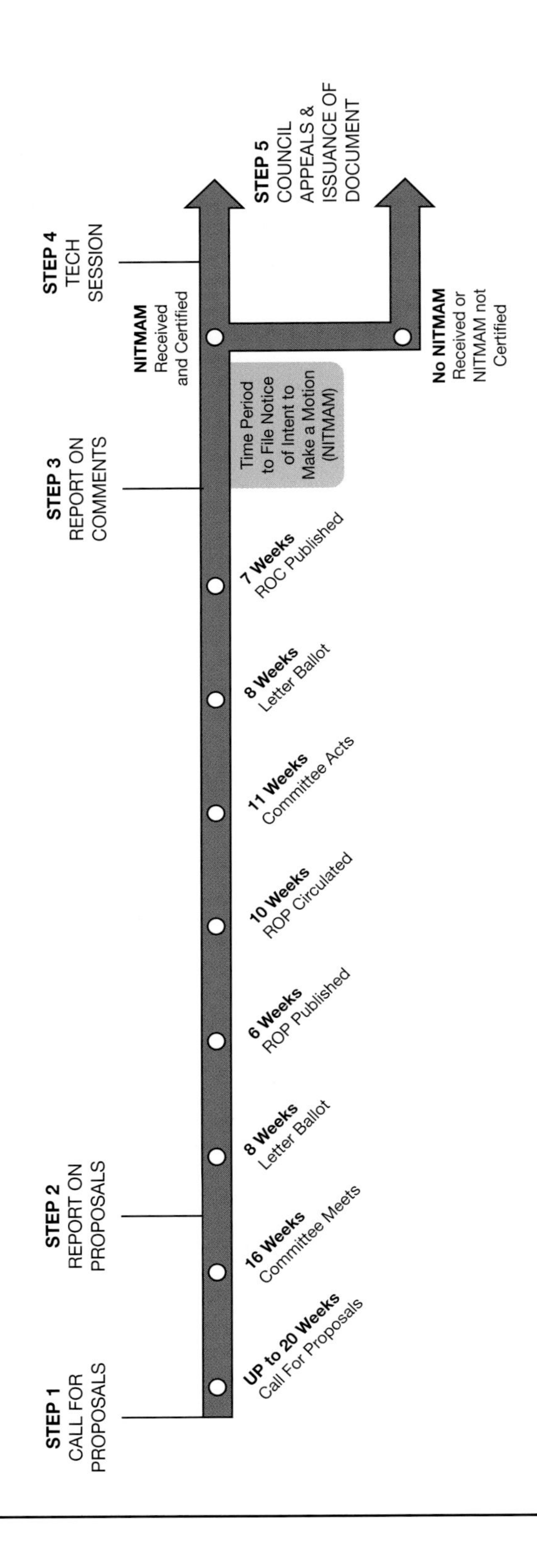

Appendix B

IBEW Code-Making Panel Members

IBEW Code-Making Panel Members

TECHNICAL CORRELATING COMMITTEE
Palmer L. Hickman, [Principal]
James T. Dollard, [Principal]

CODE–MAKING PANEL NO. 1
Palmer L. Hickman, [Principal]
Mark Christian, [Alternate]

CODE–MAKING PANEL NO. 2
Donald M. King, [Principal]
Jacob G. Benninger, [Alternate]

CODE–MAKING PANEL NO. 3
Paul J. Casparro, [Chair]
Marty L. Riesberg, [Alternate]

CODE–MAKING PANEL NO. 4
Todd W. Stafford, [Principal]
Brian L. Crise, [Alternate]

CODE–MAKING PANEL NO. 5
Dan Hammel, [Principal]
Paul J. LeVasseur, [Alternate]

CODE–MAKING PANEL NO. 6
William F. Laidler, [Principal]
James R. Weimer, [Alternate]

CODE–MAKING PANEL NO. 7
Samuel R. La Dart, [Principal]
Keith Owensby, [Alternate]

CODE–MAKING PANEL NO. 8
Joseph Dabe, [Principal]
Gary W. Pemble, [Alternate]

CODE–MAKING PANEL NO. 9
Rodney D. Belisle, [Principal]
Rhett A. Roe, [Alternate]

CODE–MAKING PANEL NO. 10
James T. Dollard, [Principal]
Richard E. Lofton, II, [Alternate]

CODE–MAKING PANEL NO. 11
James M. Fahey, [Principal]
Jebediah J. Novak, [Alternate]

CODE–MAKING PANEL NO. 12
David R. Quave, [Principal]
Jeffrey L. Holmes, [Alternate]

CODE–MAKING PANEL NO. 13
Linda J. Little, [Principal]
James T. Dollard, Jr., [Alternate]

CODE–MAKING PANEL NO. 14
John L. Simmons, [Principal]
Thomas E. Dunne, [Alternate]

CODE–MAKING PANEL NO. 15
Stephen M. Lipster, [Principal]
Gary A. Beckstrand, [Alternate]

CODE–MAKING PANEL NO. 16
Harold C. Ohde, [Principal]
Terry C. Coleman, [Alternate]

CODE–MAKING PANEL NO. 17
Randy J. Yasenchak, [Principal]
Brian Myers, [Alternate]

CODE–MAKING PANEL NO. 18
Paul Costello, [Principal]
Jesse Sprinkle, [Alternate]

CODE–MAKING PANEL NO. 19
Ronald Michaelis, [Principal]
Ronald D. Weaver, Jr., [Alternate]

Appendix C

NECA Code-Making Panel Members

NECA Code-Making Panel Members

TECHNICAL CORRELATING COMMITTEE
Stanley J. Folz, [Principal]
Larry D. Cogburn, [Alternate]

CODE–MAKING PANEL NO. 1
Harry J. Sassaman, [Principal]

CODE–MAKING PANEL NO. 2
Thomas H. Wood, [Principal]

CODE–MAKING PANEL NO. 3
Stanley D. Kahn, [Principal]

CODE–MAKING PANEL NO. 4
Ronald J. Toomer, [Chair]
Larry D. Cogburn, [Alternate]

CODE–MAKING PANEL NO. 5
Michael J. Johnston, [Chair]
Larry D. Cogburn, [Alternate]

CODE–MAKING PANEL NO. 6
Scott Cline, [Chair]
Phillip J. Huff, [Alternate]

CODE–MAKING PANEL NO. 7
Michael W. Smith, [Chair]
Wesley L. Wheeler, [Alternate]

CODE–MAKING PANEL NO. 8
Stephen P. Poholski, [Principal]

CODE–MAKING PANEL NO. 9
Monte Szendre, [Principal]

CODE–MAKING PANEL NO. 10
Richard Sobel, [Principal]

CODE–MAKING PANEL NO. 11
Wayne Brinkmeyer, [Chair]
Stanley J. Folz, [Alternate]

CODE–MAKING PANEL NO. 12
Thomas L. Hedges, [Principal]
Charles M. Trout, [Alternate]

CODE–MAKING PANEL NO. 13
Martin D. Adams, [Principal]

CODE–MAKING PANEL NO. 14
Marc J. Bernsen, [Principal]

CODE–MAKING PANEL NO. 15
Bruce D. Shelly, [Principal]

CODE–MAKING PANEL NO. 16
W. Douglas Pirkle, [Principal]

CODE–MAKING PANEL NO. 17
Don W. Jhonson, [Chair]
Bobby J. Gray, [Alternate]

CODE–MAKING PANEL NO. 18
Charles M. Trout, [Principal]

CODE–MAKING PANEL NO. 19
Howard D. Hughes, [Principal]

Appendix D

The Scope of 2011 *NEC* Code-Making Panels

2011 *NEC* CODE-MAKING PANELS	
***NEC* CODE-MAKING PANEL**	**ARTICLES, ANNEX AND CHAPTER 9 MATERIAL WITHIN THE SCOPE OF THE CODE-MAKING PANEL**
1	90, 100, 110, Annex A, Annex G
2	210, 215, 220, Annex D Examples 1 – 6
3	300, 509, 720, 725, 727, 760, Chapter 9 Tables 11(A) and (B), Tables 12(A) and (B)
4	225 & 230
5	200, 250, 280, 285
6	310, 400, 402, Chapter 9 Tables 5 – 9, Annex B
7	320, 322, 324, 326, 328, 330, 332, 334, 336, 338, 340, 382, 394, 396, 398
8	342, 344, 348, 350, 352, 353, 354, 355, 356, 358, 360, 360, 362, 366, 368, 370, 372, 374, 376, 378, 380, 384, 386, 388, 390, 392, Chapter 9 Tabels 1 – 4, Annex C
9	312, 314, 404, 408, 450, 490
10	240
11	409, 430, 440, 460, 470, Annex D, Example D8
12	610, 620, 625, 626, 630, 640, 645, 647, 650, 660, 665,668, 669, 670, 685, Annex D, Examples D9 and D10
13	445, 455, 480, 690, 692, 695, 700, 701, 702, 705
14	500, 501, 502, 503, 504, 505, 506, 510, 511, 513, 514, 515, 516
15	517, 518, 520, 522, 525, 530, 540
16	770, 800, 810, 820, 830
17	422, 424, 426, 427, 680, 682
18	406, 410, 411, 600, 605
19	545, 547, 550, 551, 552, 553, 555, 604, 675, Annex D Examples D11 and D12

Appendix E

2014 *NEC* Code-Making Schedule

NEC SCHEDULE FOR 2014		
DATE	NO. OF WEEKS BETWEEN EACH EVENT	EVENT
Nov. 4, 2011	--	Closing Date for Proposals
Jan. 9-21, 2012	10 (+1)	Code-Making Panel Meetings (ROP)
Jan. 27, 2012	1	Mail Ballots to CMPs
Feb. 24, 2012	4	Receipt of Initial Ballots
April 23-27, 2012	8	Correlating Committee Meeting
June 4-7, 2012	--	NFPA Annual Meeting - Las Vegas
June 15, 2012	7	NEC ROP to Printer
July 13, 2012	4	NEC ROP to Mailing House
Oct. 17, 2012	14	Closing Date for Comments
Nov. 28 - Dec. 8, 2012	5	Code-Making Panel Meetings (ROC)
Dec. 14, 2012	1	Mail Ballots to CMPs
Jan. 11, 2013	3	Receipt of Initial Ballots
Feb. 18-22, 2013	5	Correlating Committee Meeting
March 1, 2013	1	NEC ROC to Printer
March 22, 2013	3	NEC ROC to Mailing House
May 3, 2013	6	Notice of Intent to Make a Motion (NITMAM)
May 17, 2013	2	Posting of Certified NITMAMs
June 2-6, 2013	2	NFPA Annual Meeting - To Be Determined
July 2013		Standards Council Issuance
Sept. 2013		Release of 2014 NEC

Appendix F

NEC Technical Correlating Committee

2011 *NEC* TECHNICAL CORRELATING COMMITTEE

James W. Carpenter - Chair
International Association of Electrical Inspectors [E]

Mark W. Earley - Secrectary (non voting)
National Fire Protection Association

Jean A. O'Connor - Recording Secretary (non voting)
National Fire Protection Association

Principals

James E. Brunssen
Alliance for Telecommunications Industry Solutions [UT]

Merton W. Bunker, Jr.
U.S. Department of State [U]

James M. Daly
National Electrical Manufacturers Association [M]

William R. Drake
Marinco [M]

William T. Fiske
Intertek Testing Services [RT]

Stanley J. Folz
National Electrical Contractors Association [IM]

Palmer L. Hickman
International Brotherhood of Electrical Workers [L]

David L. Hittinger
Independent Electrical Contractors, Inc. [IM]

John R. Kovacik
Underwriters Laboratories Inc. [RT]

Neil F. LaBrake Jr.
Electric Light & Power Group/EEI [UT]

Danny Liggett
American Chemistry Council [U]

Alternates

Thomas L. Adams, Electric Light & Power Group/EEI [UT] (Alt. to N.F. LaBrake, Jr.)

Lawrence S. Ayer, Independent Electrical Contractors, Inc. [IM] (Alt. to D.L. Hittinger)

Jeffery Boksiner, Telcordia Technologies, Inc., NJ [UT] (Alt. to J.E. Brunssen)

Larry D. Cogburn, National Electrical Contractors Association [IM] (Alt. to S.J. Folz)

James T. Dollard, International Brotherhood of Electrical Workers [L] (Alt. to P.L. Hickman)

Ernest J. Gallo, Alliance for Telecommunications Industry Solutions [UT] (Alt. to J.E. Brunssen)

Michael E. McNeil, American Chemistry Council [U] (Alt. to D. Liggett)

Daniel J. Kissane, National Electrical Manufacturers Association [U] (Alt. to J.M. Daly)

Mark C. Ode, Underwriters Laboratories Inc. [RT] (Alt. to J.R. Kovacik)

Richard P. Owen, International Association of Electrical Inspectors [E] (Alt. to J.W. Carpenter)

Nonvoting

Richard G. Biermann, Biermann Electric Company, Inc. [IM] (Member Emeritus)

D. Harold Ware, Libra Electric Company [IM] (Member Emeritus)